Nonlinear Filters

Nonlinear Filters

Theory and Applications

Peyman Setoodeh
McMaster University
Ontario, Canada

Saeid Habibi
McMaster University
Ontario, Canada

Simon Haykin
McMaster University
Ontario, Canada

WILEY

This edition first published 2022

Registered Office
John Wiley & Sons, Inc., 111 River Street, Hoboken, NJ 07030, USA

Editorial Office
111 River Street, Hoboken, NJ 07030, USA

For details of our global editorial offices, customer services, and more information about Wiley products visit us at www.wiley.com.

Wiley also publishes its books in a variety of electronic formats and by print-on-demand. Some content that appears in standard print versions of this book may not be available in other formats.

Library of Congress Cataloging-in-Publication Data Applied for:
ISBN: 9781118835814

Cover Design: Wiley
Cover Image: © Emrah Turudu/Getty Images

Set in 9.5/12.5pt STIXTwoText by Straive, Chennai, India

10 9 8 7 6 5 4 3 2 1

To the memory of Rudolf
Emil Kalman

Contents

List of Figures

List of Table

Preface

Taking an algorithmic approach, this book provides a step towards bridging the gap between control theory, statistical signal processing, and machine learning regarding the state/parameter estimation problem. State estimation is an important concept that has profoundly influenced different branches of science and engineering. State of a system refers to a minimal record of the past history, which is required for predicting the future behavior. In this regard, a dynamic system can be described from the state perspective by selecting a set of independent variables as state variables. It is often desirable to know the state variables, and in control applications, to force them to follow desired trajectories in the state space. State estimation refers to the process of reconstructing the hidden or latent state variables, which cannot be directly measured, from system inputs and outputs in the minimum possible length of time. Filtering algorithms, which are deployed for state estimation, aim at minimizing the error between the estimated and the true values of the state variables.

The first part of the book is dedicated to classic estimation algorithms. A thorough presentation of the notion of observability, which refers to the ability to reconstruct the state variables from measurements, is followed by covering a number of observers as state estimators for deterministic systems. Regarding stochastic systems, optimal Bayesian filtering is presented that provides a conceptual solution for the general state estimation problem. Different Bayesian filtering algorithms have been developed based on computationally tractable approximations of the conceptual Bayesian solution. For the special case of linear systems with Gaussian noise, Kalman filter provides the optimal Bayesian solution. To extend the application of Kalman filter to nonlinear systems, two main approaches have been proposed to provide suboptimal solutions: using power series to approximate the nonlinear functions and approximating the probability distributions. While extended Kalman filter, extended information filter, and divided-difference filter approximate the nonlinear functions, unscented Kalman filter, cubature Kalman filter, and particle filter approximate the probability distributions. Other Kalman filter variants include Gaussian-sum filter, which handles non-Gaussianity, and generalized PID filter. Among the mentioned filters, particle filter is capable of handling nonlinear and non-Gaussian systems. Smooth variable-structure filter, which has been derived based on a stability theorem, is able to handle model uncertainties. Moreover, it benefits from using a secondary set of performance indicators in addition to the innovation vector.

The second part of the book is dedicated to machine learning-based filtering algorithms. Basic learning algorithms, deep learning architectures, and variational inference are reviewed to lay the groundwork for such algorithms. Different deep learning-based filters have been developed, which deploy supervised or unsupervised learning. These filters include deep Kalman filter, backpropagation Kalman filter, differentiable particle filter, deep Rao–Blackwellized particle filter, deep variational Bayes filter, Kalman variational autoencoder, and deep variational information bottleneck. Wasserstein distributionally robust Kalman filter and hierarchical invertible neural transport are presented in addition to the mentioned filtering algorithms. Expectation maximization allows for joint state and parameter estimation. Different variants of expectation maximization algorithm are implemented using particles, Gaussian mixture models, deep neural networks, relational deep neural networks, variational filters, and amortized variational filters. Variational inference and reinforcement learning can be viewed as instances of a generic expectation maximization problem. As a result, (deep) reinforcement learning methods can be used to develop novel filtering algorithms. Finally, the book covers nonparametric Bayesian models. In addition to reviewing measure-theoretic probability concepts and the notions of exchangeability, posterior computability, and algorithmic sufficiency, guidelines are provided for constructing nonparametric Bayesian models from parametric ones.

This book reviews a wide range of applications of classic and machine learning-based filtering algorithms regarding COVID-19 pandemic, influenza incidence, prediction of drug effect, robotics, information fusion, augmented reality, battery state-of-charge estimation for electric vehicles, autonomous driving, target tracking, urban traffic network, cybersecurity and optimal power flow in power systems, single-molecule fluorescence microscopy, and finance.

Hamilton, Ontario, Canada
January 2022

P. Setoodeh, S. Habibi, and S. Haykin

Acknowledgments

We would like to express our deepest gratitude to several colleagues, who helped us in one form or another while writing this book: Dr. Mehdi Fatemi, Dr. Pouya Dehghani Tafti, Dr. Ehsan Taghavi, Dr. Andrew Gadsden, Dr. Hamed Hossein Afshari, Dr. Mina Attari, Dr. Dhafar Al-Ani, Dr. Ulaş Güntürkün, Dr. Yanbo Xue, Dr. Ienkaran Arasaratnam, Dr. Mohammad Al-Shabi, Dr. Alireza Khayatian, Dr. Ali Akbar Safavi, Dr. Ebrahim Farjah, Dr. Paknoosh Karimaghaee, Dr. Mohammad Ali Masnadi-Shirazi, Dr. Mohammad Eghtesad, Dr. Majid Rostami-Shahrbabaki, Dr. Zahra Kazemi, Dr. Farshid Naseri, Dr. Zahra Khatami, Dr. Mohsen Mohammadi, Dr. Thiagalingam Kirubarajan, Dr. Stergios Roumeliotis, Dr. Magnus Norgaard, Dr. Eric Foxlin, Dr. Maryam Dehghani, Dr. Mohammad Mehdi Arefi, Dr. Mohammad Hassan Asemani, Dr. Mohammad Mohammadi, Dr. Mehdi Allahbakhshi, Dr. Haidar Samet, Dr. Mohammad Rastegar, Dr. Behrooz Zaker, Dr. Ali Reza Seifi, Dr. Mahdi Raoofat, Dr. Jun Luo, and Dr. Steven Hockema.

Last but by no means least, we would like to thank our families. Their endless support, encouragement, and love have always been a source of energy for us.

P. Setoodeh, S. Habibi, and S. Haykin

Acronyms

Backprop KF	backpropagation Kalman filter
BMS	battery management systems
CKF	cubature Kalman filter
CNN	convolutional neural network
CRLB	Cramér–Rao lower bound
DDF	divided-difference filter
DKF	deep Kalman filter
DRBPF	deep Rao–Blackwellized particle filter
DVBF	deep variational Bayes filter
DVIB	deep variational information bottleneck
EKF	extended Kalman filter
ELBO	evidence lower bound
EM	expectation maximization
FATE	fairness, accountability, transparency, and ethics
GAN	generative adversarial network
GRU	gated recurrent unit
HINT	hierarchical invertible neural transport
IB	information bottleneck
IMM	interacting multiple model
IS	importance sampling
KF	Kalman filter
KLD	Kullback–Leibler divergence
KVAE	Kalman variational autoencoder
LSTM	long short-term memory
LTI	linear time-invariant
LTV	linear time-varying
MAP	maximum a posteriori
MCMC	Markov chain Monte Carlo
MDP	Markov decision process

ML	maximum likelihood
MMSE	minimum mean square error
MPF	marginalized particle filter
N-EM	neural expectation maximization
NIB	nonlinear information bottleneck
NLL	negative log likelihood
PCRLB	posterior Cramér–Rao lower bound
PDF	probability distribution function
P-EM	particle expectation maximization
PF	particle filter
PID	proportional-integral-derivative
POMDP	partially-observable Markov decision process
RBPF	Rao-Blackwellised particle filter
ReLU	rectified linear unit
R-NEM	relational neural expectation maximization
RNN	recurrent neural network
SGVB	stochastic gradient variational Bayes
SIR	sampling importance resampling
SLAM	simultaneous localization and mapping
SMAUG	single molecule analysis by unsupervised Gibbs sampling
SMC	sequential Monte Carlo
SoC	state of charge
SoH	state of health
SVSF	smooth variable-structure filter
TD learning	temporal-difference learning
UIO	unknown-input observer
UKF	unscented Kalman filter
VAE	variational autoencoder
VFEM	variational filtering expectation maximization
VSC	variable-structure control
wILI	weighted influenza-like illness

1

Introduction

1.1 State of a Dynamic System

In many branches of science and engineering, deriving a probabilistic model for sequential data plays a key role. System theory provides guidelines for studying the underlying dynamics of sequential data (time series). In describing a dynamic system, the notion of *state* is a key concept [1]:

Definition 1.1 ***State of a dynamic system*** *is the smallest collection of variables that must be specified at a time instant k_0 in order to be able to predict the behavior of the system for any time instant $k \geq k_0$. To be more precise, the state is the minimal record of the past history, which is required to predict the future behavior.*

According to the *principle of causality*, any dynamic system may be described from the state perspective. Deploying a state-transition model allows for determining the future state of a system, $\mathbf{x}_k \in \mathbb{R}^{n_x}$, at any time instant $k \geq k_0$, given its initial state, $\mathbf{x}_{k_0}$, at time instant k_0 as well as the inputs to the system, $\mathbf{u}_k \in \mathbb{R}^{n_u}$, for $k \geq k_0$. The output of the system, $\mathbf{y}_k \in \mathbb{R}^{n_y}$, is a function of the state, which can be computed using a measurement model. In this regard, state-space models are powerful tools for analysis and control of dynamic systems.

1.2 State Estimation

Observability is a key concept in system theory, which refers to the ability to reconstruct the *hidden* or *latent* state variables that cannot be directly measured, from

Nonlinear Filters: Theory and Applications, First Edition. Peyman Setoodeh, Saeid Habibi, and Simon Haykin.

the measured variables in the minimum possible length of time [1]. In building state-space models, two key questions deserve special attention [2]:

(i) Is it possible to identify the governing dynamics from data?
(ii) Is it possible to perform inference from observables to the latent state variables?

At time instant k, the inference problem to be solved is to find the estimate of $\mathbf{x}_{k+\alpha}$ in the presence of noise, which is denoted by $\hat{\mathbf{x}}_{k+\alpha}$. Depending of the value of α, estimation algorithms are categorized into three groups [3]:

(i) *Prediction*: $\alpha > 0$,
(ii) *Filtering*: $\alpha = 0$,
(iii) *Smoothing*: $\alpha < 0$.

Regarding the mentioned two challenging questions, in order to improve performance, sophisticated representations can be deployed for the system under study. However, the corresponding inference algorithms may become computationally demanding. Hence, for designing efficient data-driven inference algorithms, the following points must be taken into account [2]:

(i) The underlying assumptions for building a state-space model must allow for reliable system identification and plausible long-term prediction of the system behavior.
(ii) The inference mechanism must be able to capture rich dependencies.
(iii) The algorithm must be able to inherit the merit of *learning machines* to be trainable on raw data such as sensory inputs in a control system.
(iv) The algorithm must be scalable to big data regarding the optimization of model parameters based on the *stochastic gradient descent* method.

Regarding the important role of computation in inference problems, Section 1.3 provides a brief account of the foundations of computing.

1.3 Construals of Computing

According to [4], a comprehensive *theory of computing* must meet three criteria:

(i) *Empirical criterion*: Doing justice to practice by keeping the analysis grounded in real-world examples.
(ii) *Conceptual criterion*: Being understandable in terms of what it says, where it comes from, and what it costs.
(iii) *Cognitive criterion*: Providing an intelligible foundation for the computational theory of mind that underlies both artificial intelligence and cognitive science.

Following this line of thinking, it was proposed in [4] to distinguish the following construals of computation:

1. *Formal symbol manipulation* is rooted in formal logic and metamathematics. The idea is to build machines that are capable of manipulating symbolic or meaningful expressions regardless of their interpretation or semantic content.
2. *Effective computability* deals with the question of what can be done, and how hard it is to do it mechanically.
3. *Execution of an algorithm or rule following* focuses on what is involved in following a set of rules or instructions, and what behavior would be produced.
4. *Calculation of a function* considers the behavior of producing the value of a mathematical function as output, when a set of arguments is given as input.
5. *Digital state machine* is based on the idea of a finite-state automaton.
6. *Information processing* focuses on what is involved in storing, manipulating, displaying, and trafficking of information.
7. *Physical symbol systems* is based on the idea that the way computers interact with symbols depends on their mutual physical embodiment. In this regard, computers may be assumed to be made of symbols.
8. *Dynamics* must be taken into account in terms of the roles that nonlinear elements, attractors, criticality, and emergence play in computing.
9. *Interactive agents* are capable of interacting and communicating with other agents and even people.
10. *Self-organizing or complex adaptive systems* are capable of adjusting their organization or structure in response to changes in their environment in order to survive and improve their performance.
11. *Physical implementation* emphasizes on the occurrence of computational practice in real-world systems.

1.4 Statistical Modeling

Statistical modeling aims at extracting information about the underlying data mechanism that allows for making predictions. Then, such predictions can be used to make decisions. There are two cultures in deploying statistical models for data analysis [5]:

- *Data modeling culture* is based on the idea that a given stochastic model generates the data.
- *Algorithmic modeling culture* uses algorithmic models to deal with an unknown data mechanism.

An algorithmic approach has the advantage of being able to handle large complex datasets. Moreover, it can avoid irrelevant theories or questionable conclusions.

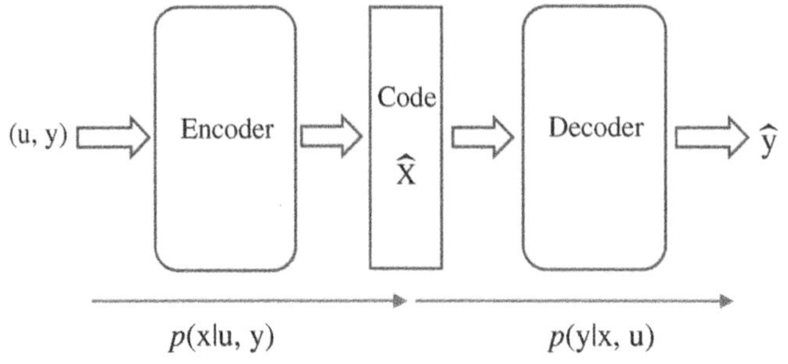

Figure 1.1 The encoder of an asymmetric autoencoder plays the role of a nonlinear filter.

Taking an algorithmic approach, in *machine learning*, statistical models can be classified as [6]:

(i) *Generative models* predict visible effects from hidden causes, $p(\mathbf{y}|\mathbf{x}, \mathbf{u})$.
(ii) *Discriminative models* infer hidden causes from visible effects, $p(\mathbf{x}|\mathbf{u}, \mathbf{y})$.

While the former is associated with the measurement process in a state-space model, the latter is associated with the state estimation or filtering problem. Deploying machine learning, a wide range of filtering algorithms can be developed that are able to learn the corresponding state-space models. For instance, an asymmetric autoencoder can be designed by combining a generative model and a discriminative model as shown in Figure 1.1 [7]. Deep neural networks can be used to implement both the encoder and the decoder. Then, the resulting autoencoder can be trained in an unsupervised manner. After training, the encoder can be used as a filter, which estimates the latent state variables.

1.5 Vision for the Book

This book provides an algorithmic perspective on the nonlinear state/parameter estimation problem for discrete-time systems, where measurements are available at discrete sampling times and estimators are implemented using digital processors. In Chapter 2, guidelines are provided for discretizing continuous-time linear and nonlinear state-space models. The rest of the book is organized as follows:

- Chapter 2 presents the notion of observability for deterministic and stochastic systems.

Chapters 3–7 cover classic estimation algorithms:

- Chapter 3 is dedicated to observers as state estimators for deterministic systems.

- Chapter 4 presents the general formulation of the optimal Bayesian filtering for stochastic systems.
- Chapter 5 covers the Kalman filter as the optimal Bayesian filter in the sense of minimizing the mean-square estimation error for linear systems with Gaussian noise. Moreover, Kalman filter variants are presented that extend its applicability to nonlinear or non-Gaussian cases.
- Chapter 6 covers the particle filter, which handles severe nonlinearity and non-Gaussianity by approximating the corresponding distributions using a set of particles (random samples).
- Chapter 7 covers the smooth variable-structure filter, which provides robustness against bounded uncertainties and noise. In addition to the innovation vector, this filter benefits from a secondary set of performance indicators.

Chapters 8–11 cover learning-based estimation algorithms:

- Chapter 8 covers the basics of deep learning.
- Chapter 9 covers deep-learning-based filtering algorithms using supervised and unsupervised learning.
- Chapter 10 presents the expectation maximization algorithm and its variants, which are used for joint state and parameter estimation.
- Chapter 11 presents the reinforcement learning-based filter, which is built on viewing variational inference and reinforcement learning as instances of a generic expectation maximization problem.

The last chapter is dedicated to nonparametric Bayesian models:

- Chapter 12 covers measure-theoretic probability concepts as well as the notions of exchangeability, posterior computability, and algorithmic sufficiency. Furthermore, it provides guidelines for constructing nonparametric Bayesian models from finite parametric Bayesian models.

In each chapter, selected applications of the presented filtering algorithms are reviewed, which cover a wide range of problems. Moreover, the last section of each chapter usually refers to a few topics for further study.

2 Observability

2.1 Introduction

In many branches of science and engineering, it is common to deal with sequential data, which is generated by dynamic systems. In different applications, it is often desirable to predict future observations based on the collected data up to a certain time instant. Since the future is always uncertain, it is preferred to have a measure that shows our confidence about the predictions. A probability distribution over possible future outcomes can provide this information [8]. A great deal of what we know about a system cannot be presented in terms of quantities that can be directly measured. In such cases, we try to build a model for the system that helps to explain the cause behind what we observe via the measurement process. This leads to the notions of *state* and *state-space* model of a dynamic system. Chapters 3–7 and 9–11 are dedicated to different methods for reconstructing (estimating) the state of dynamic systems from inputs and measurements. Each estimation algorithm has its own advantages and limitations that should be taken into account, when we want to choose an estimator for a specific application. However, before trying to choose a proper estimation algorithm among different candidates, we need to know if for a given model of the dynamic system under study, it is possible to estimate the state of the system from inputs and measurements [9]. This critical question leads to the concept of *observability*, which is the focus of this chapter.

2.2 State-Space Model

The behavioral approach for studying dynamic systems is based on an abstract model of the system of interest, which determines the relationship between its input and output. In this abstract model, the input of the system, denoted

Nonlinear Filters: Theory and Applications, First Edition. Peyman Setoodeh, Saeid Habibi, and Simon Haykin.

by $\mathbf{u} \in \mathbb{R}^{n_u}$, represents the effect of the external events on the system, and its output, denoted by $\mathbf{y} \in \mathbb{R}^{n_y}$, represents any change it causes to the surrounding environment. The output can be directly measured [10]. State of the system, denoted by $\mathbf{x} \in \mathbb{R}^{n_x}$, is defined as the minimum amount of information required at each time instant to uniquely determine the future behavior of the system provided that we know inputs to the system as well as the system's parameters. Parameter values reflect the underlying physical characteristics based on which the model of the system was built [9]. State variables may not be directly accessible for measurement; hence, the reason for calling them *hidden* or *latent* variables. Regarding the abstract nature of the state variables, they may not even represent physical quantities. However, these variables help us to improve a model's ability to capture the causal structure of the system under study [11].

A state-space model includes the corresponding mappings from input to state and from state to output. This model also describes evolution of the system's state over time [12]. In other words, any state-space model has three constituents [8]:

- A prior, $p(\mathbf{x}_0)$, which is associated with the initial state $\mathbf{x}_0$.
- A state-transition function, $p(\mathbf{x}_{k+1}|\mathbf{x}_k, \mathbf{u}_k)$.
- An observation function, $p(\mathbf{y}_k|\mathbf{x}_k, \mathbf{u}_k)$.

For controlled systems, the state-transition function depends on control inputs as well. To be able to model active perception (sensing), the observation function must be allowed to depend on inputs too.

The state-space representation is based on the assumption that the model is a first-order Markov process, which means that value of the state vector at cycle $k+1$ depends only on its value at cycle k, but not on its values in previous cycles. In other words, the state vector at cycle k contains all the information about the system from the initial cycle till cycle k. In a sense, the concept of state inherently represents the memory of the system [13]. The first-order Markov-model assumption can be shown mathematically as follows:

$$p(\mathbf{x}_{k+1}|\mathbf{x}_{0:k}) = p(\mathbf{x}_{k+1}|\mathbf{x}_k). \tag{2.1}$$

It should be noted that if a model is not a first-order Markov process, it would be possible to build a corresponding first-order Markov model based on an augmented state vector, which includes the state vector at current cycle as well as the state vectors in previous cycles. The order of the Markov process determines that the state vectors from how many previous cycles must be included in the augmented state vector. For instance, if the system is an nth-order Markov process with state vector $\mathbf{x}_k$, the corresponding first-order Markov model is built based on the augmented state vector:

$$\begin{bmatrix} \mathbf{x}_{k-n+1} \\ \vdots \\ \mathbf{x}_{k-1} \\ \mathbf{x}_k \end{bmatrix}. \tag{2.2}$$

Moreover, if model parameters are time-varying, they can be treated as random variables by including them in the augmented state vector as well.

2.3 The Concept of Observability

Observability and *controllability* are two basic properties of dynamic systems. These two concepts were first introduced by Kalman in 1960 for analyzing control systems based on linear state-space models [1]. While observability is concerned with how the state vector influences the output vector, controllability is concerned with how the input vector influences the state vector. If a state has no effect on the output, it is *unobservable*; otherwise, it is *observable*. To be more precise, starting from an unobservable initial state $\mathbf{x}_0$, system's output will be $\mathbf{y}_k = 0, k \geq 0$, in the absence of an input, $\mathbf{u}_k = 0, k \geq 0$ [14]. Another interpretation would be that unobservable systems allow for the existence of *indistinguishable* states, which means that if an input is applied to the system at any one of the indistinguishable states, then the output will be the same. On the contrary, observability implies that an observer would be able to distinguish between different initial states based on inputs and measurements. In other words, an observer would be able to uniquely determine observable initial states from inputs and measurements [13, 15]. In a general case, the state vector may be divided into two parts including observable and unobservable states.

Definition 2.1 ***(State observability)*** *A dynamic system is state observable if for any time $k_f > 0$, the initial state $\mathbf{x}_0$ can be uniquely determined from the time history of the input $\mathbf{u}_k$ and the output $\mathbf{y}_k$ for $k \in [0, k_f]$; otherwise, the system is unobservable.*

Unlike linear systems, there is not a universal definition for observability of nonlinear systems. Hence, different definitions have been proposed in the literature, which take two questions into consideration:

- How to check the observability of a nonlinear system?
- How to design an observer for such a system?

While for linear systems, observability is a global property, for nonlinear systems, observability is usually studied locally [9].

Definition 2.2 ***(State detectability)*** *If all unstable modes of a system are observable, then the system is state detectable.*

A system with undetectable modes is said to have hidden unstable modes [16, 17]. Sections 2.4–2.7 provide observability tests for different classes of systems, whether they be linear or nonlinear, continuous-time or discrete-time.

2.4 Observability of Linear Time-Invariant Systems

If the system matrices in the state-space model of a linear system are constant, then, the model represents a *linear time-invariant* (LTI) system.

2.4.1 Continuous-Time LTI Systems

The state-space model of a continuous-time LTI system is represented by the following algebraic and differential equations:

$$\dot{\mathbf{x}}(t) = \mathbf{A}\mathbf{x}(t) + \mathbf{B}\mathbf{u}(t), \tag{2.3}$$

$$\mathbf{y}(t) = \mathbf{C}\mathbf{x}(t) + \mathbf{D}\mathbf{u}(t), \tag{2.4}$$

where $\mathbf{x} \in \mathbb{R}^{n_x}$, $\mathbf{u} \in \mathbb{R}^{n_u}$, and $\mathbf{y} \in \mathbb{R}^{n_y}$ are the state, the input, and the output vectors, respectively, and $\mathbf{A} \in \mathbb{R}^{n_x \times n_x}$, $\mathbf{C} \in \mathbb{R}^{n_y \times n_x}$, $\mathbf{B} \in \mathbb{R}^{n_x \times n_u}$, and $\mathbf{D} \in \mathbb{R}^{n_y \times n_u}$ are the system matrices. Here, we need to find out when an initial state vector $\mathbf{x}(t_0)$ can be uniquely reconstructed from nonzero initial system output vector and its successive derivatives. We start by writing the system output vector and its successive derivatives based on the state vector as well as the input vector and its successive derivatives as follows:

$$\begin{aligned}
\mathbf{y}(t) &= \mathbf{C}\mathbf{x}(t) + \mathbf{D}\mathbf{u}(t),\\
\dot{\mathbf{y}}(t) &= \mathbf{C}\mathbf{A}\mathbf{x}(t) + \mathbf{C}\mathbf{B}\mathbf{u}(t) + \mathbf{D}\dot{\mathbf{u}}(t),\\
\ddot{\mathbf{y}}(t) &= \mathbf{C}\mathbf{A}^2\mathbf{x}(t) + \mathbf{C}\mathbf{A}\mathbf{B}\mathbf{u}(t) + \mathbf{C}\mathbf{B}\dot{\mathbf{u}}(t) + \mathbf{D}\ddot{\mathbf{u}}(t),\\
\vdots &= \vdots\\
\mathbf{y}^{(n-1)}(t) &= \mathbf{C}\mathbf{A}^{n-1}\mathbf{x}(t) + \mathbf{C}\mathbf{A}^{n-2}\mathbf{B}\mathbf{u}(t) + \cdots + \mathbf{C}\mathbf{B}\mathbf{u}^{(n-2)}(t) + \mathbf{D}\mathbf{u}^{(n-1)}(t),
\end{aligned} \tag{2.5}$$

where the superscript in the parentheses denotes the order of the derivative. The aforementioned equations can be rewritten in the following compact form:

$$\mathcal{O}\mathbf{x}(t) = \mathcal{Y}(t), \tag{2.6}$$

where

$$\mathcal{O} = \begin{bmatrix} \mathbf{C} \\ \mathbf{CA} \\ \vdots \\ \mathbf{CA}^{n-1} \end{bmatrix} \tag{2.7}$$

and

$$\mathcal{Y}(t) = \begin{bmatrix} \mathbf{y}(t) - \mathbf{Du}(t) \\ \dot{\mathbf{y}}(t) - \mathbf{CBu}(t) - \mathbf{D}\dot{\mathbf{u}}(t) \\ \vdots \\ \mathbf{y}^{(n-1)}(t) - \mathbf{CA}^{n-2}\mathbf{Bu}(t) - \cdots - \mathbf{CBu}^{(n-2)}(t) - \mathbf{Du}^{(n-1)}(t) \end{bmatrix}. \tag{2.8}$$

Initially we have

$$\mathcal{O}\mathbf{x}(t_0) = \mathcal{Y}(t_0). \tag{2.9}$$

The continuous-time system is observable, if and only if the *observability matrix* $\mathcal{O}$ is nonsingular (it is full rank), therefore the initial state can be found as:

$$\mathbf{x}(t_0) = \mathcal{O}^{-1}\mathcal{Y}(t_0). \tag{2.10}$$

The observable subspace of the linear system, denoted by $\mathbf{T}^{O}$, is composed of the basis vectors of the range of $\mathcal{O}$, and the unobservable subspace of the linear system, denoted by $\mathbf{T}^{\bar{O}}$, is composed of the basis vectors of the null space of $\mathcal{O}$. These two subspaces can be combined to form the following nonsingular transformation matrix:

$$\mathbf{T} = \begin{bmatrix} \mathbf{T}^{O} \\ \mathbf{T}^{\bar{O}} \end{bmatrix}. \tag{2.11}$$

If we apply the aforementioned transformation to the state vector $\mathbf{x}$ such that:

$$\mathbf{z}(t) = \mathbf{Tx}(t), \tag{2.12}$$

the transformed state vector $\mathbf{z}$ will be partitioned to observable modes, $\mathbf{z}^{o}$, and unobservable modes, $\mathbf{z}^{\bar{o}}$:

$$\mathbf{z}(t) = \begin{bmatrix} \mathbf{z}^{o}(t) \\ \mathbf{z}^{\bar{o}}(t) \end{bmatrix}. \tag{2.13}$$

Then, the state-space model of (2.3) and (2.4) can be rewritten based on the transformed state vector, $\mathbf{z}$, as follows:

$$\dot{\mathbf{z}}(t) = \mathbf{TAT}^{-1}\mathbf{z}(t) + \mathbf{TBu}(t), \tag{2.14}$$

$$\mathbf{y}(t) = \mathbf{CT}^{-1}\mathbf{z}(t) + \mathbf{Du}(t), \tag{2.15}$$

or equivalently as:

$$\begin{bmatrix} \dot{\mathbf{z}}^{o}(t) \\ \dot{\mathbf{z}}^{\bar{o}}(t) \end{bmatrix} = \begin{bmatrix} \widetilde{\mathbf{A}}^{o} & 0 \\ \widetilde{\mathbf{A}} & \widetilde{\mathbf{A}}^{\bar{o}} \end{bmatrix} \begin{bmatrix} \mathbf{z}^{o}(t) \\ \mathbf{z}^{\bar{o}}(t) \end{bmatrix} + \begin{bmatrix} \widetilde{\mathbf{B}}^{o} \\ \widetilde{\mathbf{B}}^{\bar{o}} \end{bmatrix} \mathbf{u}(t), \tag{2.16}$$

$$\mathbf{y}(t) = \begin{bmatrix} \widetilde{\mathbf{C}}^{o} & 0 \end{bmatrix} \begin{bmatrix} \mathbf{z}^{o}(t) \\ \mathbf{z}^{\bar{o}}(t) \end{bmatrix} + \mathbf{D}\mathbf{u}(t). \tag{2.17}$$

Any pair of equations (2.14) and (2.15) or (2.16) and (2.17) is called the state-space model of the system in the *observable canonical* form. For the system to be detectable (to have stable unobservable modes), the eigenvalues of $\widetilde{\mathbf{A}}^{\bar{o}}$ must have negative real parts ($\widetilde{\mathbf{A}}^{\bar{o}}$ must be *Hurwitz*).

2.4.2 Discrete-Time LTI Systems

The state-space model of a discrete-time LTI system is represented by the following algebraic and difference equations:

$$\mathbf{x}_{k+1} = \mathbf{A}\mathbf{x}_k + \mathbf{B}\mathbf{u}_k, \tag{2.18}$$

$$\mathbf{y}_k = \mathbf{C}\mathbf{x}_k + \mathbf{D}\mathbf{u}_k, \tag{2.19}$$

where $\mathbf{x} \in \mathbb{R}^{n_x}$, $\mathbf{u} \in \mathbb{R}^{n_u}$, and $\mathbf{y} \in \mathbb{R}^{n_y}$ are the state, the input, and the output vectors, respectively, and $\mathbf{A} \in \mathbb{R}^{n_x \times n_x}$, $\mathbf{C} \in \mathbb{R}^{n_y \times n_x}$, $\mathbf{B} \in \mathbb{R}^{n_x \times n_u}$, and $\mathbf{D} \in \mathbb{R}^{n_y \times n_u}$ are the system matrices. Starting from the initial cycle, the system output vector at successive cycles up to cycle $k = n - 1$ can be written based on the initial state vector $\mathbf{x}_0$ and input vectors $\mathbf{u}_{k:k+n-1}$ as follows:

$$\begin{aligned} \mathbf{y}_k &= \mathbf{C}\mathbf{x}_k + \mathbf{D}\mathbf{u}_k, \\ \mathbf{y}_{k+1} &= \mathbf{C}\mathbf{A}\mathbf{x}_k + \mathbf{C}\mathbf{B}\mathbf{u}_k + \mathbf{D}\mathbf{u}_{k+1}, \\ \mathbf{y}_{k+2} &= \mathbf{C}\mathbf{A}^2\mathbf{x}_k + \mathbf{C}\mathbf{A}\mathbf{B}\mathbf{u}_k + \mathbf{C}\mathbf{B}\mathbf{u}_{k+1} + \mathbf{D}\mathbf{u}_{k+2}, \\ \vdots &= \vdots \\ \mathbf{y}_{k+n-1} &= \mathbf{C}\mathbf{A}^{n-1}\mathbf{x}_k + \mathbf{C}\mathbf{A}^{n-2}\mathbf{B}\mathbf{u}_k + \cdots + \mathbf{C}\mathbf{B}\mathbf{u}_{k+n-2} + \mathbf{D}\mathbf{u}_{k+n-1}. \end{aligned} \tag{2.20}$$

The aforementioned equations can be rewritten in the following compact form:

$$\mathcal{O}\mathbf{x}_k = \mathcal{Y}_k, \tag{2.21}$$

where

$$\mathcal{O} = \begin{bmatrix} \mathbf{C} \\ \mathbf{C}\mathbf{A} \\ \vdots \\ \mathbf{C}\mathbf{A}^{n-1} \end{bmatrix} \tag{2.22}$$

and

$$\mathcal{Y}_k = \begin{bmatrix} \mathbf{y}_k - \mathbf{D}\mathbf{u}_k \\ \mathbf{y}_{k+1} - \mathbf{C}\mathbf{B}\mathbf{u}_k - \mathbf{D}\mathbf{u}_{k+1} \\ \vdots \\ \mathbf{y}_{k+n-1} - \mathbf{C}\mathbf{A}^{n-2}\mathbf{B}\mathbf{u}_k - \cdots - \mathbf{C}\mathbf{B}\mathbf{u}_{k+n-2} - \mathbf{D}\mathbf{u}_{k+n-1} \end{bmatrix}. \tag{2.23}$$

It is obvious from the linear set of equations (2.21) that in order to be able to uniquely determine the initial state $\mathbf{x}_0$, matrix $\mathcal{O}$ must be full-rank, provided that inputs and outputs are known. In other words, if the matrix $\mathcal{O}$ is full rank, the linear system is observable or reconstructable, hence, the reason for calling $\mathcal{O}$ the *observability matrix*. The reverse is true as well, if the system is observable, then the observability matrix will be full-rank. In this case, the initial state vector can be calculated as:

$$\mathbf{x}_0 = \mathcal{O}^{-1}\mathcal{Y}_0. \tag{2.24}$$

Since $\mathcal{O}$ depends only on matrices $\mathbf{A}$ and $\mathbf{C}$, for an observable system, it is equivalently said that the pair $(\mathbf{A}, \mathbf{C})$ is observable. Any initial state that has a component in the null space of $\mathcal{O}$ cannot be uniquely determined from measurements; therefore, the null space of $\mathcal{O}$ is called the *unobservable subspace* of the system. As mentioned before, the system is detectable if the unobservable subspace does not include unstable modes of $\mathbf{A}$, which are associated with the eigenvalues that are outside the unit circle.

While the observable subspace of the linear system, denoted by $\mathbf{T}^{O}$, is composed of the basis vectors of the range of $\mathcal{O}$, the unobservable subspace of the linear system, denoted by $\mathbf{T}^{\bar{O}}$, is composed of the basis vectors of the null space of $\mathcal{O}$. These two subspaces can be combined to form the following nonsingular transformation matrix:

$$\mathbf{T} = \begin{bmatrix} \mathbf{T}^{O} \\ \mathbf{T}^{\bar{O}} \end{bmatrix}. \tag{2.25}$$

If we apply this transformation to the state vector $\mathbf{x}$ such that:

$$\mathbf{z}_k = \mathbf{T}\mathbf{x}_k, \tag{2.26}$$

the transformed state vector $\mathbf{z}$ will be partitioned to observable modes, $\mathbf{z}^{o}$, and unobservable modes, $\mathbf{z}^{\bar{o}}$:

$$\mathbf{z}_k = \begin{bmatrix} \mathbf{z}_k^{o} \\ \mathbf{z}_k^{\bar{o}} \end{bmatrix}. \tag{2.27}$$

Then, the state-space model of (2.18) and (2.19) can be rewritten based on the transformed state vector, $\mathbf{z}$, as follows:

$$\mathbf{z}_{k+1} = \mathbf{T}\mathbf{A}\mathbf{T}^{-1}\mathbf{z}_k + \mathbf{T}\mathbf{B}\mathbf{u}_k, \tag{2.28}$$

$$\mathbf{y}_k = \mathbf{C}\mathbf{T}^{-1}\mathbf{z}_k + \mathbf{D}\mathbf{u}_k, \tag{2.29}$$

or equivalently as:

$$\begin{bmatrix} \mathbf{z}^{o}_{k+1} \\ \mathbf{z}^{\bar{o}}_{k+1} \end{bmatrix} = \begin{bmatrix} \widetilde{\mathbf{A}}^{o} & 0 \\ \widetilde{\mathbf{A}} & \widetilde{\mathbf{A}}^{\bar{o}} \end{bmatrix} \begin{bmatrix} \mathbf{z}^{o}_{k} \\ \mathbf{z}^{\bar{o}}_{k} \end{bmatrix} + \begin{bmatrix} \widetilde{\mathbf{B}}^{o} \\ \widetilde{\mathbf{B}}^{\bar{o}} \end{bmatrix} \mathbf{u}_k, \tag{2.30}$$

$$\mathbf{y}_k = \begin{bmatrix} \widetilde{\mathbf{C}}^{o} & 0 \end{bmatrix} \begin{bmatrix} \mathbf{z}^{o}_{k} \\ \mathbf{z}^{\bar{o}}_{k} \end{bmatrix} + \mathbf{D}\mathbf{u}_k. \tag{2.31}$$

Any pair of equations (2.28) and (2.29) or (2.30) and (2.31) is called the state-space model of the system in the *observable canonical* form.

2.4.3 Discretization of LTI Systems

When a continuous-time system is connected to a computer via analog-to-digital and digital-to-analog converters at input and output, respectively, we need to find a discrete-time equivalent of the continuous-time system that describes the relationship between the system's input and its output at certain time instants (sampling times t_k for $k = 0,1,\ldots$). This process is called *sampling* the continuous-time system. Using *zero-order-hold* sampling, where the corresponding analog signals are kept constant over the sampling period, we will have the following discrete-time equivalent for the continuous-time system of (2.3) and (2.4) [18]:

$$\mathbf{x}(t_{k+1}) = \mathbf{\Phi}(t_{k+1}, t_k)\mathbf{x}(t_k) + \mathbf{\Gamma}(t_{k+1}, t_k)\mathbf{u}(t_k), \tag{2.32}$$

$$\mathbf{y}(t_k) = \mathbf{C}\mathbf{x}(t_k) + \mathbf{D}\mathbf{u}(t_k), \tag{2.33}$$

where

$$\mathbf{\Phi}(t_{k+1}, t_k) = \mathrm{e}^{\mathbf{A}(t_{k+1}-t_k)}, \tag{2.34}$$

$$\mathbf{\Gamma}(t_{k+1}, t_k) = \int_0^{t_{k+1}-t_k} \mathrm{e}^{\mathbf{A}\tau}\, \mathrm{d}\tau \mathbf{B}. \tag{2.35}$$

2.5 Observability of Linear Time-Varying Systems

If the system matrices in the state-space model of a linear system change with time, then, the model represents a *linear time-varying* (LTV) system. Obviously, the observability condition would be more complicated for LTV systems compared to LTI systems.

2.5.1 Continuous-Time LTV Systems

The state-space model of a continuous-time LTV system is represented by the following algebraic and differential equations:

$$\dot{\mathbf{x}}(t) = \mathbf{A}(t)\mathbf{x}(t) + \mathbf{B}(t)\mathbf{u}(t), \tag{2.36}$$

$$\mathbf{y}(t) = \mathbf{C}(t)\mathbf{x}(t) + \mathbf{D}(t)\mathbf{u}(t). \tag{2.37}$$

In order to determine the *relative observability* of different state variables, we investigate their contributions to the energy of the system output. Knowing the input, we can eliminate its contribution to the energy of the output. Therefore, without loss of generality, we can assume that the input is zero. Without an input, evolution of the state vector is governed by the following unforced differential equation:

$$\dot{\mathbf{x}}(t) = \mathbf{A}(t)\mathbf{x}(t), \tag{2.38}$$

whose solution is:

$$\mathbf{x}(t) = \mathbf{\Phi}(t, t_0)\mathbf{x}(t_0), \tag{2.39}$$

where $\mathbf{\Phi}(t, t_0)$ is called the continuous-time *state-transition matrix*, which is itself the solution of the following differential equation:

$$\frac{\mathrm{d}\mathbf{\Phi}(t, t_0)}{\mathrm{d}t} = \mathbf{A}(t)\mathbf{\Phi}(t, t_0), \tag{2.40}$$

with the initial condition:

$$\mathbf{\Phi}(t_0, t_0) = \mathbf{I}. \tag{2.41}$$

Note that for an LTI system, where the matrix $\mathbf{A}$ is constant, the state transition matrix will be:

$$\mathbf{\Phi}(t, t_0) = \mathrm{e}^{\mathbf{A}(t-t_0)}. \tag{2.42}$$

Without an input, output of the unforced system is obtained from (2.37) as follows:

$$\mathbf{y}(t) = \mathbf{C}(t)\mathbf{x}(t). \tag{2.43}$$

Replacing for $\mathbf{x}(t)$ from (2.39) in the aforementioned equation, we will have:

$$\mathbf{y}(t) = \mathbf{C}(t)\mathbf{\Phi}(t, t_0)\mathbf{x}(t_0), \tag{2.44}$$

whose energy is obtained from:

$$\int_{t_0}^{t} \mathbf{y}^T(\tau)\mathbf{y}(\tau)\mathrm{d}\tau = \mathbf{x}^T(t_0)\left(\int_{t_0}^{t} \mathbf{\Phi}^T(\tau, t_0)\mathbf{C}^T(\tau)\mathbf{C}(\tau)\mathbf{\Phi}(\tau, t_0)\mathrm{d}\tau\right)\mathbf{x}(t_0). \tag{2.45}$$

In the aforementioned equation, the matrix in the parentheses is called the continuous-time *observability Gramian* matrix:

$$\mathbf{W}_o(t_0, t) = \int_{t_0}^{t} \mathbf{\Phi}^T(\tau, t_0)\mathbf{C}^T(\tau)\mathbf{C}(\tau)\mathbf{\Phi}(\tau, t_0)\mathrm{d}\tau. \tag{2.46}$$

From its structure, it is obvious that the observability Gramian matrix is symmetric and nonnegative. If we apply a transformation, $\mathbf{T}$, to the state vector, $\mathbf{x}$, such that $\mathbf{z} = \mathbf{Tx}$, the output energy:

$$\int_{t_0}^{t} \mathbf{y}^T(\tau)\mathbf{y}(\tau)\mathrm{d}\tau = \mathbf{x}^T(t_0)\mathbf{W}_o(t_0, t)\mathbf{x}(t_0) \tag{2.47}$$

can be rewritten as:

$$\int_{t_0}^{t} \mathbf{y}^T(\tau)\mathbf{y}(\tau)\mathrm{d}\tau = \mathbf{z}^T(t_0)\left(\mathbf{T}^{-T}\mathbf{W}_o(t_0,t)\mathbf{T}^{-1}\right)\mathbf{z}(t_0). \tag{2.48}$$

If the transformation $\mathbf{T}$ is chosen in a way that the transformed observability Gramian matrix, $\mathbf{T}^{-T}\mathbf{W}_o(t_0,t)\mathbf{T}^{-1}$, is diagonal, then, its diagonal elements can be viewed as the contribution of different state variables in the initial state vector $\mathbf{z}(t_0)$ to the energy of the output. The continuous-time LTV system of (2.36) and (2.37) is observable, if and only if the observability Gramian matrix $\mathbf{W}_o(t_0,t)$ is nonsingular [9, 14].

2.5.2 Discrete-Time LTV Systems

The state-space model of a discrete-time LTV system is represented by the following algebraic and difference equations:

$$\mathbf{x}_{k+1} = \mathbf{A}_k\mathbf{x}_k + \mathbf{B}_k\mathbf{u}_k, \tag{2.49}$$

$$\mathbf{y}_k = \mathbf{C}_k\mathbf{x}_k + \mathbf{D}_k\mathbf{u}_k. \tag{2.50}$$

Before proceeding with a discussion on the observability condition, we need to define the discrete-time *state-transition matrix*, $\mathbf{\Phi}(k,j)$, as the solution of the following difference equation:

$$\mathbf{\Phi}(k+1,j) = \mathbf{A}_k\mathbf{\Phi}(k,j) \tag{2.51}$$

with the initial condition:

$$\mathbf{\Phi}(j,j) = \mathbf{I}. \tag{2.52}$$

The reason that $\mathbf{\Phi}(k,j)$ is called the state-transition matrix is that it describes the dynamic behavior of the following autonomous system (a system with no input):

$$\mathbf{x}_{k+1} = \mathbf{A}_k\mathbf{x}_k \tag{2.53}$$

with $\mathbf{x}_k$ being obtained from

$$\mathbf{x}_k = \mathbf{\Phi}(k,0)\mathbf{x}_0. \tag{2.54}$$

Following a discussion on energy of the system output similar to the continuous-time case, we reach the following definition for the discrete-time *observability Gramian* matrix:

$$\mathbf{W}_o(j,k) = \sum_{i=j+1}^{k} \mathbf{\Phi}^T(i,j+1)\mathbf{C}_i^T\mathbf{C}_i\mathbf{\Phi}(i,j+1). \tag{2.55}$$

As before, the system (2.49) and (2.50) is observable, if and only if the observability Gramian matrix $\mathbf{W}_o(j,k)$ is full-rank (nonsingular) [9].

2.5.3 Discretization of LTV Systems

This section generalizes the method presented before for discretization of continuous-time LTI systems and describes how the continuous-time LTV system (2.36) and (2.37) can be discretized. Solving the differential equation in (2.36), we obtain:

$$\mathbf{x}(t) = \mathbf{\Phi}(t, t_0)\mathbf{x}(t_0) + \int_{t_0}^{t} \mathbf{\Phi}(t, \tau)\mathbf{B}(\tau)\mathbf{u}(\tau)\mathrm{d}\tau, \tag{2.56}$$

where $\mathbf{\Phi}(t, \tau)$ is the state-transition matrix as described in (2.40) and (2.41). Using zero-order-hold sampling results in a piecewise-constant input $\mathbf{u}(t)$, which remains constant at $\mathbf{u}(t_k)$ over the time interval $t \in [t_k, t_{k+1})$. Setting $t_0 = t_k$ and $t = t_{k+1}$, from (2.56), we will have:

$$\mathbf{x}(t_{k+1}) = \mathbf{\Phi}(t_{k+1}, t_k)\mathbf{x}(t_k) + \left(\int_{t_k}^{t_{k+1}} \mathbf{\Phi}(t_{k+1}, \tau)\mathbf{B}(\tau)\mathrm{d}\tau\right)\mathbf{u}(\tau). \tag{2.57}$$

Therefore, dynamics of the discrete-time equivalent of the continuous-time system in (2.36) and (2.37) will be governed by the following state-space model [19]:

$$\mathbf{x}(t_{k+1}) = \mathbf{\Phi}(t_{k+1}, t_k)\mathbf{x}(t_k) + \mathbf{\Gamma}(t_{k+1}, t_k)\mathbf{u}(t_k), \tag{2.58}$$

$$\mathbf{y}(t_k) = \mathbf{C}(t_k)\mathbf{x}(t_k) + \mathbf{D}(t_k)\mathbf{u}(t_k), \tag{2.59}$$

where

$$\mathbf{\Gamma}(t_{k+1}, t_k) = \int_{t_k}^{t_{k+1}} \mathbf{\Phi}(t_{k+1}, \tau)\mathbf{B}(\tau)\mathrm{d}\tau. \tag{2.60}$$

2.6 Observability of Nonlinear Systems

As mentioned before, observability is a global property for linear systems. However, for nonlinear systems, a weaker form of observability is defined, in which an initial state must be *distinguishable* only from its neighboring points. Two states $\mathbf{x}_a(t_0)$ and $\mathbf{x}_b(t_0)$ are *indistinguishable*, if their corresponding outputs are equal: $\mathbf{y}_a(t) = \mathbf{y}_b(t)$ for $t_0 < t < T$, where T is finite. If the set of states in the neighborhood of a particular initial state $\mathbf{x}(t_0)$ that are indistinguishable from it includes only $\mathbf{x}(t_0)$, then, the nonlinear system is said to be *weakly observable* at that initial state. A nonlinear system is called to be weakly observable if it is weakly observable at all $\mathbf{x}_0 \in \mathbb{R}^{n_x}$. If the state and the output trajectories of a weakly observable nonlinear system remain close to the corresponding initial conditions, then the system that satisfies this additional constraint is called *locally weakly observable* [13, 20].

There is another difference between linear and nonlinear systems regarding observability and that is the role of inputs in nonlinear observability. While inputs do not affect the observability of a linear system, in nonlinear systems, some initial states may be distinguishable for some inputs and indistinguishable for others. This leads to the concept of *uniform observability*, which is the property of a class of systems, for which initial states are distinguishable for all inputs [12, 20]. Furthermore, in nonlinear systems, distinction between time-invariant and time-varying systems is not critical because by adding time as an extra state such as $\mathbf{x}_{n+1} = t - t_0$, the latter can be converted to the former [21].

2.6.1 Continuous-Time Nonlinear Systems

The state-space model of a continuous-time nonlinear system is represented by the following system of nonlinear equations:

$$\dot{\mathbf{x}}(t) = \mathbf{f}(\mathbf{x}(t), \mathbf{u}(t)), \tag{2.61}$$

$$\mathbf{y}(t) = \mathbf{g}(\mathbf{x}(t), \mathbf{u}(t)), \tag{2.62}$$

where $\mathbf{f} : \mathbb{R}^{n_x} \times \mathbb{R}^{n_u} \to \mathbb{R}^{n_x}$ is the system function, and $\mathbf{g} : \mathbb{R}^{n_x} \times \mathbb{R}^{n_u} \to \mathbb{R}^{n_y}$ is the measurement function. It is common practice to deploy a control law that uses state feedback. In such cases, the control input $\mathbf{u}$ is itself a function of the state vector $\mathbf{x}$. Before proceeding, we need to recall the concept of *Lie derivative* from differential geometry [22, 23]. Assuming that $\mathbf{f}$ and $\mathbf{g}$ are smooth vector functions (they have derivatives of all orders or they can be differentiated infinitely many times), the Lie derivative of $\mathbf{g}_i$ (the ith element of $\mathbf{g}$) with respect to $\mathbf{f}$ is a scalar function defined as:

$$L_{\mathbf{f}}\, \mathbf{g}_i = \nabla \mathbf{g}_i\, \mathbf{f}, \tag{2.63}$$

where ∇ denotes the gradient with respect to $\mathbf{x}$. For simplicity, arguments of the functions have not been shown. Repeated Lie derivatives are defined as:

$$L_{\mathbf{f}}^{j}\, \mathbf{g}_i = L_{\mathbf{f}}(L_{\mathbf{f}}^{j-1}\mathbf{g}_i) = \nabla(L_{\mathbf{f}}^{j-1}\mathbf{g}_i)\, \mathbf{f}, \tag{2.64}$$

with

$$L_{\mathbf{f}}^{0}\, \mathbf{g}_i = \mathbf{g}_i. \tag{2.65}$$

Relevance of Lie derivatives to observability of nonlinear systems becomes clear, when we consider successive derivatives of the output vector as follows:

$$\begin{aligned}
\mathbf{y} &= \mathbf{g} = \begin{bmatrix} L_{\mathbf{f}}^{0}\mathbf{g}_1 & \cdots & L_{\mathbf{f}}^{0}\mathbf{g}_i & \cdots & L_{\mathbf{f}}^{0}\mathbf{g}_{n_y} \end{bmatrix}^T, \\
\dot{\mathbf{y}} &= \frac{\partial \mathbf{g}}{\partial \mathbf{x}}\, \dot{\mathbf{x}} = \begin{bmatrix} \nabla \mathbf{g}_1\, \mathbf{f} & \cdots & \nabla \mathbf{g}_i\, \mathbf{f} & \cdots & \nabla \mathbf{g}_{n_y}\, \mathbf{f} \end{bmatrix}^T \\
&= \begin{bmatrix} L_{\mathbf{f}}^{1}\mathbf{g}_1 & \cdots & L_{\mathbf{f}}^{1}\mathbf{g}_i & \cdots & L_{\mathbf{f}}^{1}\mathbf{g}_{n_y} \end{bmatrix}^T, \\
\ddot{\mathbf{y}} &= \frac{\partial}{\partial \mathbf{x}} \begin{bmatrix} L_{\mathbf{f}}\mathbf{g}_1 & \cdots & L_{\mathbf{f}}\mathbf{g}_i & \cdots & L_{\mathbf{f}}\mathbf{g}_{n_y} \end{bmatrix}^T \dot{\mathbf{x}}
\end{aligned}$$

$$= \left[L_{\mathbf{f}}^{2}\mathbf{g}_1 \ \cdots \ L_{\mathbf{f}}^{2}\mathbf{g}_i \ \cdots \ L_{\mathbf{f}}^{2}\mathbf{g}_{n_y} \right]^T,$$

$$\vdots = \vdots$$

$$\mathbf{y}^{(n-1)} = \frac{\partial}{\partial \mathbf{x}} \left[L_{\mathbf{f}}^{n-2}\mathbf{g}_1 \ \cdots \ L_{\mathbf{f}}^{n-2}\mathbf{g}_i \ \cdots \ L_{\mathbf{f}}^{n-2}\mathbf{g}_{n_y} \right]^T \dot{\mathbf{x}}$$

$$= \left[L_{\mathbf{f}}^{n-1}\mathbf{g}_1 \ \cdots \ L_{\mathbf{f}}^{n-1}\mathbf{g}_i \ \cdots \ L_{\mathbf{f}}^{n-1}\mathbf{g}_{n_y} \right]^T. \tag{2.66}$$

The aforementioned differential equations can be rewritten in the following compact form:

$$\mathcal{Y}(t) = \mathcal{L}\left(\mathbf{x}(t), \mathbf{u}(t)\right), \tag{2.67}$$

where

$$\mathcal{Y}(t) = \begin{bmatrix} \mathbf{y}(t) \\ \dot{\mathbf{y}}(t) \\ \vdots \\ \mathbf{y}^{(n-1)}(t) \end{bmatrix} \tag{2.68}$$

and

$$\mathcal{L}\left(\mathbf{x}(t), \mathbf{u}(t)\right) = \begin{bmatrix} \begin{bmatrix} L_{\mathbf{f}}^{0}\mathbf{g}_1 \\ \vdots \\ L_{\mathbf{f}}^{0}\mathbf{g}_{n_y} \end{bmatrix} \\ \begin{bmatrix} L_{\mathbf{f}}^{1}\mathbf{g}_1 \\ \vdots \\ L_{\mathbf{f}}^{1}\mathbf{g}_{n_y} \end{bmatrix} \\ \vdots \\ \begin{bmatrix} L_{\mathbf{f}}^{n-1}\mathbf{g}_1 \\ \vdots \\ L_{\mathbf{f}}^{n-1}\mathbf{g}_{n_y} \end{bmatrix} \end{bmatrix}. \tag{2.69}$$

The system of nonlinear differential equations in (2.67) can be linearized about an initial state $\mathbf{x}_0$ to develop a test for local observability of the nonlinear system (2.61) and (2.62) at this specific initial state, where the linearized test would be similar to the observability test for linear systems. Writing the Taylor series expansion of the function $\mathcal{L}\left(\mathbf{x}(t), \mathbf{u}(t)\right)$ about $\mathbf{x}_0$ and ignoring higher-order terms, we will have:

$$\mathcal{Y}(t) \approx \mathcal{L}\left(\mathbf{x}(t), \mathbf{u}(t)\right)\Big|_{\mathbf{x}(t_0)} + \nabla \mathcal{L}\left(\mathbf{x}(t), \mathbf{u}(t)\right)\Big|_{\mathbf{x}(t_0)} \left(\mathbf{x}(t) - \mathbf{x}(t_0)\right). \tag{2.70}$$

Using Cartan's formula:

$$\nabla\left(L_{\mathbf{f}}\,\mathbf{g}_i\right) = L_{\mathbf{f}}\left(\nabla \mathbf{g}_i\right), \tag{2.71}$$

we obtain:

$$\nabla\mathcal{L}(\mathbf{x}(t),\mathbf{u}(t)) = \begin{bmatrix} \begin{bmatrix} L_{\mathbf{f}}^{0}(\nabla\mathbf{g}_1) \\ \vdots \\ L_{\mathbf{f}}^{0}(\nabla\mathbf{g}_{n_y}) \end{bmatrix} \\ \begin{bmatrix} L_{\mathbf{f}}^{1}(\nabla\mathbf{g}_1) \\ \vdots \\ L_{\mathbf{f}}^{1}(\nabla\mathbf{g}_{n_y}) \end{bmatrix} \\ \vdots \\ \begin{bmatrix} L_{\mathbf{f}}^{n-1}(\nabla\mathbf{g}_1) \\ \vdots \\ L_{\mathbf{f}}^{n-1}(\nabla\mathbf{g}_{n_y}) \end{bmatrix} \end{bmatrix}. \tag{2.72}$$

Now, we can proceed with deriving the local observability test for nonlinear systems based on the aforementioned linearized system of equations. The nonlinear system in (2.61) and (2.62) is observable at $\mathbf{x}(t_0)$, if there exists a neighborhood of $\mathbf{x}(t_0)$ and an n_y-tuple of integers $(k_1, k_2, \dots, k_{n_y})$ called *observability indices* such that [9, 24]:

1. $\sum_{i=1}^{n_y} k_i = n_x$ for $k_1 \geq k_2 \geq \cdots \geq k_{n_y} \geq 0$.
2. The n_x row vectors of $\{L_f^{j-1}(\nabla\mathbf{g}_i) | \ i = 1, \dots, n_y; \ j = 1, \dots, k_i\}$ are linearly independent.

From the row vectors $L_f^{j-1}(\nabla\mathbf{g}_i)$, an observability matrix can be constructed for the continuous-time nonlinear system in (2.61) and (2.62) as follows:

$$\mathcal{O} = \left[L_f^{j-1}(\nabla\mathbf{g}_i)\right], \quad i = 1, \dots, n_y; \ j = 1, \dots, k_i. \tag{2.73}$$

If $\mathcal{O}$ is full-rank, then the nonlinear system in (2.61) and (2.62) is locally weakly observable. It is worth noting that the observability matrix for continuous-time linear systems (2.7) is a special case of the observability matrix for continuous-time nonlinear systems (2.73). In other words, if $\mathbf{f}$ and $\mathbf{g}$ are linear functions, then (2.73) will be reduced to (2.7) [9, 24].

The nonlinear system in (2.61) and (2.62) can be linearized about $\mathbf{x}(t_0)$. Using Taylor series expansion and ignoring higher-order terms, we will have the following linearized system:

$$\dot{\mathbf{x}}(t) \approx \mathbf{f}(\mathbf{x}(t),\mathbf{u}(t))\Big|_{\mathbf{x}(t_0)} + \frac{\partial\mathbf{f}(\mathbf{x}(t),\mathbf{u}(t))}{\partial\mathbf{x}}\Big|_{\mathbf{x}(t_0)} \left(\mathbf{x}(t) - \mathbf{x}(t_0)\right), \tag{2.74}$$

$$\mathbf{y}(t) \approx \mathbf{g}(\mathbf{x}(t),\mathbf{u}(t))\Big|_{\mathbf{x}(t_0)} + \frac{\partial\mathbf{g}(\mathbf{x}(t),\mathbf{u}(t))}{\partial\mathbf{x}}\Big|_{\mathbf{x}(t_0)} \left(\mathbf{x}(t) - \mathbf{x}(t_0)\right). \tag{2.75}$$

Then, the observability test for linear systems can be applied to the following linearized system matrices:

$$\mathbf{A}(t) = \frac{\partial \mathbf{f}\left(\mathbf{x}(t), \mathbf{u}(t)\right)}{\partial \mathbf{x}}\bigg|_{\mathbf{x}(t_0)}, \tag{2.76}$$

$$\mathbf{C}(t) = \frac{\partial \mathbf{g}\left(\mathbf{x}(t), \mathbf{u}(t)\right)}{\partial \mathbf{x}}\bigg|_{\mathbf{x}(t_0)}. \tag{2.77}$$

In this way, the nonlinear observability matrix in (2.73) can be approximated by the observability matrix, which is constructed using $\mathbf{A}(t)$ and $\mathbf{C}(t)$ in (2.76) and (2.77). Although this approach may seem simpler, observability of the linearized system may not imply the observability of the original nonlinear system [9].

2.6.2 Discrete-Time Nonlinear Systems

The state-space model of a discrete-time nonlinear system is represented by the following system of nonlinear equations:

$$\mathbf{x}_{k+1} = \mathbf{f}(\mathbf{x}_k, \mathbf{u}_k), \tag{2.78}$$

$$\mathbf{y}_k = \mathbf{g}(\mathbf{x}_k, \mathbf{u}_k), \tag{2.79}$$

where $\mathbf{f} : \mathbb{R}^{n_x} \times \mathbb{R}^{n_u} \to \mathbb{R}^{n_x}$ is the system function, and $\mathbf{g} : \mathbb{R}^{n_x} \times \mathbb{R}^{n_u} \to \mathbb{R}^{n_y}$ is the measurement function. Similar to the discrete-time linear case, starting from the initial cycle, system's output vectors at successive cycles till cycle $k = n - 1$ can be written based on the initial state $\mathbf{x}_0$ and input vectors $\mathbf{u}_{0:n-1}$ as follows:

$$\begin{aligned}
\mathbf{y}_0 &= \mathbf{g}(\mathbf{x}_0, \mathbf{u}_0), \\
\mathbf{y}_1 &= \mathbf{g}(\mathbf{x}_1, \mathbf{u}_1) = \mathbf{g}\left(\mathbf{f}(\mathbf{x}_0, \mathbf{u}_0), \mathbf{u}_1\right), \\
\mathbf{y}_2 &= \mathbf{g}(\mathbf{x}_2, \mathbf{u}_2) = \mathbf{g}\left(\mathbf{f}(\mathbf{x}_1, \mathbf{u}_1), \mathbf{u}_2\right) = \mathbf{g}\left(\mathbf{f}\left(\mathbf{f}(\mathbf{x}_0, \mathbf{u}_0), \mathbf{u}_1\right), \mathbf{u}_2\right), \\
\vdots \; &= \; \vdots \\
\mathbf{y}_{n-1} &= \mathbf{g}(\mathbf{x}_{n-1}, \mathbf{u}_{n-1}) = \mathbf{g}\left(\mathbf{f} \cdots \left(\mathbf{f}(\mathbf{x}_0, \mathbf{u}_0)\right), \ldots, \mathbf{u}_{n-1}\right).
\end{aligned} \tag{2.80}$$

Functional powers of the system function $\mathbf{f}$ can be used to simplify the notation in the aforementioned equations. Functional powers are obtained by repeated composition of a function with itself:

$$\mathbf{f}^n = \mathbf{f} \circ \mathbf{f}^{n-1} = \mathbf{f}^{n-1} \circ \mathbf{f}, \quad n \in \mathbb{N}, \tag{2.81}$$

where $\circ$ denotes the function-composition operator: $\mathbf{f} \circ \mathbf{f}(\mathbf{x}) = \mathbf{f}(\mathbf{f}(\mathbf{x}))$, and $\mathbf{f}^0$ is the identity map. Alternatively, the difference equations in (2.80) can be rewritten as:

$$\begin{bmatrix} \mathbf{y}_0 \\ \mathbf{y}_1 \\ \mathbf{y}_2 \\ \vdots \\ \mathbf{y}_{n-1} \end{bmatrix} = \begin{bmatrix} \mathbf{g} \\ \mathbf{g} \circ \mathbf{f} \\ \mathbf{g} \circ \mathbf{f}^2 \\ \vdots \\ \mathbf{g} \circ \mathbf{f}^{n-1} \end{bmatrix}_{\mathbf{x}_0, \mathbf{u}_{0:n-1}} = \begin{bmatrix} \mathbf{g} \circ \mathbf{f}^0 \\ \mathbf{g} \circ \mathbf{f}^1 \\ \mathbf{g} \circ \mathbf{f}^2 \\ \vdots \\ \mathbf{g} \circ \mathbf{f}^{n-1} \end{bmatrix}_{\mathbf{x}_0, \mathbf{u}_{0:n-1}}. \tag{2.82}$$

Similar to the continuous-time case, the system of nonlinear difference equations in (2.82) can be linearized about the initial state $\mathbf{x}_0$ based on the Taylor series expansion to develop a linearized test for weak local observability of the nonlinear discrete-time system (2.78) and (2.79). The nonlinear system in (2.78) and (2.79) is locally weakly observable at $\mathbf{x}_0$, if there exist a neighborhood of $\mathbf{x}_0$ and an n_y-tuple of integers $(k_1, k_2, \dots, k_{n_y})$ such that [9, 25]:

1. $\sum_{i=1}^{n_y} k_i = n_x$ for $k_1 \geq k_2 \geq \cdots \geq k_{n_y} \geq 0$.
2. The following observability matrix is full rank:

$$\mathcal{O} = \begin{bmatrix} \mathcal{O}_1 \\ \vdots \\ \mathcal{O}_{n_y} \end{bmatrix}, \tag{2.83}$$

where

$$\mathcal{O}_i = \frac{\partial}{\partial \mathbf{x}} \begin{bmatrix} \mathbf{g}_i(\mathbf{x}_0) \\ \mathbf{g}_i\left(\mathbf{f}(\mathbf{x}_0)\right) \\ \vdots \\ \mathbf{g}_i\left(\mathbf{f}^{k_i-1}(\mathbf{x}_0)\right) \end{bmatrix}. \tag{2.84}$$

The observability matrix for discrete-time linear systems (2.22) is a special case of the observability matrix for discrete-time nonlinear systems (2.83). In other words, if $\mathbf{f}$ and $\mathbf{g}$ are linear functions, then (2.83) will be reduced to (2.22) [9, 25].

2.6.3 Discretization of Nonlinear Systems

Unlike linear systems, there is not a general functional representation for discrete-time equivalents of continuous-time nonlinear systems. One approach is to find a discrete-time equivalent for the perturbed state-space model of the nonlinear system under study [19]. In this approach, first, we need to linearize the nonlinear system in (2.61) and (2.62) about nominal values of state and input vectors, denoted by $\bar{\mathbf{x}}(t)$ and $\bar{\mathbf{u}}(t)$, respectively. The perturbation terms, denoted by $\delta\mathbf{x}(t)$, $\delta\mathbf{u}(t)$, and $\delta\mathbf{y}(t)$, are defined as the difference between the actual and the nominal values of state, input, and output vectors, respectively:

$$\delta\mathbf{x}(t) = \mathbf{x}(t) - \bar{\mathbf{x}}(t), \tag{2.85}$$

$$\delta\mathbf{u}(t) = \mathbf{u}(t) - \bar{\mathbf{u}}(t), \tag{2.86}$$

$$\delta\mathbf{y}(t) = \mathbf{y}(t) - \bar{\mathbf{y}}(t). \tag{2.87}$$

Since input is usually derived from a feedback control law, it may be a function of the state, $\mathbf{u}(\mathbf{x}(t))$. In such cases, a difference between the actual and the nominal values of the state (a perturbation in the state) leads to a difference between the actual and the nominal values of the input (a perturbation in the input), and in effect therefore, $\delta\mathbf{u}(t) \neq 0$. Otherwise, $\delta\mathbf{u}(t)$ can be zero. Using the Taylor series expansion and neglecting the higher-order terms, we obtain the following perturbation state-space model:

$$\delta\dot{\mathbf{x}}(t) = \mathbf{f}_{\mathbf{x}}\left(\overline{\mathbf{x}}(t), \overline{\mathbf{u}}(t)\right)\delta\mathbf{x}(t) + \mathbf{f}_{\mathbf{u}}\left(\overline{\mathbf{x}}(t), \overline{\mathbf{u}}(t)\right)\delta\mathbf{u}(t), \tag{2.88}$$

$$\delta\mathbf{y}(t) = \mathbf{g}_{\mathbf{x}}\left(\overline{\mathbf{x}}(t), \overline{\mathbf{u}}(t)\right)\delta\mathbf{x}(t) + \mathbf{g}_{\mathbf{u}}\left(\overline{\mathbf{x}}(t), \overline{\mathbf{u}}(t)\right)\delta\mathbf{u}(t), \tag{2.89}$$

where $\mathbf{f}_{\mathbf{x}}$ and $\mathbf{f}_{\mathbf{u}}$, respectively, denote the Jacobian matrices obtained by taking the derivatives of $\mathbf{f}$ with respect to $\mathbf{x}$ and $\mathbf{u}$. Similarly, $\mathbf{g}_{\mathbf{x}}$ and $\mathbf{g}_{\mathbf{u}}$ are Jacobians of $\mathbf{g}$ with respect to $\mathbf{x}$ and $\mathbf{u}$.

Now, the continuous-time state-space model in (2.88) and (2.89) can be treated as an LTV system. The discrete-time equivalent of (2.88) and (2.89) is obtained as:

$$\delta\mathbf{x}(t_{k+1}) = \overline{\mathbf{\Phi}}(t_{k+1}, t_k)\delta\mathbf{x}(t_k) + \overline{\mathbf{\Gamma}}(t_{k+1}, t_k)\delta\mathbf{u}(t_k), \tag{2.90}$$

$$\delta\mathbf{y}(t_k) = \mathbf{g}_{\mathbf{x}}\left(\overline{\mathbf{x}}(t_k), \overline{\mathbf{u}}(t_k)\right)\delta\mathbf{x}(t_k) + \mathbf{g}_{\mathbf{u}}\left(\overline{\mathbf{x}}(t_k), \overline{\mathbf{u}}(t_k)\right)\delta\mathbf{u}(t_k), \tag{2.91}$$

where $\overline{\mathbf{\Phi}}(t_{k+1}, t_k)$ is the solution to

$$\frac{\mathrm{d}\overline{\mathbf{\Phi}}(t, t_0)}{\mathrm{d}t} = \mathbf{f}_{\mathbf{x}}\left(\overline{\mathbf{x}}(t), \overline{\mathbf{u}}(t)\right)\overline{\mathbf{\Phi}}(t, t_0), \tag{2.92}$$

with the initial condition:

$$\overline{\mathbf{\Phi}}(t_0, t_0) = \mathbf{I}, \tag{2.93}$$

when we set $t_0 = t_k$ and $t = t_{k+1}$. As before, $\overline{\mathbf{\Gamma}}(t_{k+1}, t_k)$ is given by:

$$\overline{\mathbf{\Gamma}}(t_{k+1}, t_k) = \int_{t_k}^{t_{k+1}} \overline{\mathbf{\Phi}}(t_{k+1}, \tau)\mathbf{f}_{\mathbf{u}}\left(\overline{\mathbf{x}}(\tau), \overline{\mathbf{u}}(\tau)\right)\mathrm{d}\tau. \tag{2.94}$$

So far, this chapter has been focused on studying the observability of deterministic systems. Section 2.7 discusses the observability of stochastic systems.

2.7 Observability of Stochastic Systems

Before proceeding with defining observability for stochastic systems, we need to recall a few concepts from information theory [26]:

Definition 2.3 ***Entropy*** *is a measure of our uncertainty about an event in Shannon's information theory. Specifically, the entropy of a discrete random vector* $\mathbf{X}$ *with* *alphabet $\mathcal{X}$ is defined as:*

$$H(\mathbf{X}) = -\sum_{\mathbf{x}\in\mathcal{X}} p(\mathbf{x}) \log p(\mathbf{x}), \tag{2.95}$$

and correspondingly, for a continuous random vector $\mathbf{X}$*, we have:*

$$H(\mathbf{X}) = -\int_{\mathbb{R}} \log p(\mathbf{x}) dp(\mathbf{x}). \tag{2.96}$$

Entropy can also be interpreted as the expected value of the term $1/\log p(\mathbf{x})$:

$$H(\mathbf{X}) = \mathbb{E}\left[\frac{1}{\log p(\mathbf{x})}\right], \tag{2.97}$$

where $\mathbb{E}$ is the expectation operator and $p(\mathbf{x})$ is the probability density function (PDF) of $\mathbf{x}$. Definition of Shannon's entropy, H, shows that it is a function of the corresponding PDF. It will be insightful to examine the way that this information measure is affected by the shape of the PDF. A relatively broad and flat PDF, which is associated with lack of predictability, has high entropy. On the other hand, if the PDF is relatively narrow and has sharp slopes around a specific value of $\mathbf{x}$, which is associated with bias toward that particular value of $\mathbf{x}$, then the PDF has low entropy. A rearrangement of the tuples $(\mathbf{x}_i, p(\mathbf{x}_i))$ may change the shape of the PDF curve significantly but it does not affect the value of the summation or integral in (2.95) or (2.96), because summation and integration can be calculated in any order. Since H is not affected by local changes in the PDF curve, it can be considered as a global measure of the behavior of the corresponding PDF [27].

Definition 2.4 ***Joint entropy*** *is defined for a pair of random vectors* $(\mathbf{X}, \mathbf{Y})$ *based on their joint distribution* $p(\mathbf{x}, \mathbf{y})$ *as:*

$$H(\mathbf{X}, \mathbf{Y}) = \mathbb{E}\left[\frac{1}{\log p(\mathbf{x}, \mathbf{y})}\right]. \tag{2.98}$$

Definition 2.5 ***Conditional entropy*** *is defined as the entropy of a random variable (state vector) conditional on the knowledge of another random variable (measurement vector):*

$$H(\mathbf{X}|\mathbf{Y}) = H(\mathbf{X}, \mathbf{Y}) - H(\mathbf{Y}). \tag{2.99}$$

It can also be expressed as:

$$H(\mathbf{X}|\mathbf{Y}) = \mathbb{E}\left[\frac{1}{\log p(\mathbf{x}|\mathbf{y})}\right]. \tag{2.100}$$

Definition 2.6 ***Mutual information*** *between two random variables is a measure of the amount of information that one contains about the other. It can also be interpreted as the reduction in the uncertainty about one random variable due to knowledge about the other one. Mathematically it is defined as:*

$$I(\mathbf{X};\mathbf{Y}) = H(\mathbf{X}) - H(\mathbf{X}|\mathbf{Y}). \tag{2.101}$$

Substituting for $H(\mathbf{X}|\mathbf{Y})$ from (2.99) into the aforementioned equation, we will have:

$$I(\mathbf{X};\mathbf{Y}) = H(\mathbf{X}) + H(\mathbf{Y}) - H(\mathbf{X},\mathbf{Y}). \tag{2.102}$$

Therefore, mutual information is symmetric with respect to $\mathbf{X}$ and $\mathbf{Y}$. It can also be viewed as a measure of dependence between the two random vectors. Mutual information is nonnegative; being equal to zero, if and only if $\mathbf{X}$ and $\mathbf{Y}$ are independent. The notion of observability for stochastic systems can be defined based on the concept of mutual information.

Definition 2.7 ***(Stochastic observability)*** *The random vector $\mathbf{X}$ (state) is unobservable from the random vector $\mathbf{Y}$ (measurement), if they are independent or equivalently $I(\mathbf{X};\mathbf{Y}) = 0$. Otherwise, $\mathbf{X}$ is observable from $\mathbf{Y}$.*

Since mutual information is nonnegative, (2.101) leads to the following conclusion: if either $H(\mathbf{X}) = 0$ or $H(\mathbf{X}|\mathbf{Y}) < H(\mathbf{X})$, then $\mathbf{X}$ is observable from $\mathbf{Y}$ [28].

2.8 Degree of Observability

Instead of considering the notion of observability as a yes/no question, it will be helpful in practice to pose the question of how observable a system may be [29]. Knowing the answer to this question, we can select the best set of variables, which can be directly measured, as outputs to improve observability [30]. With this in mind and building on Section 2.7, mutual information can be used as a measure for the degree of observability [31].

An alternative approach aiming at providing insight into the observability of the system of interest in filtering applications uses eigenvalues of the estimation error covariance matrix. The largest eigenvalue of the covariance matrix is the variance of the state or a function of states, which is poorly observable. Hence, its corresponding eigenvector provides the direction of poor observability. On the other hand, states or functions of states that are highly observable are associated with smaller eigenvalues, where their corresponding eigenvectors provide the directions of good observability [30].

A deterministic system is either observable or unobservable, but for stochastic systems, the degree of observability can be defined as [32]:

$$\rho(\mathbf{X}, \mathbf{Y}) = \frac{I(\mathbf{X}; \mathbf{Y})}{\max\left(H(\mathbf{X}), H(\mathbf{Y})\right)}, \tag{2.103}$$

which is a time-dependent non-decreasing function that varies between 0 and 1. Before starting the measurement process, $H(\mathbf{X}|\mathbf{Y}) = H(\mathbf{X})$ and therefore, $I(\mathbf{X}; \mathbf{Y}) = 0$, which makes $\rho(\mathbf{X}, \mathbf{Y}) = 0$. As more measurements become available, $H(\mathbf{X}|\mathbf{Y})$ may reduce and therefore, $I(\mathbf{X}; \mathbf{Y})$ may increase, which leads to the growth of $\rho(\mathbf{X}, \mathbf{Y})$ up to 1 [33].

2.9 Invertibility

Observability can be studied regarding the characteristics of the information set, $\mathbf{I}_k = \{\mathbf{U}_k, \mathbf{Y}_k\} = \{\mathbf{u}_{0:k}, \mathbf{y}_{0:k}\}$ as well as the invertibility property of the corresponding maps. From this viewpoint, *information sets* can be divided into three categories [34]:

- *Instantaneously invertible*: In this case, full information can be recovered and it would not be necessary to use a filter.
- *Asymptotically invertible*: A filter can provide a solution that converges to the true values of states. In this case, the deployed filter may recover information with fewer observables compared with the instantaneously invertible case.
- *Noninvertible*: In this case, even a filter cannot replicate full information.

For *systems with unknown inputs*, a subset of the inputs to the system may be unknown. For such systems, assuming that the initial state is known, it is often desired to reconstruct the unknown inputs through *dynamic system inversion*. Since a number of faults and attacks can be modeled as unknown inputs to the system, the concept of invertibility is important in fault diagnosis and cybersecurity. To be more precise, the following definitions are recalled from [35].

Definition 2.8 ***(ℓ-delay inverse)*** *The system described by the state-space model (2.18) and (2.19) has an ℓ-delay inverse, if its input at time step k, $\mathbf{u}_k$, can be uniquely recovered from the outputs up to time step $k + \ell$, $\mathbf{y}_{k:k+\ell}$, for some nonnegative integer ℓ, assuming that the initial state $\mathbf{x}_0$ is known.*

Definition 2.9 ***(Invertibility)*** *The system with the state-space model (2.18) and (2.19) is invertible, if it has an ℓ-delay inverse for some finite ℓ, where the inherent delay of the system is the least integer ℓ for which an ℓ-delay inverse exists.*

The topic of invertibility will be further discussed in Section 3.5, which covers the *unknown input observers*.

2.10 Concluding Remarks

Observability is a key property of dynamic systems, which deals with the question of whether the state of a system can be uniquely determined in a finite time interval from inputs and measured outputs provided that the system dynamic model is known:

- For linear systems, observability is a global property and there is a universal definition for it. An LTI (LTV) system is observable, if and only if its observability matrix (observability Gramian matrix) is full-rank, and the state can be reconstructed from inputs and measured outputs using the inverse of the observability matrix (observability Gramian matrix).
- For nonlinear systems, a unique definition of observability does not exist and locally weak observability is considered in a neighborhood of the initial state. A nonlinear system is locally weakly observable if its Jacobian matrix about that particular state has full rank. Then, the initial state can be reconstructed from inputs and measured outputs using the inverse of the Jacobian.
- For stochastic systems, mutual information between states and outputs can be used as a measure for the degree of observability, which helps to reconfigure the sensory (perceptual) part of the system in a way to improve the observability [33].

3

Observers

3.1 Introduction

State of a system refers to the minimum amount of information, which is required at the current time instant to uniquely describe the dynamic behavior of the system in the future, given the inputs and parameters. *Parameters* reflect the physical properties used in a model as a description of the system under study. *Inputs* or *actions* are manipulated variables that act on the dynamic system as forcing functions. *Outputs* or *observations* are variables that can be directly measured. In many practical situations, the full state of a dynamic system cannot be directly measured. Hence, the current state of the system must be reconstructed from the known inputs and the measured outputs. State estimation is deployed as a process to determine the state from inputs and outputs given a dynamic model of the system. For linear systems, reconstruction of system state can be performed by deploying the well-established optimal linear estimation theory. However, for nonlinear systems, we need to settle for sub-optimal methods, which are computationally tractable and can be implemented in real-time applications. Such methods rely on simplifications of or approximations to the underlying nonlinear system in the presence of uncertainty [9]. In this chapter, starting from deterministic linear state estimation, the stage will be set for nonlinear methods, and then, unknown inputs such as faults and attacks will be discussed. Since state estimators are usually implemented using digital processors, emphasis will be put on methods in which measurements are available at discrete sampling times. There are two classes of observers regarding the order of the observer as a dynamic system and the order of the system under study whose state is to be estimated:

- *Full-order observers*: The order of the observer is equal to the order of the main system.
- *Reduced-order observers*: The order of the observer is less than the order of the main system.

Nonlinear Filters: Theory and Applications, First Edition. Peyman Setoodeh, Saeid Habibi, and Simon Haykin.

In the following, the symbol " ^ " denotes the estimated variable. Since such variable is determined based on the measured outputs, for clarification, two time indices will be used: one for the estimate and the other for the measurement. For discrete-time systems, $\hat{\mathbf{x}}_{k_1|k_2}$ denotes the estimate of the state, $\mathbf{x}$, at time instant k_1, given output measurements up to time instant k_2. Similarly, for continuous-time systems, $\hat{\mathbf{x}}(t_1 | t_2)$ denotes the estimate of the state, $\mathbf{x}$, at time instant t_1, given output measurements up to time instant t_2. The estimate $\hat{\mathbf{x}}_{k_1|k_2}$ is

- a *smoothed* state estimate, if $k_1 < k_2$,
- a *filtered* state estimate, if $k_1 = k_2$, and
- a *predicted* state estimate, if $k_1 > k_2$.

For control applications, usually the control law uses the filtered state estimate, which is the current estimate based on all available measurements. The predicted estimate relies on the state transition model to extrapolate the filtered estimate into the future. Such estimates are usually used for computing the objective or cost function in model predictive control. Smoothed estimates are computed in an offline manner based on both past and future measurements. They provide a more accurate estimate than the filtered ones. They are usually used for process analysis and diagnosis [9].

3.2 Luenberger Observer

A deterministic discrete-time linear system is described by the following state-space model:

$$\mathbf{x}_{k+1} = \mathbf{A}\mathbf{x}_k + \mathbf{B}\mathbf{u}_k, \tag{3.1}$$

$$\mathbf{y}_k = \mathbf{C}\mathbf{x}_k + \mathbf{D}\mathbf{u}_k, \tag{3.2}$$

where $\mathbf{x}_k \in \mathbb{R}^{n_x}$, $\mathbf{u}_k \in \mathbb{R}^{n_u}$, and $\mathbf{y}_k \in \mathbb{R}^{n_y}$ denote state, input, and output vectors, respectively, and $(\mathbf{A}, \mathbf{B}, \mathbf{C}, \mathbf{D})$ are the model parameters, which are matrices with appropriate dimensions. *Luenberger observer* is a *sequential* or *recursive* state estimator, which needs the information of only the previous sample time to reconstruct the state as:

$$\hat{\mathbf{x}}_{k|k} = \hat{\mathbf{x}}_{k|k-1} + \mathbf{L}\left(\mathbf{y}_k - \mathbf{C}\hat{\mathbf{x}}_{k|k-1} - \mathbf{D}\mathbf{u}_k\right), \tag{3.3}$$

where $\mathbf{L}$ is a constant gain matrix, which is determined in a way that the closed-loop system achieves some desired performance criteria. The predicted estimate, $\hat{\mathbf{x}}_{k|k-1}$, is obtained from (3.1) as:

$$\hat{\mathbf{x}}_{k|k-1} = \mathbf{A}\hat{\mathbf{x}}_{k-1|k-1} + \mathbf{B}\mathbf{u}_{k-1}, \tag{3.4}$$

with the initial condition $\hat{\mathbf{x}}_{0|0} = \mathbf{x}_0$.

The dynamic response of the state reconstruction error, $\mathbf{e}_k$, from an initial nonzero value is governed by:

$$\mathbf{e}_{k+1} = (\mathbf{A} - \mathbf{ALC})\,\mathbf{e}_k. \tag{3.5}$$

The gain matrix, **L**, is determined by choosing the closed-loop observer poles, which are the eigenvalues of $(\mathbf{A} - \mathbf{ALC})$. Using the *pole placement* method to design the Luenberger observer requires the system observability. In order to have a stable observer, moduli of the eigenvalues of $(\mathbf{A} - \mathbf{ALC})$ must be strictly less than one. A *deadbeat* observer is obtained, if all the eigenvalues are zero. The Luenberger observer is designed based on a compromise between rapid decay of the reconstruction error and sensitivity to modeling error and measurement noise [9]. Section 3.3 provides an extension of the Luenberger observer for nonlinear systems.

3.3 Extended Luenberger-Type Observer

A class of nonlinear observers is designed based on observer error linearization. State estimation by such observers involves multiplication of a matrix gain by the difference between predicted and measured outputs. After designing the observer (selecting the observer gain), this calculation can be performed fairly quickly, which is an advantage from the computational complexity perspective. However, the domain of applicability of such observers may be restricted to a specific class of nonlinear systems. A discrete-time nonlinear Luenberger-type observer has been proposed in [36]. A deterministic discrete-time nonlinear system is described by the following state-space model:

$$\mathbf{x}_{k+1} = \mathbf{f}(\mathbf{x}_k, \mathbf{u}_k), \tag{3.6}$$

$$\mathbf{y}_k = \mathbf{g}(\mathbf{x}_k, \mathbf{u}_k), \tag{3.7}$$

where **f** and **g** are nonlinear vector functions. At time instant k, let $\mathbf{u}_{k-n_x+1:k}$ denote the sequence of the past n_x inputs:

$$\begin{bmatrix} \mathbf{u}_{k-n_x+1} \\ \mathbf{u}_{k-n_x+2} \\ \vdots \\ \mathbf{u}_{k-2} \\ \mathbf{u}_{k-1} \\ \mathbf{u}_k \end{bmatrix}, \tag{3.8}$$

where n_x is the dimension of the state vector. Given an initial state and a sequence of previous inputs, a series of predicted outputs can be generated using the state equation (3.6) as:

$$\begin{bmatrix} \vdots \\ \hat{\mathbf{y}}_{k|k-3} \\ \hat{\mathbf{y}}_{k|k-2} \\ \hat{\mathbf{y}}_{k|k-1} \end{bmatrix} = \begin{bmatrix} \vdots \\ \mathbf{g}\left(\mathbf{f}\left(\mathbf{f}(\mathbf{x}_{k-2}, \mathbf{u}_{k-2}), \mathbf{u}_{k-1}\right), \mathbf{u}_k\right) \\ \mathbf{g}\left(\mathbf{f}(\mathbf{x}_{k-1}, \mathbf{u}_{k-1}), \mathbf{u}_k\right) \\ \mathbf{g}(\mathbf{x}_k, \mathbf{u}_k) \end{bmatrix}. \tag{3.9}$$

Let $\mathbf{f}^i$ denote the ith composite of the system function $\mathbf{f}$ as:

$$\mathbf{f}^i = \underbrace{\mathbf{f} \circ \mathbf{f} \circ \cdots \circ \mathbf{f}}_{i \text{ times}}. \tag{3.10}$$

Given an initial state and the sequence $\mathbf{u}_{k-n_x+1:k}$, the corresponding sequence of predicted outputs is obtained as:

$$\begin{aligned}\boldsymbol{\Phi}(\mathbf{x}_{k-n_x+1:k}, \mathbf{u}_{k-n_x+1:k}) &= \begin{bmatrix} \hat{\mathbf{y}}_{k|k-n_x} \\ \hat{\mathbf{y}}_{k|k-n_x+1} \\ \vdots \\ \hat{\mathbf{y}}_{k|k-3} \\ \hat{\mathbf{y}}_{k|k-2} \\ \hat{\mathbf{y}}_{k|k-1} \end{bmatrix} \\ &= \begin{bmatrix} \mathbf{g}\left(\mathbf{f}^{n_x-1}(\mathbf{x}_{k-n_x+1}, \mathbf{u}_{k-n_x+1:k-1}), \mathbf{u}_k\right) \\ \mathbf{g}\left(\mathbf{f}^{n_x-2}(\mathbf{x}_{k-n_x+2}, \mathbf{u}_{k-n_x+2:k-1}), \mathbf{u}_k\right) \\ \vdots \\ \mathbf{g}\left(\mathbf{f}^2(\mathbf{x}_{k-2}, \mathbf{u}_{k-2:k-1}), \mathbf{u}_k\right) \\ \mathbf{g}\left(\mathbf{f}(\mathbf{x}_{k-1}, \mathbf{u}_{k-1}), \mathbf{u}_k\right) \\ \mathbf{g}(\mathbf{x}_k, \mathbf{u}_k) \end{bmatrix}. \end{aligned} \tag{3.11}$$

To simplify the notation, let us define $\mathbf{F}$ such that

$$\mathbf{g}\left(\mathbf{F}^i(\mathbf{x}_{k-i}, \mathbf{u}_{k-i:k})\right) = \mathbf{g}\left(\mathbf{f}^i(\mathbf{x}_{k-i}, \mathbf{u}_{k-i:k-1}), \mathbf{u}_k\right), \tag{3.12}$$

then we have:

$$\boldsymbol{\Phi}(\mathbf{x}_{k-n_x+1:k}, \mathbf{u}_{k-n_x+1:k}) = \begin{bmatrix} \mathbf{g}\left(\mathbf{F}^{n_x-1}(\mathbf{x}_{k-n_x+1}, \mathbf{u}_{k-n_x+1:k})\right) \\ \mathbf{g}\left(\mathbf{F}^{n_x-2}(\mathbf{x}_{k-n_x+2}, \mathbf{u}_{k-n_x+2:k})\right) \\ \vdots \\ \mathbf{g}\left(\mathbf{F}^2(\mathbf{x}_{k-2}, \mathbf{u}_{k-2:k})\right) \\ \mathbf{g}\left(\mathbf{F}(\mathbf{x}_{k-1}, \mathbf{u}_{k-1:k})\right) \\ \mathbf{g}(\mathbf{x}_k, \mathbf{u}_k) \end{bmatrix}.$$

Defining

$$\mathbf{z}_{k-n_x+i|k} = \mathbf{F}^{i-1}(\mathbf{z}_{k-n_x+1|k}, \mathbf{u}_{k-n_x+i-1:i-1}), \tag{3.13}$$

the filtered state estimate is computed from the following discrete-time system [9]:

$$\mathbf{z}_{k-n_x+2|k+1} = \mathbf{z}_{k-n_x+2|k} + \mathcal{O}^{-1}(\mathbf{z}_{k-n_x+1|k}, \mathbf{u}_{k-n_x+1:k-1}) \times \left(\mathbf{B}\left(\mathbf{y}_{k+1} - \mathbf{g}(\mathbf{z}_{k+1|k})\right) + \mathbf{L}\left(\mathbf{y}_{k-n_x+1} - \mathbf{g}(\mathbf{z}_{k-n_x+1|k})\right)\right), \quad (3.14)$$

$$\hat{x}_{k|k} = \mathbf{F}^{n_x-1}(\mathbf{z}_{k-n_x+1|k}, \mathbf{u}_{k-n_x+1:k}), \quad (3.15)$$

where $\mathbf{L}$ is the observer gain, $\mathbf{B} = [1\ 0\ \cdots\ 0]^T$, and $\mathcal{O}$ is the discrete nonlinear observability matrix, which is computed at sample time k as:

$$\mathcal{O}(\mathbf{x}, \mathbf{u}_{k-n_x+2:k}) = \frac{\partial \mathbf{\Phi}(\mathbf{x}, \mathbf{u}_{k-n_x+2:k})}{\partial \mathbf{x}}. \quad (3.16)$$

The observer gain is determined in a way to guarantee local stability of the perturbed linear system for the reconstruction error in $\mathbf{z}$:

$$\mathbf{e}_{k-n_x+2|k+1} = (\mathbf{A} - \mathbf{LC})\mathbf{e}_{k-n_x+1|k} + O\left(\| \mathbf{e}_{k-n_x+1|k} \|^2\right), \quad (3.17)$$

where $\mathbf{C} = \begin{bmatrix} 0 & \cdots & 0 & 1 \end{bmatrix}$ and

$$\mathbf{A} = \begin{bmatrix} 0 & 0 & \cdots & 0 & 0 \\ 1 & 0 & \cdots & 0 & 0 \\ 0 & 1 & \cdots & 0 & 0 \\ \vdots & \vdots & \ddots & \vdots & \vdots \\ 0 & 0 & \cdots & 1 & 0 \end{bmatrix}.$$

This approach for designing nonlinear observers is applicable to systems that are observable for every bounded input, $\| \mathbf{u}_k \| \leq M$, with $\mathbf{g}(\mathbf{x}, .)$ and $\mathbf{F}^n\left(\mathbf{\Phi}^{-1}(\mathbf{x}, .)\right)$ being uniformly *Lipschitz* continuous functions of the state:

$$\sup_{\|\mathbf{u}_k\| \leq M} \| \mathbf{g}(\mathbf{x}_1, .) - \mathbf{g}(\mathbf{x}_2, .) \| \leq L_1 \| \mathbf{x}_1 - \mathbf{x}_2 \|, \quad (3.18)$$

$$\sup_{\|\mathbf{u}_k\| \leq M} \| \mathbf{F}^n\left(\mathbf{\Phi}^{-1}(\mathbf{x}_1, .)\right) - \mathbf{F}^n\left(\mathbf{\Phi}^{-1}(\mathbf{x}_2, .)\right) \| \leq L_2 \| \mathbf{x}_1 - \mathbf{x}_2 \|, \quad (3.19)$$

where L_1 and L_2 denote the corresponding Lipschitz constants. However, convergence is guaranteed only for a neighborhood around the true state [36].

3.4 Sliding-Mode Observer

The equivalent control approach allows for designing the discrete-time sliding-mode realization of a reduced-order asymptotic observer [37]. Let us consider the following discrete-time state-space model:

$$\mathbf{x}_{k+1} = \mathbf{A}\mathbf{x}_k + \mathbf{B}\mathbf{u}_k, \quad (3.20)$$

$$\mathbf{y}_k = \mathbf{C}\mathbf{x}_k, \quad (3.21)$$

where $\mathbf{x}_k \in \mathbb{R}^{n_x}$, $\mathbf{u}_k \in \mathbb{R}^{n_u}$, and $\mathbf{y}_k \in \mathbb{R}^{n_y}$. It is assumed that the pair $(\mathbf{A}, \mathbf{C})$ is observable, and $\mathbf{C}$ is full rank. The goal is to reconstruct the state vector from the input and the output vectors using the discrete-time sliding-mode framework. Using a nonsingular transformation matrix, $\mathbf{T}$, the state vector is transformed into a partitioned form:

$$\mathbf{T}\mathbf{x}_k = \begin{bmatrix} \mathbf{z}_k^u \\ \mathbf{z}_k^l \end{bmatrix}, \tag{3.22}$$

such that the upper partition has the identity relationship with the output vector:

$$\mathbf{y}_k = \mathbf{z}_k^u. \tag{3.23}$$

Then, the system given in (3.20) and (3.21) is transformed to the following canonical form:

$$\begin{bmatrix} \mathbf{y}_{k+1} \\ \mathbf{z}_{k+1}^l \end{bmatrix} = \begin{bmatrix} \mathbf{\Phi}_{11} & \mathbf{\Phi}_{12} \\ \mathbf{\Phi}_{21} & \mathbf{\Phi}_{22} \end{bmatrix} \begin{bmatrix} \mathbf{y}_k \\ \mathbf{z}_k^l \end{bmatrix} + \begin{bmatrix} \mathbf{G}_1 \\ \mathbf{G}_2 \end{bmatrix} \mathbf{u}_k, \tag{3.24}$$

where

$$\mathbf{\Phi} = \mathbf{T}\mathbf{A}\mathbf{T}^{-1} = \begin{bmatrix} \mathbf{\Phi}_{11} & \mathbf{\Phi}_{12} \\ \mathbf{\Phi}_{21} & \mathbf{\Phi}_{22} \end{bmatrix}, \tag{3.25}$$

$$\mathbf{G} = \mathbf{T}\mathbf{B} = \begin{bmatrix} \mathbf{G}_1 \\ \mathbf{G}_2 \end{bmatrix}. \tag{3.26}$$

The corresponding sliding-mode observer is obtained as:

$$\begin{bmatrix} \hat{\mathbf{y}}_{k+1} \\ \hat{\mathbf{z}}_{k+1}^l \end{bmatrix} = \begin{bmatrix} \mathbf{\Phi}_{11} & \mathbf{\Phi}_{12} \\ \mathbf{\Phi}_{21} & \mathbf{\Phi}_{22} \end{bmatrix} \begin{bmatrix} \hat{\mathbf{y}}_k \\ \hat{\mathbf{z}}_k^l \end{bmatrix} + \begin{bmatrix} \mathbf{G}_1 \\ \mathbf{G}_2 \end{bmatrix} \mathbf{u}_k + \begin{bmatrix} -\mathbf{v}_k \\ \mathbf{L}\mathbf{v}_k \end{bmatrix}, \tag{3.27}$$

where $\mathbf{v}_k \in \mathbb{R}^{n_y}$, and $\mathbf{L} \in \mathbb{R}^{(n_x - n_y) \times n_y}$ is the observer gain. Defining the estimation errors as:

$$\begin{bmatrix} \mathbf{e}_{\mathbf{y}_k} \\ \mathbf{e}_{\mathbf{z}_k^l} \end{bmatrix} = \begin{bmatrix} \mathbf{y}_k - \hat{\mathbf{y}}_k \\ \mathbf{z}_k^l - \hat{\mathbf{z}}_k^l \end{bmatrix}, \tag{3.28}$$

and subtracting (3.27) from (3.24), we obtain the following error dynamics:

$$\begin{bmatrix} \mathbf{e}_{\mathbf{y}_{k+1}} \\ \mathbf{e}_{\mathbf{z}_{k+1}^l} \end{bmatrix} = \begin{bmatrix} \mathbf{\Phi}_{11} & \mathbf{\Phi}_{12} \\ \mathbf{\Phi}_{21} & \mathbf{\Phi}_{22} \end{bmatrix} \begin{bmatrix} \mathbf{e}_{\mathbf{y}_k} \\ \mathbf{e}_{\mathbf{z}_k^l} \end{bmatrix} + \begin{bmatrix} \mathbf{v}_k \\ -\mathbf{L}\mathbf{v}_k \end{bmatrix}. \tag{3.29}$$

According to (3.29), the state estimation problem can be viewed as finding an auxiliary observer input $\mathbf{v}_k$ in terms of the available quantities such that the observer estimation errors $\mathbf{e}_{\mathbf{y}_k}$ and $\mathbf{e}_{\mathbf{z}_k^l}$ are steered to zero in a finite number of steps. To design the discrete-time sliding-mode observer, the sliding manifold is defined as $\{\mathbf{e}_\mathbf{x} | \mathbf{e}_\mathbf{y} = 0\}$. Hence, by putting $\mathbf{e}_{\mathbf{y}_{k+1}} = 0$ in (3.29), the equivalent control input, $\mathbf{v}_k^{eq}$, can be found as [37, 38]:

$$\mathbf{v}_k^{eq} = -\mathbf{\Phi}_{11}\mathbf{e}_{\mathbf{y}_k} - \mathbf{\Phi}_{12}\mathbf{e}_{\mathbf{z}_k^l}. \tag{3.30}$$

Applying the equivalent control, $\mathbf{v}_k^{eq}$, sliding mode occurs at the next step, and the governing dynamics of $\mathbf{e}_{\mathbf{z}_k^l}$ will be described by:

$$\mathbf{e}_{\mathbf{z}_{k+1}^l} = \left(\mathbf{\Phi}_{22} + \mathbf{L}\mathbf{\Phi}_{12}\right)\mathbf{e}_{\mathbf{z}_k^l}, \tag{3.31}$$

which converges to zero by properly choosing the eigenvalues of $\left(\mathbf{\Phi}_{22} + \mathbf{L}\mathbf{\Phi}_{12}\right)$ through designing the observer gain matrix, $\mathbf{L}$. In order to place the eigenvalues of $\left(\mathbf{\Phi}_{22} + \mathbf{L}\mathbf{\Phi}_{12}\right)$ at the desired locations, the pair $\left(\mathbf{\Phi}_{22}, \mathbf{\Phi}_{12}\right)$ must be observable. This condition is satisfied if the pair $(\mathbf{\Phi}, \mathbf{C})$ is observable. This leads to a sliding-mode realization of the standard reduced order asymptotic observer [37].

In (3.30), $\mathbf{e}_{\mathbf{z}_k^l}$ is unknown. Therefore, the following recursive equation is obtained from (3.29) to compute $\mathbf{e}_{\mathbf{z}_k^l}$:

$$\mathbf{e}_{\mathbf{z}_k^l} = \left(\mathbf{\Phi}_{22} + \mathbf{L}\mathbf{\Phi}_{12}\right)\mathbf{e}_{\mathbf{z}_{k-1}^l} + \left(\mathbf{\Phi}_{21} + \mathbf{L}\mathbf{\Phi}_{11}\right)\mathbf{e}_{\mathbf{y}_{k-1}} - \mathbf{L}\mathbf{e}_{\mathbf{y}_k}. \tag{3.32}$$

Now, the equivalent observer auxiliary input can be calculated as:

$$\begin{aligned}\mathbf{v}_k^{eq} = &-\left(\mathbf{\Phi}_{11} + \mathbf{\Phi}_{12}\mathbf{L}\right)\mathbf{e}_{\mathbf{y}_k} - \mathbf{\Phi}_{12}\left(\mathbf{\Phi}_{21} + \mathbf{L}\mathbf{\Phi}_{11}\right)\mathbf{e}_{\mathbf{y}_{k-1}} \\ &- \mathbf{\Phi}_{12}\left(\mathbf{\Phi}_{22} + \mathbf{L}\mathbf{\Phi}_{12}\right)\mathbf{e}_{\mathbf{z}_{k-1}^l},\end{aligned} \tag{3.33}$$

where $\mathbf{e}_{\mathbf{z}_{k-1}^l}$ is obtained from (3.32). Given $\mathbf{v}_k^{eq}$, the discrete-time sliding-mode observer provides the state estimate as [37]:

$$\hat{\mathbf{x}}_{k+1} = \mathbf{A}\hat{\mathbf{x}}_k + \mathbf{B}\mathbf{u}_k - \mathbf{T}^{-1}\begin{bmatrix}\mathbf{I}_{n_y} \\ -\mathbf{L}\end{bmatrix}\mathbf{v}_k^{eq}. \tag{3.34}$$

3.5 Unknown-Input Observer

The *unknown-input observer* (UIO) aims at estimating the state of uncertain systems in the presence of unknown inputs or uncertain disturbances and faults. The UIO is very useful in diagnosing system faults and detecting cyber-attacks [35, 39]. Let us consider the following discrete-time linear system:

$$\mathbf{x}_{k+1} = \mathbf{A}\mathbf{x}_k + \mathbf{B}\mathbf{u}_k, \tag{3.35}$$

$$\mathbf{y}_k = \mathbf{C}\mathbf{x}_k + \mathbf{D}\mathbf{u}_k, \tag{3.36}$$

where $\mathbf{x}_k \in \mathbb{R}^{n_x}$, $\mathbf{u}_k \in \mathbb{R}^{n_u}$, and $\mathbf{y}_k \in \mathbb{R}^{n_y}$. It is assumed that the matrix $\begin{bmatrix}\mathbf{B} \\ \mathbf{D}\end{bmatrix}$ has full column rank, which can be achieved using an appropriate transformation. Response of the system (3.35) and (3.36) over $k+1$ time steps is given by [35]:

$$\underbrace{\begin{bmatrix} \mathbf{y}_0 \\ \mathbf{y}_1 \\ \mathbf{y}_2 \\ \vdots \\ \mathbf{y}_\ell \end{bmatrix}}_{\mathbf{y}_{0:\ell}} = \underbrace{\begin{bmatrix} \mathbf{C} \\ \mathbf{CA} \\ \mathbf{CA}^2 \\ \vdots \\ \mathbf{CA}^\ell \end{bmatrix}}_{\mathcal{O}_\ell} \mathbf{x}_0 + \underbrace{\begin{bmatrix} \mathbf{D} & \mathbf{0} & \mathbf{0} & \cdots & \mathbf{0} \\ \mathbf{CB} & \mathbf{D} & \mathbf{0} & \cdots & \mathbf{0} \\ \mathbf{CAB} & \mathbf{CB} & \mathbf{D} & \cdots & \mathbf{0} \\ \vdots & \vdots & \vdots & \ddots & \vdots \\ \mathbf{CA}^{\ell-1}\mathbf{B} & \mathbf{CA}^{\ell-2}\mathbf{B} & \mathbf{CA}^{\ell-3}\mathbf{B} & \cdots & \mathbf{D} \end{bmatrix}}_{\mathcal{J}_\ell} \underbrace{\begin{bmatrix} \mathbf{u}_0 \\ \mathbf{u}_1 \\ \mathbf{u}_2 \\ \vdots \\ \mathbf{u}_\ell \end{bmatrix}}_{\mathbf{u}_{0:\ell}}. \tag{3.37}$$

The matrix $\mathcal{O}_\ell$ is the *observability matrix* for the pair $(\mathbf{A}, \mathbf{C})$, and $\mathcal{J}_\ell$ is the *invertibility matrix* for the tuple $(\mathbf{A}, \mathbf{B}, \mathbf{C}, \mathbf{D})$. The matrices $\mathcal{O}_\ell$ and $\mathcal{J}_\ell$ can also be expressed as [35]:

$$\mathcal{O}_\ell = \begin{bmatrix} \mathbf{C} \\ \mathcal{O}_{\ell-1}\mathbf{A} \end{bmatrix}, \tag{3.38}$$

$$\mathcal{J}_\ell = \begin{bmatrix} \mathbf{D} & \mathbf{0} \\ \mathcal{O}_{\ell-1}\mathbf{B} & \mathcal{J}_{\ell-1} \end{bmatrix}. \tag{3.39}$$

Equation (3.37) can be rewritten in the following compact form:

$$\mathbf{y}_{k:k+\ell} = \mathcal{O}_\ell \mathbf{x}_k + \mathcal{J}_\ell \mathbf{u}_{k:k+\ell}. \tag{3.40}$$

Then, the dynamic system

$$\hat{\mathbf{x}}_{k+1} = \mathbf{E}\hat{\mathbf{x}}_k + \mathbf{F}\mathbf{y}_{k:k+\ell} \tag{3.41}$$

is a UIO with delay ℓ, if

$$\lim_{k\to\infty} \mathbf{x}_k - \hat{\mathbf{x}}_k = \mathbf{0}, \tag{3.42}$$

regardless of the values of $\mathbf{u}_k$. Since the input is unknown, the observer equation (3.41) does not depend on the input. Moreover, the system outputs up to time step $k+\ell$ are used to estimate the state at time step k. Hence, the observer given by (3.41) is a *delayed state estimator*. Alternatively, it can be said that at time step k, the observer estimates the state at time step $k-\ell$ [35].

In order to design the observer in (3.41), the matrices $\mathbf{E}$ and $\mathbf{F}$ are chosen regarding the state estimation error:

$$\begin{aligned} \mathbf{e}_{k+1} &= \mathbf{x}_{k+1} - \hat{\mathbf{x}}_{k+1} \\ &= \mathbf{A}\mathbf{x}_k + \mathbf{B}\mathbf{u}_k - \mathbf{E}\hat{\mathbf{x}}_k - \mathbf{F}\mathbf{y}_{k:k+\ell} \\ &= (\mathbf{A}-\mathbf{E})\mathbf{x}_k + \mathbf{B}\mathbf{u}_k + \mathbf{E}\mathbf{e}_k - \mathbf{F}\mathbf{y}_{k:k+\ell}. \end{aligned} \tag{3.43}$$

Using (3.40), the state estimation error can be rewritten as:

$$\mathbf{e}_{k+1} = \mathbf{E}\mathbf{e}_k + \left(\mathbf{A} - \mathbf{E} - \mathbf{F}\mathcal{O}_\ell\right)\mathbf{x}_k + \mathbf{B}\mathbf{u}_k - \mathbf{F}\mathcal{J}_\ell \mathbf{u}_{k:k+\ell}. \tag{3.44}$$

To force $\mathbf{e}_k$ to go to zero, regardless of the values of $\mathbf{x}_k$ and $\mathbf{u}_{k:k+\ell}$, $\mathbf{E}$ must be a Hurwitz matrix (its eigenvalues must be in the left-half of the complex plane), and $\mathbf{F}$ must simultaneously satisfy the following conditions:

$$\mathbf{F}\mathcal{J}_\ell = \begin{bmatrix}\mathbf{B} & \mathbf{0} & \cdots & \mathbf{0}\end{bmatrix}, \tag{3.45}$$

$$\mathbf{E} = \mathbf{A} - \mathbf{F}\mathcal{O}_\ell. \tag{3.46}$$

Existence of a matrix $\mathbf{F}$ that satisfies condition (3.45) is guaranteed by the following theorem [35].

Theorem 3.1 *There exists a matrix* $\mathbf{F}$ *that satisfies (3.45), if and only if*

$$rank(\mathcal{J}_\ell) - rank(\mathcal{J}_{\ell-1}) = n_u. \tag{3.47}$$

Equation (3.47) can be interpreted as the inversion condition of the inputs with a known initial state and delay ℓ, which is a fairly strict condition. In the design phase, starting from $\ell = 0$, the delay is increased until a value is found that satisfies (3.47). However, n_x is an upper bound for ℓ. To be more precise, if (3.47) is not satisfied for $\ell = n_x$, then asymptotic state estimation will not be possible using the observer in (3.41).

In order to satisfy condition (3.45), matrix $\mathbf{F}$ must be in the left nullspace of the last $\ell \times n_u$ columns of $\mathcal{J}_\ell$ given by $\begin{bmatrix}\mathbf{0} \\ \mathcal{J}_{\ell-1}\end{bmatrix}$. Let $\overline{\mathbf{N}}$ be a matrix whose rows form a basis for the left nullspace of $\mathcal{J}_{\ell-1}$:

$$\overline{\mathbf{N}}\mathcal{J}_{\ell-1} = \mathbf{0}, \tag{3.48}$$

then we have:

$$\begin{bmatrix}\mathbf{I}_{n_y} & \mathbf{0} \\ \mathbf{0} & \overline{\mathbf{N}}\end{bmatrix}\begin{bmatrix}\mathbf{0} \\ \mathcal{J}_{\ell-1}\end{bmatrix} = \mathbf{0}. \tag{3.49}$$

Let us define:

$$\mathbf{N} = \mathbf{W}\begin{bmatrix}\mathbf{I}_{n_y} & \mathbf{0} \\ \mathbf{0} & \overline{\mathbf{N}}\end{bmatrix}, \tag{3.50}$$

where $\mathbf{W}$ is an invertible matrix. Then, we have:

$$\mathbf{N}\begin{bmatrix}\mathbf{0} \\ \mathcal{J}_{\ell-1}\end{bmatrix} = \mathbf{0}. \tag{3.51}$$

To choose $\mathbf{W}$, note that:

$$\begin{aligned}\mathbf{N}\mathcal{J}_\ell &= \mathbf{N}\begin{bmatrix}\mathbf{D} & \mathbf{0} \\ \mathcal{O}_{\ell-1}\mathbf{B} & \mathcal{J}_{\ell-1}\end{bmatrix} \\ &= \mathbf{W}\begin{bmatrix}\mathbf{D} & \mathbf{0} \\ \overline{\mathbf{N}}\mathcal{O}_{\ell-1}\mathbf{B} & \mathbf{0}\end{bmatrix}.\end{aligned} \tag{3.52}$$

From Theorem 3.1, the first n_u columns of $\mathcal{J}_\ell$ must be linearly independent of each other and of the other $\ell \times n_u$ columns. Now, **W** is chosen such that:

$$\mathbf{N}\mathcal{J}_\ell = \begin{bmatrix} \mathbf{0} & \mathbf{0} \\ \mathbf{I}_{n_u} & \mathbf{0} \end{bmatrix}. \tag{3.53}$$

Regarding (3.45), **F** can be expressed as:

$$\begin{aligned} \mathbf{F} &= \hat{\mathbf{F}}\mathbf{N} \\ &= \begin{bmatrix} \hat{\mathbf{F}}_1 & \hat{\mathbf{F}}_2 \end{bmatrix} \mathbf{N}, \end{aligned} \tag{3.54}$$

where $\hat{\mathbf{F}}_2$ has n_y columns. Then, equation (3.45) leads to:

$$\begin{bmatrix} \hat{\mathbf{F}}_1 & \hat{\mathbf{F}}_2 \end{bmatrix} \begin{bmatrix} \mathbf{0} & \mathbf{0} \\ \mathbf{I}_{n_u} & \mathbf{0} \end{bmatrix} = \begin{bmatrix} \mathbf{B} & \mathbf{0} \end{bmatrix}. \tag{3.55}$$

Hence, $\hat{\mathbf{F}}_2 = \mathbf{B}$ and $\hat{\mathbf{F}}_1$ is a free matrix. According to (3.46), we have:

$$\mathbf{E} = \mathbf{A} - \mathbf{F}\mathcal{O}_\ell \tag{3.56}$$

$$= \mathbf{A} - \begin{bmatrix} \hat{\mathbf{F}}_1 & \mathbf{B} \end{bmatrix} \mathbf{N}\mathcal{O}_\ell. \tag{3.57}$$

Defining $\mathbf{N}\mathcal{O}_\ell = \begin{bmatrix} \mathbf{S}_1 \\ \mathbf{S}_2 \end{bmatrix}$, where $\mathbf{S}_2$ has n_u rows, we obtain:

$$\mathbf{E} = (\mathbf{A} - \mathbf{B}\mathbf{S}_2) - \hat{\mathbf{F}}_1\mathbf{S}_1. \tag{3.58}$$

Since **E** is required to be a stable matrix, the pair $(\mathbf{A} - \mathbf{B}\mathbf{S}_2, \mathbf{S}_1)$ must be detectable [35].

From (3.35) and (3.36), we have:

$$\begin{bmatrix} \mathbf{x}_{k+1} - \mathbf{A}\mathbf{x}_k \\ \mathbf{y}_k - \mathbf{C}\mathbf{x}_k \end{bmatrix} = \begin{bmatrix} \mathbf{B} \\ \mathbf{D} \end{bmatrix} \mathbf{u}_k. \tag{3.59}$$

Assuming that $\begin{bmatrix} \mathbf{B} \\ \mathbf{D} \end{bmatrix}$ has full column rank, there exists a matrix **G** such that:

$$\mathbf{G} \begin{bmatrix} \mathbf{B} \\ \mathbf{D} \end{bmatrix} = \mathbf{I}_{n_u}. \tag{3.60}$$

Left-multiplying both sides of the equation (3.59) by **G**, and then using (3.60), the input vector can be estimated based on the state-vector estimate as:

$$\hat{\mathbf{u}}_k = \mathbf{G} \begin{bmatrix} \hat{\mathbf{x}}_{k+1} - \mathbf{A}\hat{\mathbf{x}}_k \\ \mathbf{y}_k - \mathbf{C}\hat{\mathbf{x}}_k \end{bmatrix}. \tag{3.61}$$

Regarding (3.42), this estimate asymptotically approaches the true value of the input [35].

3.6 Concluding Remarks

Observers are dynamic processes, which are used to estimate the states or the unknown inputs of linear as well as nonlinear dynamic systems. This chapter covered the Luenberger observer, the extended Luenberger-type observer, the sliding-mode observer, and the UIO. In addition to the mentioned observers, *high-gain observers* have been proposed in the literature to handle uncertainty. Although the deployed high gains in high-gain observers allow for fast convergence and performance recovery, they amplify the effect of measurement noise [40]. Hence, there is a trade-off between fast state reconstruction under uncertainty and measurement noise attenuation. Due to this trade-off, in the transient and steady-state periods, relatively high and low gains are used, respectively. However, stochastic approximation allows for an implementation of the high-gain observer, which is able to cope with measurement noise [41]. Alternatively, the *bounding observer* or *interval observer* provides two simultaneous state estimations, which play the role of an upper bound and a lower bound on the true value of the state. The true value of the state is guaranteed to remain within these two bounds [42].

4

Bayesian Paradigm and Optimal Nonlinear Filtering

4.1 Introduction

Immanuel Kant proposed the two concepts of the *noumenal* world and the *phenomenal* world. While the former is the world of things as they are, which is independent of our modes of perception and thought, the latter is the world of things as they appear to us, which depends on how we perceive things. According to Kant, everything about the noumenal world is *transcendental* that means it exists but is not prone to concept formation by us [43].

Following this line of thinking, statistics will aim at interpretation rather than explanation. In this framework, *statistical inference* is built on *probabilistic modeling* of the observed phenomenon. A probabilistic model must include the available information about the phenomenon of interest as well as the uncertainty associated with this information. The purpose of statistical inference is to solve an inverse problem aimed at retrieving the causes, which are presented by states and/or parameters of the developed probabilistic model, from the effects, which are summarized in the observations. On the other hand, probabilistic modeling describes the behavior of the system and allows us to predict what will be observed in the future conditional on states and/or parameters [44].

Bayesian paradigm provides a mathematical framework in which degrees of belief are quantified by probabilities. It is the method of choice for dealing with uncertainty in measurements. Using the Bayesian approach, probability of an event of interest (state) can be calculated based on the probability of other events (observations or measurements) that are logically connected to and therefore, stochastically dependent on the event of interest. Moreover, the Bayesian method allows us to iteratively update probability of the state when new measurements become available [45]. This chapter reviews the Bayesian paradigm and presents the formulation of the optimal nonlinear filtering problem.

Nonlinear Filters: Theory and Applications, First Edition. Peyman Setoodeh, Saeid Habibi, and Simon Haykin.

4.2 Bayes' Rule

Bayes' theorem describes the inversion of probabilities. Let us consider two events A and B. Provided that $P(B) \neq 0$, we have the following relationship between the conditional probabilities $P(A|B)$ and $P(B|A)$:

$$P(A|B) = \frac{P(B|A)P(A)}{P(B)}. \tag{4.1}$$

Considering two random variables $\mathbf{x}$ and $\mathbf{y}$ with conditional distribution $p(\mathbf{x}|\mathbf{y})$ and marginal distribution $p(\mathbf{y})$, the continuous version of Bayes' rule is as follows:

$$p(\mathbf{x}|\mathbf{y}) = \frac{p(\mathbf{y}|\mathbf{x})p(\mathbf{x})}{\int p(\mathbf{y}|\mathbf{x})p(\mathbf{x})\mathrm{d}\mathbf{x}}, \tag{4.2}$$

where $p(\mathbf{x})$ is the *prior* distribution, $p(\mathbf{x}|\mathbf{y})$ is the *posterior* distribution, and $p(\mathbf{y}|\mathbf{x})$ is the *likelihood* function, which is also denoted by $\ell(\mathbf{x}|\mathbf{y})$. This formula captures the essence of Bayesian statistical modeling, where $\mathbf{y}$ denotes observations, and $\mathbf{x}$ represents states or parameters. In order to build a Bayesian model, we need a parametric statistical model described by the likelihood function $\ell(\mathbf{x}|\mathbf{y})$. Furthermore, we need to incorporate our knowledge about the system under study and the uncertainty about this information, which is represented by the prior distribution $p(\mathbf{x})$ [44].

4.3 Optimal Nonlinear Filtering

The following discrete-time stochastic state-space model describes the behavior of a discrete-time nonlinear system:

$$\mathbf{x}_{k+1} = \mathbf{f}(\mathbf{x}_k, \mathbf{u}_k, \mathbf{v}_k), \tag{4.3}$$

$$\mathbf{y}_k = \mathbf{g}(\mathbf{x}_k, \mathbf{u}_k, \mathbf{w}_k), \tag{4.4}$$

where $\mathbf{x} \in \mathbb{R}^{n_x}$, $\mathbf{u} \in \mathbb{R}^{n_u}$, and $\mathbf{y} \in \mathbb{R}^{n_y}$ denote the state, the input, and the output vectors, respectively. Compared with the deterministic nonlinear discrete-time model in (2.78) and (2.79), two additional variables are included in the above stochastic model, which are the process noise, $\mathbf{v}_k$, and the measurement noise, $\mathbf{w}_k$. These two random variables take account of model inaccuracies and other sources of uncertainty. The more accurate the model is, the smaller the contribution of noise terms will be. These two noise sequences are assumed to be white, independent of each other, and independent from the initial state. The probabilistic model of the state evolution in (4.3) is assumed to be a first-order Markov process, and therefore, can be rewritten as the following state-transition probability density function (PDF) [46]:

$$p(\mathbf{x}_{k+1}|\mathbf{x}_k, \mathbf{u}_k). \tag{4.5}$$

Similarly, the measurement model in (4.4) can be represented by the following PDF:

$$p(\mathbf{y}_k|\mathbf{x}_k, \mathbf{u}_k). \tag{4.6}$$

The input sequence and the available measurement sequence at time instant k are denoted by $\mathbf{U}_k = \{\mathbf{u}_i|i = 0, \dots, k\} \equiv \mathbf{u}_{0:k}$ and $\mathbf{Y}_k = \{\mathbf{y}_i|i = 0, \dots, k\} \equiv \mathbf{y}_{0:k}$, respectively. These two sequences form the available information at time k, hence the union of these two sets is called the *information set*, $\mathbf{I}_k = \{\mathbf{U}_k, \mathbf{Y}_k\} = \{\mathbf{u}_{0:k}, \mathbf{y}_{0:k}\}$ [47].

A filter uses the inputs and available observations up to time instant k, to estimate the state at k, $\hat{\mathbf{x}}_{k|k}$. In other words, a filter tries to solve an inverse problem to infer the states (cause) from the observations (effect). Due to uncertainties, different values of the state could have led to the obtained measurement sequence, $\mathbf{y}_{0:k}$. The Bayesian framework allows us to associate a degree of belief to these possibly valid values of state. The main idea here is to start from an initial density for the state vector, $p(\mathbf{x}_0)$, and recursively calculate the posterior PDF, $p(\mathbf{x}_k|\mathbf{u}_{0:k}, \mathbf{y}_{0:k})$ based on the measurements. This can be done by a filtering algorithm that includes two-stages of *prediction* and *update* [46].

In the prediction stage, the *Chapman–Kolmogorov* equation can be used to calculate the prediction density, $p(\mathbf{x}_{k+1}|\mathbf{u}_{0:k}, \mathbf{y}_{0:k})$, from $p(\mathbf{x}_k|\mathbf{u}_{0:k}, \mathbf{y}_{0:k})$, which is provided by the update stage of the previous iteration. Using (4.5), the prediction density can be computed as [46, 47]:

$$p(\mathbf{x}_{k+1}|\mathbf{u}_{0:k}, \mathbf{y}_{0:k}) = \int p(\mathbf{x}_{k+1}|\mathbf{x}_k, \mathbf{u}_k) p(\mathbf{x}_k|\mathbf{u}_{0:k}, \mathbf{y}_{0:k}) d\mathbf{x}_k. \tag{4.7}$$

When a new measurement $\mathbf{y}_{k+1}$ is obtained, the prediction stage is followed by the update stage, where the above prediction density will play the role of the prior. Bayes' rule is used to compute the posterior density of the state as [46, 47]:

$$\begin{aligned}
p(\mathbf{x}_{k+1}|\mathbf{u}_{0:k+1}, \mathbf{y}_{0:k+1}) &= p(\mathbf{x}_{k+1}|\mathbf{u}_{0:k+1}, \mathbf{y}_{k+1}, \mathbf{y}_{0:k}) \\
&= \frac{p(\mathbf{x}_{k+1}, \mathbf{y}_{k+1}|\mathbf{u}_{0:k+1}, \mathbf{y}_{0:k})}{p(\mathbf{y}_{k+1}|\mathbf{u}_{0:k+1}, \mathbf{y}_{0:k})} \\
&= \frac{p(\mathbf{y}_{k+1}|\mathbf{x}_{k+1}, \mathbf{u}_{0:k+1}, \mathbf{y}_{0:k}) p(\mathbf{x}_{k+1}|\mathbf{u}_{0:k}, \mathbf{y}_{0:k})}{p(\mathbf{y}_{k+1}|\mathbf{u}_{0:k+1}, \mathbf{y}_{0:k})} \\
&= \frac{p(\mathbf{y}_{k+1}|\mathbf{x}_{k+1}, \mathbf{u}_{k+1}) p(\mathbf{x}_{k+1}|\mathbf{u}_{0:k}, \mathbf{y}_{0:k})}{p(\mathbf{y}_{k+1}|\mathbf{u}_{0:k+1}, \mathbf{y}_{0:k})},
\end{aligned} \tag{4.8}$$

where the normalization constant in the denominator is obtained as:

$$p(\mathbf{y}_{k+1}|\mathbf{u}_{0:k+1}, \mathbf{y}_{0:k}) = \int p(\mathbf{y}_{k+1}|\mathbf{x}_{k+1}, \mathbf{u}_{k+1}) p(\mathbf{x}_{k+1}|\mathbf{u}_{0:k}, \mathbf{y}_{0:k}) d\mathbf{x}_{k+1}. \tag{4.9}$$

The recursive propagation of the state posterior density according to equations (4.7) and (4.8) provides the basis of the *Bayesian solution*. Having the posterior,

$p(\mathbf{x}_{k+1}|\mathbf{u}_{0:k+1}, \mathbf{y}_{0:k+1})$, the optimal state estimate $\hat{\mathbf{x}}_{k+1:k+1}$ can be obtained based on a specific criterion. Depending on the chosen criterion, different estimators are obtained [46–48]:

- *Minimum mean-square error* (MMSE) estimator

$$\hat{\mathbf{x}}_{k|k}^{MMSE} \triangleq \arg\min_{\hat{\mathbf{x}}_{k|k}} \mathbb{E}[(\mathbf{x}_k - \hat{\mathbf{x}}_{k|k})^2|\mathbf{u}_{0:k}, \mathbf{y}_{0:k}]. \tag{4.10}$$

  This is equivalent to minimizing the trace (sum of the diagonal elements) of the estimation-error covariance matrix. The MMSE estimate is the conditional mean of $\mathbf{x}_k$:

$$\hat{\mathbf{x}}_{k|k}^{MMSE} = \mathbb{E}\left[\mathbf{x}_k|\mathbf{u}_{0:k}, \mathbf{y}_{0:k}\right], \tag{4.11}$$

  where the expectation is taken with respect to the posterior, $p(\mathbf{x}_k|\mathbf{u}_{0:k}, \mathbf{y}_{0:k})$.
- *Risk-sensitive* (RS) estimator

$$\hat{\mathbf{x}}_{k|k}^{RS} \triangleq \arg\min_{\hat{\mathbf{x}}_{k|k}} \mathbb{E}\left[e^{(\mathbf{x}_k - \hat{\mathbf{x}}_{k|k})^2}|\mathbf{u}_{0:k}, \mathbf{y}_{0:k}\right]. \tag{4.12}$$

  Compared to the MMSE estimator, the RS estimator is less sensitive to uncertainties. In other words, it is a more robust estimator [49].
- *Maximum a posteriori* (MAP) estimator

$$\hat{\mathbf{x}}_{k|k}^{MAP} \triangleq \arg\max_{\hat{\mathbf{x}}_{k|k}} p(\mathbf{x}_k|\mathbf{u}_{0:k}, \mathbf{y}_{0:k}). \tag{4.13}$$

- *Minimax* estimator

$$\hat{\mathbf{x}}_{k|k}^{Minimax} \triangleq \arg\min_{\hat{\mathbf{x}}_{k|k}} \max |\mathbf{x}_k - \hat{\mathbf{x}}_{k|k}|. \tag{4.14}$$

  The minimax estimate is the medium of the posterior, $p(\mathbf{x}_k|\mathbf{u}_{0:k}, \mathbf{y}_{0:k})$. The minimax technique is used to achieve optimal performance under the worst-case condition [50].
- *The most probable* (MP) estimator

$$\hat{\mathbf{x}}_{k|k}^{MP} \triangleq \arg\max_{\hat{\mathbf{x}}_{k|k}} P(\hat{\mathbf{x}}_{k|k} = \mathbf{x}). \tag{4.15}$$

  MP estimate is the mode of the posterior, $p(\mathbf{x}_k|\mathbf{u}_{0:k}, \mathbf{y}_{0:k})$. For a uniform prior, this estimate will be identical to the *maximum likelihood* (ML) estimate.

$$\hat{\mathbf{x}}_{k|k}^{ML} \triangleq \arg\max_{\hat{\mathbf{x}}_{k|k}} p(\mathbf{y}_k|\mathbf{x}_k, \mathbf{u}_{0:k}, \mathbf{y}_{0:k}). \tag{4.16}$$

In general, there may not exist simple analytic forms for the corresponding PDFs. Without an analytic form, the PDF for a single variable will be equivalent to an infinite-dimensional vector that must be stored for performing the required computations. In such cases, obtaining the Bayesian solution will be computationally intractable. In other words, the Bayesian solution except for special cases, is a conceptual solution, and generally speaking, it cannot be determined analytically. In many practical situations, we will have to use some sort of approximation, and therefore, settle for a suboptimal Bayesian solution [46]. Different approximation methods lead to different filtering algorithms.

4.4 Fisher Information

The relevant portion of the data obtained by measurement can be interpreted as information. In this line of thinking, a summary of the amount of information with regard to the variables of interest is provided by the *Fisher information matrix* [51]. To be more specific, Fisher information plays two basic roles:

1. It is a measure of the ability to estimate a quantity of interest.
2. It is a measure of the state of disorder in a system or phenomenon of interest.

The first role implies that the Fisher information matrix has a close connection to the estimation-error covariance matrix and can be used to calculate the confidence region of estimates. The second role implies that the Fisher information has a close connection to Shannon's entropy.

Let us consider the PDF $p_\theta(\mathbf{x})$, which is parameterized by the set of parameters θ. The Fisher information matrix is defined as:

$$\mathbf{F}_\theta = \mathbb{E}_\mathbf{x}\left[\left(\nabla_\theta \log p_\theta(\mathbf{x})\right)\left(\nabla_\theta \log p_\theta(\mathbf{x})\right)^T\right]. \tag{4.17}$$

This definition is based on the outer product of the gradient of $\log p_\theta(\mathbf{x})$ with itself, where the gradient is a column vector denoted by ∇_θ. There is an equivalent definition based on the second derivative of $\log p_\theta(\mathbf{x})$ as:

$$\mathbf{F}_\theta = -\mathbb{E}_\mathbf{x}\left[\frac{\partial^2 \log p_\theta(\mathbf{x})}{\partial^2 \theta}\right]. \tag{4.18}$$

From the definition of $\mathbf{F}_\theta$, it is obvious that Fisher information is a function of the corresponding PDF. A relatively broad and flat PDF, which is associated with lack of predictability and high entropy, has small gradient contents and, in effect therefore, low Fisher information. On the other hand, if the PDF is relatively narrow and has sharp slopes around a specific value of $\mathbf{x}$, which is associated with bias toward that particular value of $\mathbf{x}$ and low entropy, it has large gradient contents and therefore high Fisher information. In summary, there is a duality between Shannon's entropy and Fisher information. However, a closer look at their mathematical definitions reveals an important difference [27]:

- A rearrangement of the tuples $(\mathbf{x}_i, p(\mathbf{x}_i))$ may change the shape of the PDF curve significantly, but it does not affect the value of the summation in (2.95) or integration in (2.96), because the summation and integration can be calculated in

any order. Since H is not affected by local changes in the PDF curve, it can be considered as a global measure of the behavior of the corresponding PDF.
- On the other hand, such a rearrangement of points changes the slope, and therefore gradient of the PDF curve, which, in turn, changes the Fisher information significantly. Hence, the Fisher information is sensitive to local rearrangement of points and can be considered as a local measure of the behavior of the corresponding PDF.

Both entropy (as a global measure of smoothness in the PDF) and Fisher information (as a local measure of smoothness in the PDF) can be used in a variational principle to infer about the PDF that describes the phenomenon under consideration. However, the local measure may be preferred in general [27]. This leads to another performance metric, which is discussed in Section 4.5.

4.5 Posterior Cramér–Rao Lower Bound

To assess the performance of an estimator, a lower bound is always desirable. Such a bound is a measure of performance limitation that determines whether or not the design criterion is realistic and implementable. The *Cramér–Rao lower bound* (CRLB) is a lower bound that represents the lowest possible mean-square error in the estimation of deterministic parameters for all unbiased estimators. It can be computed as the inverse of the Fisher information matrix. For random variables, a similar version of the CRLB, namely, the posterior Cramér–Rao lower bound (PCRLB) was derived in [52] as:

$$\mathbf{P}_{k|k} = \mathbb{E}[(\mathbf{x}_k - \hat{\mathbf{x}}_k)\,(\mathbf{x}_k - \hat{\mathbf{x}}_k)^T] \geq \mathbf{F}_k^{-1}, \tag{4.19}$$

where $\mathbf{F}_k^{-1}$ denotes the inverse of Fisher information matrix at time instant k. This bound is also referred to as the Bayesian CRLB [53, 54]. To compute it in an online manner, an iterative version of the PCRLB for nonlinear filtering using state-space models was proposed in [55], where the posterior information matrix of the hidden state vector is decomposed for each discrete-time instant by virtue of the factorization of the joint PDF of the state variables. In this way, an iterative structure is obtained for evolution of the information matrix. For a nonlinear system with the following state-space model with zero-mean additive Gaussian noise:

$$\mathbf{x}_{k+1} = \mathbf{f}_k(\mathbf{x}_k) + \mathbf{v}_k, \tag{4.20}$$

$$\mathbf{y}_k = \mathbf{g}_k(\mathbf{x}_k) + \mathbf{w}_k, \tag{4.21}$$

the sequence of posterior information matrices, $\mathbf{F}_k$, for estimating state vectors, $\mathbf{x}_k$, can be computed as [55]:

$$\mathbf{F}_{k+1} = \mathbf{D}_k^{22} - \mathbf{D}_k^{21}(\mathbf{F}_k + \mathbf{D}_k^{11})^{-1}\mathbf{D}_k^{12}, \tag{4.22}$$

where

$$\mathbf{D}_k^{11} = \mathbb{E}\left[\nabla_{\mathbf{x}_k}\mathbf{f}_k(\mathbf{x}_k)\,\mathbf{Q}_k^{-1}\,\nabla_{\mathbf{x}_k}^T\mathbf{f}_k(\mathbf{x}_k)\right], \tag{4.23}$$

$$\mathbf{D}_k^{12} = -\mathbb{E}\left[\nabla_{\mathbf{x}_k}\mathbf{f}_k(\mathbf{x}_k)\right]\,\mathbf{Q}_k^{-1}, \tag{4.24}$$

$$\mathbf{D}_k^{21} = \left(\mathbf{D}_k^{12}\right)^T, \tag{4.25}$$

$$\mathbf{D}_k^{22} = \mathbb{E}\left[\nabla_{\mathbf{x}_{k+1}}\mathbf{g}_{k+1}(\mathbf{x}_{k+1})\,\mathbf{R}_{k+1}^{-1}\,\nabla_{\mathbf{x}_{k+1}}^T\mathbf{g}_{k+1}(\mathbf{x}_{k+1})\right] + \mathbf{Q}_k^{-1}, \tag{4.26}$$

where $\mathbf{Q}_k$ and $\mathbf{R}_k$ are the process and measurement noise covariance matrices, respectively.

4.6 Concluding Remarks

The general formulation of the optimal nonlinear Bayesian filtering leads to a computationally intractable problem; hence, the Bayesian solution is a conceptual solution. Settling for computationally tractable suboptimal solutions through deploying different approximation methods has led to a wide range of classic as well as machine learning-based filtering algorithms. Such algorithms have their own advantages, restrictions, and domains of applicability. To assess and compare such filtering algorithms, several performance metrics can be used including entropy, Fisher information, and PCRLB. Furthermore, the Fisher information matrix is used to define the *natural gradient*, which is helpful in machine learning.

5

Kalman Filter

5.1 Introduction

Chapter 4 presented the general formulation of the optimal Bayesian filter. Furthermore, it was mentioned that except for special cases, a closed-form solution for this problem does not exist. Perhaps the most famous special case is a linear dynamic system perturbed by additive white Gaussian noise, for which a recursive estimator that is optimal in the sense of minimizing the mean-square of estimation error has an analytic solution [56]. Such an optimal estimator is the celebrated Kalman filter (KF), which is considered as the most useful and widely applied result of the state-space approach of modern control theory [57]. Since the publication of Rudolf Emil Kalman's seminal paper in 1960 [58], KF found immediate applications in almost all practical control systems, and it has become a necessary tool in any control engineer's toolbox [3, 47, 56, 59–63].

However, almost all real systems are nonlinear in nature, and therefore, do not satisfy the basic assumptions for derivation of KF. In order to extend application of KF to nonlinear estimation problems, a number of filtering algorithms have been proposed over decades. These algorithms can be categorized into two main groups:

- Filters that use power series to approximate the nonlinear functions in the state and/or measurement equations. *Extended Kalman filter* (EKF), which uses Taylor series expansion [3], and *divided-difference filter* (DDF), which uses Stirling's interpolation formula [64], belong to this category. The main idea is to linearize the corresponding nonlinear functions about the estimated state trajectory by keeping the linear term and ignoring the higher-order terms of the chosen power series. Then, the linearized functions are used to build a linear state-space model. This linear model, is a crude approximation of the original nonlinear model, which is only locally legitimate around the current state estimate. Having a linear state-space model, we can use the

Nonlinear Filters: Theory and Applications, First Edition. Peyman Setoodeh, Saeid Habibi, and Simon Haykin.

original Kalman filter algorithm for state estimation. In each iteration, the corresponding nonlinear functions are linearized around the new estimate to obtain the linear model and the algorithm carries on from one iteration to the next. However, this local linearization around the current state estimate will be a valid approximation only for mild nonlinearities. In case of severe nonlinearities, higher-order terms may not be ignorable, and in effect therefore, the crude linear approximation is not accurate enough for estimation purposes. To remedy this problem, second-order versions of EKF [65] and DDF [64] have been proposed.

- An alternative way to handle sever nonlinearities is the idea that approximating a probability distribution may be easier than approximating a nonlinear function. In the second category of filters, instead of analytical approximation of nonlinear functions, numerical methods are deployed to approximate the corresponding distributions. *Unscented Kalman filter* (UKF), which uses the *unscented transform* [66] and *cubature Kalman filter* (CKF), which uses the *cubature rule* [67] for approximating integrals, belong to this category.

It should be noted that none of the mentioned algorithms clearly outperform the others in all applications that involve nonlinear functions. In order to select a filter for a specific application, attention must be paid to several factors such as estimation accuracy, computational complexity, ease of implementation, and numerical robustness [64].

5.2 Kalman Filter

The state-space model of a discrete-time stochastic linear dynamic system is represented by:

$$\mathbf{x}_{k+1} = \mathbf{A}\mathbf{x}_k + \mathbf{B}\mathbf{u}_k + \mathbf{v}_k, \tag{5.1}$$

$$\mathbf{y}_k = \mathbf{C}\mathbf{x}_k + \mathbf{D}\mathbf{u}_k + \mathbf{w}_k. \tag{5.2}$$

Compared with the deterministic linear discrete-time model in (2.18) and (2.19), two additional variables are included in the above stochastic model, which are the process noise, $\mathbf{v}_k$, and the measurement noise, $\mathbf{w}_k$. These two random variables are assumed to be independent white Gaussian random sequences, which are independent of the initial state as well. In mathematical terms, for the independence condition, we have [48]:

$$p(\mathbf{v}_k, \mathbf{w}_k, |\mathbf{x}_k, \mathbf{u}_k, \mathbf{y}_{0:k}) = p(\mathbf{v}_k)\, p(\mathbf{w}_k). \tag{5.3}$$

Moreover, process noise and measurement noise have the following statistical characteristics:

$$\mathbb{E}[\mathbf{v}_k] = 0, \quad \text{cov}(\mathbf{v}_k) = \mathbb{E}\left[\mathbf{v}_k\mathbf{v}_k^T\right] = \mathbf{Q}_k, \tag{5.4}$$

$$\mathbb{E}[\mathbf{w}_k] = 0, \quad \text{cov}(\mathbf{w}_k) = \mathbb{E}\left[\mathbf{w}_k\mathbf{w}_k^T\right] = \mathbf{R}_k. \tag{5.5}$$

Since the process noise has Gaussian distribution, $p(\mathbf{x}_{k+1}|\mathbf{u}_{0:k+1}, \mathbf{y}_{0:k+1})$ is Gaussian as well. For the statistical characteristics of the state vector, we use the following notation [48]:

$$\mathbb{E}[\mathbf{x}_{k+1}|\mathbf{u}_{0:k}, \mathbf{y}_{0:k}] = \hat{\mathbf{x}}_{k+1|k}, \tag{5.6}$$

$$\mathbb{E}[\mathbf{x}_{k+1}|\mathbf{u}_{0:k+1}, \mathbf{y}_{0:k+1}] = \hat{\mathbf{x}}_{k+1|k+1}, \tag{5.7}$$

$$\text{cov}(\mathbf{x}_{k+1}|\mathbf{u}_{0:k}, \mathbf{y}_{0:k}) = \mathbf{P}_{k+1|k}, \tag{5.8}$$

$$\text{cov}(\mathbf{x}_{k+1}|\mathbf{u}_{0:k+1}, \mathbf{y}_{0:k+1}) = \mathbf{P}_{k+1|k+1}. \tag{5.9}$$

If the above assumptions are valid for a system, then the maximum a posteriori (MAP) estimator and the minimum mean square error (MMSE) estimator will be equivalent for that system. Therefore, Kalman filter can be derived either way. For instance, by recalling the posterior of the Bayesian solution from (4.8):

$$\begin{aligned} p(\mathbf{x}_{k+1}|\mathbf{u}_{0:k+1}, \mathbf{y}_{0:k+1}) &= p(\mathbf{x}_{k+1}|\mathbf{u}_{0:k+1}, \mathbf{y}_{k+1}, \mathbf{y}_{0:k}) \\ &= \frac{p(\mathbf{x}_{k+1}, \mathbf{y}_{k+1}|\mathbf{u}_{0:k+1}, \mathbf{y}_{0:k})}{p(\mathbf{y}_{k+1}|\mathbf{u}_{0:k+1}, \mathbf{y}_{0:k})} \\ &= \frac{p(\mathbf{y}_{k+1}|\mathbf{x}_{k+1}, \mathbf{u}_{0:k+1}, \mathbf{y}_{0:k})p(\mathbf{x}_{k+1}|\mathbf{u}_{0:k}, \mathbf{y}_{0:k})}{p(\mathbf{y}_{k+1}|\mathbf{u}_{0:k+1}, \mathbf{y}_{0:k})} \\ &= \frac{p(\mathbf{y}_{k+1}|\mathbf{x}_{k+1}, \mathbf{u}_{k+1})p(\mathbf{x}_{k+1}|\mathbf{u}_{0:k}, \mathbf{y}_{0:k})}{p(\mathbf{y}_{k+1}|\mathbf{u}_{0:k+1}, \mathbf{y}_{0:k})}, \end{aligned} \tag{5.10}$$

and maximizing it with respect to the state estimate $\hat{\mathbf{x}}_{k+1|k+1}$, we obtain the celebrated Kalman filter algorithm as summarized in Algorithm 5.1.

Since the covariance matrix must be positive semi-definite, $\mathbf{P}_k \geq \mathbf{0}$, to reduce the effect of numerical errors and avoid the *divergence phenomenon*, the *square-root* implementation of the Kalman filter has been proposed [68]. In the square-root Kalman filter, *Cholesky factorization* is used to perform covariance computations based on the square root of the estimation error covariance matrix:

$$\mathbf{P}_k = \mathbf{P}_k^{\frac{1}{2}}\left(\mathbf{P}_k^{\frac{1}{2}}\right)^T. \tag{5.11}$$

Using the **LDU** decomposition:

$$\mathbf{P}_k = \mathbf{L}_k\mathbf{D}_k\mathbf{L}_k^T, \tag{5.12}$$

Algorithm 5.1: Kalman filter

State-space model

$$\mathbf{x}_{k+1} = \mathbf{A}_k\mathbf{x}_k + \mathbf{B}_k\mathbf{u}_k + \mathbf{v}_k$$
$$\mathbf{y}_k = \mathbf{C}_k\mathbf{x}_k + \mathbf{D}_k\mathbf{u}_k + \mathbf{w}_k$$

where $\mathbf{v}_k \sim \mathcal{N}(\mathbf{0}, \mathbf{Q}_k)$ and $\mathbf{w}_k \sim \mathcal{N}(\mathbf{0}, \mathbf{R}_k)$.

Initialization

$$\hat{\mathbf{x}}_0 = \mathbb{E}[\mathbf{x}_0]$$
$$\mathbf{P}_0 = \mathbb{E}\left[\left(\mathbf{x}_0 - \mathbb{E}[\mathbf{x}_0]\right)\left(\mathbf{x}_0 - \mathbb{E}[\mathbf{x}_0]\right)^T\right]$$

for $k = 0, 1, \ldots,$ **do**

A priori state estimate

$$\hat{\mathbf{x}}_{k+1|k} = \mathbf{A}_k\hat{\mathbf{x}}_{k|k} + \mathbf{B}_k\mathbf{u}_k$$

A priori error covariance

$$\mathbf{P}_{k+1|k} = \mathbf{A}_k\mathbf{P}_{k|k}\mathbf{A}_k^T + \mathbf{Q}_k$$

A priori measurement estimate

$$\hat{\mathbf{y}}_{k+1|k} = \mathbf{C}_{k+1}\hat{\mathbf{x}}_{k+1|k} + \mathbf{D}_{k+1}\mathbf{u}_{k+1}$$

A priori output error estimate (innovation)

$$\mathbf{e}_{\mathbf{y}_{k+1|k}} = \mathbf{y}_{k+1} - \hat{\mathbf{y}}_{k+1|k}$$

Kalman gain

$$\mathbf{K}_{k+1} = \mathbf{P}_{k+1|k}\mathbf{C}_{k+1}^T\left(\mathbf{C}_{k+1}\mathbf{P}_{k|k}\mathbf{C}_{k+1}^T + \mathbf{R}_{k+1}\right)^{-1}$$

A posteriori state estimate

$$\hat{\mathbf{x}}_{k+1|k+1} = \hat{\mathbf{x}}_{k+1|k} + \mathbf{K}_{k+1}\mathbf{e}_{\mathbf{y}_{k+1|k}}$$

A posteriori error covariance

$$\mathbf{P}_{k+1|k+1} = \left(\mathbf{I} - \mathbf{K}_{k+1}\mathbf{C}_{k+1}\right)\mathbf{P}_{k+1|k}$$

end

---

the square-root of the covariance matrix is defined as:

$$\mathbf{P}_k^{\frac{1}{2}} = \mathbf{L}_k\mathbf{D}_k^{\frac{1}{2}}, \tag{5.13}$$

where $\mathbf{L}_k$ is a unit lower-triangular matrix and $\mathbf{D}_k$ is a diagonal matrix. The noise covariance matrices $\mathbf{Q}_k$ and $\mathbf{R}_k$ are similarly factorized.

5.3 Kalman Smoother

Smoothing refers to state estimation at time instant k based on the collected data up to the time instant $\ell > k$ [47]:

$$\hat{\mathbf{x}}_{k|\ell} = \mathbb{E}[\mathbf{x}_k | \mathbf{u}_{0:\ell}, \mathbf{y}_{0:\ell}]. \tag{5.14}$$

There are three main categories of smoothing [3, 47, 59]:

- *Fixed-point smoothing*: It is desired to estimate the state at a single fixed point in time, k, while the measurement process continues indefinitely ahead of this point, $\ell = k+1, k+2, \ldots$, for instance, the initial state, $\mathbf{x}_0$ can be estimated by a fixed-point smoother based on a noisy trajectory of the observations after $k = 0$.
- *Fixed-lag smoothing*: It is desired to estimate the state at a varying point in time, k, based on the measurements over a fixed-length of time ahead of this point, $\ell = k + L$, where L denotes the fixed length. Alternatively, it can be said that at time instant $k + L$, the fixed-lag smoother estimates the state at time instant k, hence, L plays the role of a fixed lag. This is the case with delayed state estimators presented in Chapter 3.
- *Fixed-interval smoothing*: It is desired to estimate the state at $k = 0,1, \ldots, N$ based on measurements for a fixed time interval (data span), $\ell = N$. The fixed-interval smoother is used for offline information processing based on noisy measurement data.

Let us consider the fixed-interval smoothing as the most common smoothing method. After performing the forward filtering iterations based on the collected dataset, the following quantities are computed and stored for $k = 0,1, \ldots, N-1$:

$$\hat{\mathbf{x}}_{k|k},$$
$$\mathbf{P}_{k|k},$$
$$\hat{\mathbf{x}}_{k+1|k},$$
$$\mathbf{P}_{k+1|k}.$$

Then, smoothing requires backward iterations. The *smoothing gain* is computed as:

$$\mathbf{K}_k^S = \mathbf{P}_{k|k}\mathbf{A}_k^T\mathbf{P}_{k+1|k}^{-1}, \tag{5.15}$$

which is used to calculate the smoothed state:

$$\hat{\mathbf{x}}_{k|N} = \hat{\mathbf{x}}_{k|k} + \mathbf{K}_k^S\left(\hat{\mathbf{x}}_{k+1|N} - \hat{\mathbf{x}}_{k+1|k}\right), \tag{5.16}$$

and its corresponding estimation error covariance matrix:

$$\mathbf{P}_{k|N} = \mathbf{P}_{k|k} + \mathbf{K}_k^S\left(\mathbf{P}_{k+1|N} - \mathbf{P}_{k+1|k}\right)\left(\mathbf{K}_k^S\right)^T, \tag{5.17}$$

for $k = N-1, N-2, \ldots, 1, 0$ [47].

5.4 Information Filter

Regarding the fact that the Fisher information matrix is the inverse of the covariance matrix, an alternative formulation of the Kalman filter is obtained, where the inverse of the estimation error covariance matrix is recursively computed for both the prediction and the update stages. This alternative formulation is known as the *information filter* [47]. Two key entities in the information filter algorithm are the information matrix:

$$\mathbf{P}_{k|k}^{-1}, \tag{5.18}$$

and the information state vector:

$$\hat{\mathbf{z}}_{k|k} = \mathbf{P}_{k|k}^{-1}\hat{\mathbf{x}}_{k|k}. \tag{5.19}$$

Algorithm 5.2 summarizes the information filter procedure [69]. For cases without any initial estimate, the initialization can be performed as follows [47]:

$$\mathbf{P}_0^{-1} = \mathbf{0}, \tag{5.20}$$

$$\hat{\mathbf{z}}_0 = \mathbf{P}_0^{-1}\hat{\mathbf{x}}_0 = \mathbf{0}. \tag{5.21}$$

5.5 Extended Kalman Filter

Assuming that the required derivatives exist and can be obtained with reasonable effort, in the EKF algorithm, the corresponding nonlinear functions are replaced with their first-order Taylor series approximations about the current state estimate in each iteration. The EKF algorithm is summarized in Algorithm 5.3.

5.6 Extended Information Filter

Similar to the EKF algorithm, application of the information filter can be extended to nonlinear systems. As shown in Algorithm 5.4, the extended information filter is derived based on the linearization of the state-space model about the current state estimate in each iteration [69].

5.7 Divided-Difference Filter

The idea behind the DDF is conceptually similar to the EKF, however, the implementation is different [64, 70]. DDF uses a multidimensional extension of the

Algorithm 5.2: Information filter

State-space model

$$\mathbf{x}_{k+1} = \mathbf{A}_k\mathbf{x}_k + \mathbf{B}_k\mathbf{u}_k + \mathbf{v}_k$$
$$\mathbf{y}_k = \mathbf{C}_k\mathbf{x}_k + \mathbf{D}_k\mathbf{u}_k + \mathbf{w}_k$$

where $\mathbf{v}_k \sim \mathcal{N}(\mathbf{0}, \mathbf{Q}_k)$ and $\mathbf{w}_k \sim \mathcal{N}(\mathbf{0}, \mathbf{R}_k)$.

Initialization

$$\hat{\mathbf{x}}_0 = \mathbb{E}[\mathbf{x}_0]$$
$$\mathbf{P}_0 = \mathbb{E}\left[\left(\mathbf{x}_0 - \mathbb{E}[\mathbf{x}_0]\right)\left(\mathbf{x}_0 - \mathbb{E}[\mathbf{x}_0]\right)^T\right]$$
$$\hat{\mathbf{z}}_0 = \mathbf{P}_0^{-1}\hat{\mathbf{x}}_0$$

for $k = 0, 1, \ldots,$ **do**

A priori information matrix

$$\mathbf{P}_{k+1|k}^{-1} = \left(\mathbf{A}_k\mathbf{P}_{k|k}\mathbf{A}_k^T + \mathbf{Q}_k\right)^{-1}$$

A priori information state estimate

$$\hat{\mathbf{z}}_{k+1|k} = \mathbf{P}_{k+1|k}^{-1}\mathbf{A}_k\mathbf{P}_{k|k}\hat{\mathbf{z}}_{k|k}$$

A posteriori information matrix

$$\mathbf{P}_{k+1|k+1}^{-1} = \mathbf{P}_{k+1|k}^{-1} + \mathbf{C}_{k+1}^T\mathbf{R}_{k+1}^{-1}\mathbf{C}_{k+1}$$

A posteriori information state estimate

$$\hat{\mathbf{z}}_{k+1|k+1} = \hat{\mathbf{z}}_{k+1|k} + \mathbf{C}_{k+1}^T\mathbf{R}_{k+1}^{-1}\mathbf{y}_{k+1}$$

A posteriori state estimate

$$\hat{\mathbf{x}}_{k+1|k+1} = \mathbf{P}_{k+1|k+1}\hat{\mathbf{z}}_{k+1|k+1}$$

end

Stirling's interpolation formula to approximate the nonlinear functions in the state and measurement equations [71, 72]. Unlike the Taylor series approximation used in the EKF, the interpolation formula used in the DDF does not require the derivatives of the corresponding nonlinear functions. It only uses function evaluations. This facilitates the estimation process even at singular points, where the derivatives do not exist. However, the DDF is more computationally demanding than the EKF. Algorithm 5.5 summarizes the DDF procedure. In the DDF algorithm, the *Householder triangularization* is used to transform rectangular matrices to square Cholesky factors [64, 70].

Algorithm 5.3: Extended Kalman filter

State-space model

$$\mathbf{x}_{k+1} = \mathbf{f}(\mathbf{x}_k, \mathbf{u}_k) + \mathbf{v}_k$$
$$\mathbf{y}_k = \mathbf{g}(\mathbf{x}_k, \mathbf{u}_k) + \mathbf{w}_k$$

where $\mathbf{v}_k \sim \mathcal{N}(\mathbf{0}, \mathbf{Q}_k)$ and $\mathbf{w}_k \sim \mathcal{N}(\mathbf{0}, \mathbf{R}_k)$.

Initialization

$$\hat{\mathbf{x}}_0 = \mathbb{E}[\mathbf{x}_0]$$
$$\mathbf{P}_0 = \mathbb{E}\left[\left(\mathbf{x}_0 - \mathbb{E}[\mathbf{x}_0]\right)\left(\mathbf{x}_0 - \mathbb{E}[\mathbf{x}_0]\right)^T\right]$$

for $k = 0, 1, \ldots,$ **do**

Linearization of the state-transition function

$$\mathbf{A}_k = \nabla_{\mathbf{x}} \mathbf{f}|_{\mathbf{x}=\hat{\mathbf{x}}_{k|k}}$$

A priori state estimate

$$\hat{\mathbf{x}}_{k+1|k} = \mathbf{f}(\hat{\mathbf{x}}_{k|k}, \mathbf{u}_k)$$

A priori error covariance

$$\mathbf{P}_{k+1|k} = \mathbf{A}_k \mathbf{P}_{k|k} \mathbf{A}_k^T + \mathbf{Q}_k$$

Linearization of the measurement function

$$\mathbf{C}_{k+1} = \nabla_{\mathbf{x}} \mathbf{g}|_{\mathbf{x}=\hat{\mathbf{x}}_{k+1|k}}$$

A priori measurement estimate

$$\hat{\mathbf{y}}_{k+1|k} = \mathbf{g}(\hat{\mathbf{x}}_{k+1|k}, \mathbf{u}_{k+1})$$

A priori output error estimate (innovation)

$$\mathbf{e}_{\mathbf{y}_{k+1|k}} = \mathbf{y}_{k+1} - \hat{\mathbf{y}}_{k+1|k}$$

Kalman gain

$$\mathbf{K}_{k+1} = \mathbf{P}_{k+1|k} \mathbf{C}_{k+1}^T \left(\mathbf{C}_{k+1} \mathbf{P}_{k|k} \mathbf{C}_{k+1}^T + \mathbf{R}_{k+1}\right)^{-1}$$

A posteriori state estimate

$$\hat{\mathbf{x}}_{k+1|k+1} = \hat{\mathbf{x}}_{k+1|k} + \mathbf{K}_{k+1} \mathbf{e}_{\mathbf{y}_{k+1|k}}$$

A posteriori error covariance

$$\mathbf{P}_{k+1|k+1} = \left(\mathbf{I} - \mathbf{K}_{k+1} \mathbf{C}_{k+1}\right) \mathbf{P}_{k+1|k}$$

end

Algorithm 5.4: Extended information filter

State-space model

$$\mathbf{x}_{k+1} = \mathbf{f}(\mathbf{x}_k, \mathbf{u}_k) + \mathbf{v}_k$$
$$\mathbf{y}_k = \mathbf{g}(\mathbf{x}_k, \mathbf{u}_k) + \mathbf{w}_k$$

where $\mathbf{v}_k \sim \mathcal{N}(\mathbf{0}, \mathbf{Q}_k)$ and $\mathbf{w}_k \sim \mathcal{N}(\mathbf{0}, \mathbf{R}_k)$.

Initialization

$$\hat{\mathbf{x}}_0 = \mathbb{E}[\mathbf{x}_0]$$
$$\mathbf{P}_0 = \mathbb{E}\left[\left(\mathbf{x}_0 - \mathbb{E}[\mathbf{x}_0]\right)\left(\mathbf{x}_0 - \mathbb{E}[\mathbf{x}_0]\right)^T\right]$$
$$\hat{\mathbf{z}}_0 = \mathbf{P}_0^{-1}\hat{\mathbf{x}}_0$$

for $k = 0, 1, \ldots,$ **do**

Linearization of the state-transition function

$$\mathbf{A}_k = \nabla_{\mathbf{x}}\mathbf{f}|_{\mathbf{x}=\hat{\mathbf{x}}_{k|k}}$$

A priori information matrix

$$\mathbf{P}_{k+1|k}^{-1} = \left(\mathbf{A}_k\mathbf{P}_{k|k}\mathbf{A}_k^T + \mathbf{Q}_k\right)^{-1}$$

A priori information state estimate

$$\hat{\mathbf{z}}_{k+1|k} = \mathbf{P}_{k+1|k}^{-1}\mathbf{f}(\hat{\mathbf{x}}_{k|k}, \mathbf{u}_k)$$

Linearization of the measurement function

$$\mathbf{C}_{k+1} = \nabla_{\mathbf{x}}\mathbf{g}|_{\mathbf{x}=\hat{\mathbf{x}}_{k+1|k}}$$

A posteriori information matrix

$$\mathbf{P}_{k+1|k+1}^{-1} = \mathbf{P}_{k+1|k}^{-1} + \mathbf{C}_{k+1}^T\mathbf{R}_{k+1}^{-1}\mathbf{C}_{k+1}$$

A posteriori information state estimate

$$\begin{aligned}\hat{\mathbf{z}}_{k+1|k+1} = {} & \hat{\mathbf{z}}_{k+1|k} \\ & + \mathbf{C}_{k+1}^T\mathbf{R}_{k+1}^{-1}\left(\mathbf{y}_{k+1} - \mathbf{g}(\hat{\mathbf{x}}_{k+1|k}, \mathbf{u}_{k+1}) + \mathbf{C}_{k+1}\hat{\mathbf{x}}_{k+1|k}\right)\end{aligned}$$

A posteriori state estimate

$$\hat{\mathbf{x}}_{k+1|k+1} = \mathbf{P}_{k+1|k+1}\hat{\mathbf{z}}_{k+1|k+1}$$

end

Algorithm 5.5: Divided-difference filter

State-space model

$$\mathbf{x}_{k+1} = \mathbf{f}(\mathbf{x}_k, \mathbf{u}_k) + \mathbf{v}_k$$
$$\mathbf{y}_k = \mathbf{g}(\mathbf{x}_k, \mathbf{u}_k) + \mathbf{w}_k$$

where $\mathbf{v}_k \sim \mathcal{N}(\mathbf{0}, \mathbf{Q}_k)$ and $\mathbf{w}_k \sim \mathcal{N}(\mathbf{0}, \mathbf{R}_k)$.

Initialization

$$\hat{\mathbf{x}}_0 = \mathbb{E}[\mathbf{x}_0]$$
$$\mathbf{P}_0 = \mathbb{E}\left[\left(\mathbf{x}_0 - \mathbb{E}[\mathbf{x}_0]\right)\left(\mathbf{x}_0 - \mathbb{E}[\mathbf{x}_0]\right)^T\right]$$

for $k = 0, 1, \ldots,$ **do**

A priori state estimate

$$\hat{\mathbf{x}}_{k+1|k} = \mathbf{f}(\hat{\mathbf{x}}_{k|k}, \mathbf{u}_k)$$

A priori error covariance

$$\left(\mathbf{S}^{\mathbf{xx}}_{k+1|k}\right)_{ij} = \frac{1}{2\alpha}\left[\mathbf{f}_i\left(\hat{\mathbf{x}}_{k|k} + \alpha\left(\mathbf{P}^{\frac{1}{2}}_{k|k}\right)_j, \mathbf{u}_k\right) - \mathbf{f}_i\left(\hat{\mathbf{x}}_{k|k} - \alpha\left(\mathbf{P}^{\frac{1}{2}}_{k|k}\right)_j, \mathbf{u}_k\right)\right]$$

$$\mathbf{P}^{\frac{1}{2}}_{k+1|k} = \text{triang}\left(\left[\mathbf{S}^{\mathbf{xx}}_{k+1|k} \quad \mathbf{Q}^{\frac{1}{2}}_k\right]\right)$$
$$\mathbf{P}_{k+1|k} = \mathbf{P}^{\frac{1}{2}}_{k+1|k}\left(\mathbf{P}^{\frac{1}{2}}_{k+1|k}\right)^T$$
$$\mathbf{Q}_k = \mathbf{Q}^{\frac{1}{2}}_k\left(\mathbf{Q}^{\frac{1}{2}}_k\right)^T$$

where triang(.) denotes the Householder triangularization, $\mathbf{f}_i(\cdot)$ denotes the ith element of $\mathbf{f}(\cdot)$, $\left(\mathbf{S}^{\mathbf{xx}}_{k+1|k}\right)_{ij}$ denotes the (i, j)th element of matrix $\mathbf{S}^{\mathbf{xx}}_{k+1|k}$, and $\left(\mathbf{P}^{\frac{1}{2}}_{k|k}\right)_j$ denotes the jth column of matrix $\mathbf{P}^{\frac{1}{2}}_{k|k}$.

A priori measurement estimate

$$\hat{\mathbf{y}}_{k+1|k} = \mathbf{g}(\hat{\mathbf{x}}_{k+1|k}, \mathbf{u}_{k+1})$$

A priori output error estimate (innovation)

$$\mathbf{e}_{\mathbf{y}_{k+1|k}} = \mathbf{y}_{k+1} - \hat{\mathbf{y}}_{k+1|k}$$

Filter gain

$$\left(\mathbf{S}^{\mathbf{xy}}_{k+1|k}\right)_{ij} = \frac{1}{2\alpha}\left[\mathbf{g}_i\left(\hat{\mathbf{x}}_{k+1|k} + \alpha\left(\mathbf{P}^{\frac{1}{2}}_{k+1|k}\right)_j, \mathbf{u}_{k+1}\right) - \mathbf{g}_i\left(\hat{\mathbf{x}}_{k+1|k} - \alpha\left(\mathbf{P}^{\frac{1}{2}}_{k+1|k}\right)_j, \mathbf{u}_{k+1}\right)\right]$$

$$\left(\mathbf{P}^{\mathbf{yy}}_{k+1|k}\right)^{\frac{1}{2}} = \text{triang}\left(\left[\mathbf{S}^{\mathbf{xy}}_{k+1|k} \quad \mathbf{R}^{\frac{1}{2}}_k\right]\right)$$

$$\mathbf{P}^{\mathbf{yy}}_{k+1|k} = \left(\mathbf{P}^{\mathbf{yy}}_{k+1|k}\right)^{\frac{1}{2}}\left(\mathbf{P}^{\mathbf{yy}}_{k+1|k}\right)^{\frac{T}{2}}$$

$$\mathbf{R}_k = \mathbf{R}^{\frac{1}{2}}_k\left(\mathbf{R}^{\frac{1}{2}}_k\right)^T$$

$$\mathbf{K}_{k+1} = \mathbf{P}^{\frac{1}{2}}_{k+1|k}\,\mathbf{S}^{\mathbf{xy}}_{k+1|k}\left(\mathbf{P}^{\mathbf{yy}}_{k+1|k}\right)^{-1}$$

where $\mathbf{g}_i(\cdot)$ denotes the ith element of $\mathbf{g}(\cdot)$, $\left(\mathbf{S}^{\mathbf{xy}}_{k+1|k}\right)_{ij}$ denotes the (i,j)th element of matrix $\mathbf{S}^{\mathbf{xy}}_{k+1|k}$, and $\left(\mathbf{P}^{\frac{1}{2}}_{k|k}\right)_j$ denotes the jth column of matrix $\mathbf{P}^{\frac{1}{2}}_{k|k}$.

A posteriori state estimate

$$\hat{\mathbf{x}}_{k+1|k+1} = \hat{\mathbf{x}}_{k+1|k} + \mathbf{K}_{k+1}\mathbf{e}_{\mathbf{y}_{k+1|k}}$$

A posteriori error covariance

$$\mathbf{P}^{\frac{1}{2}}_{k+1|k+1} = \text{triang}\left(\left[\mathbf{P}^{\frac{1}{2}}_{k+1|k} - \mathbf{K}_{k+1}\mathbf{S}^{\mathbf{xy}}_{k+1|k} \quad \mathbf{K}_{k+1}\mathbf{R}^{\frac{1}{2}}_k\right]\right)$$

end

5.8 Unscented Kalman Filter

UKF is an alternative to the EKF that has a better performance, when it comes to handling systems with severe nonlinearities. The UKF was first proposed in [73] and then further developed in [66, 74, 75]. Both EKF and UKF are suboptimal algorithms, which were developed based on the assumption that process noise and measurement noise are Gaussian. However, these two filtering algorithms use two different approaches for propagating the corresponding Gaussian random variables through the system dynamics.

In the EKF, a Gaussian distribution is considered as the approximate representation for the true distribution of the state. This approximate distribution is then propagated in an analytic manner through the linear dynamic model obtained from the first-order linearization of the nonlinear dynamic model under study. In case of severe nonlinearity, the crude first-order approximation may introduce large deviations of the posterior mean and covariance from their counterparts of the true state distribution. In the UKF, although the idea of using an approximate Gaussian representation for the state distribution is adopted again, this approximate distribution is represented by a minimal set of sample points, which are then propagated through the nonlinear dynamics. The posterior mean and covariance obtained by the UKF are accurate approximations up to the second-order in the sense of the Taylor series expansion for their counterparts of the true state distribution. While the UKF achieves second-order accuracy without the need for calculating Jacobin and Hessian of the nonlinear functions, the EKF achieves first-order accuracy with the need for Jacobian calculation. Algorithm 5.6 represents the UKF procedure [75].

5.9 Cubature Kalman Filter

CKF is another powerful approximation of the Bayesian filter [67]. The main idea for derivation of this filter is the fact that, under the Gaussian assumption, approximation of the Bayesian filter reduces to computing multidimensional integrals of a special form, where the integrand can be described as:

$$\text{(nonlinear function)} \times \text{(Gaussian function)} \tag{5.23}$$

While the UKF uses $2n_x + 1$ sigma points, the CKF uses $2n_x$ cubature points. It is worth noting that if the scaling parameter λ in the UKF is equal to zero, UKF will reduce to CKF. Algorithm 5.7 represents the CKF procedure.

Algorithm 5.6: Unscented Kalman filter

State-space model

$$\mathbf{x}_{k+1} = \mathbf{f}(\mathbf{x}_k, \mathbf{u}_k) + \mathbf{v}_k$$
$$\mathbf{y}_k = \mathbf{g}(\mathbf{x}_k, \mathbf{u}_k) + \mathbf{w}_k$$

where $\mathbf{v}_k \sim \mathcal{N}(\mathbf{0}, \mathbf{Q}_k)$ and $\mathbf{w}_k \sim \mathcal{N}(\mathbf{0}, \mathbf{R}_k)$.

Initialization

$$\hat{\mathbf{x}}_0 = \mathbb{E}[\mathbf{x}_0]$$
$$\mathbf{P}_0 = \mathbb{E}\left[\left(\mathbf{x}_0 - \mathbb{E}[\mathbf{x}_0]\right)\left(\mathbf{x}_0 - \mathbb{E}[\mathbf{x}_0]\right)^T\right]$$

$$\begin{aligned} w_0^x &= \frac{\lambda}{n_x + \lambda} \\ w_0^P &= \frac{\lambda}{n_x + \lambda} + 1 - \alpha^2 + \beta \\ w_i^x = w_i^P &= \frac{1}{2(n_x + \lambda)}, \qquad i = 1, \dots, 2n_x \\ \gamma &= \sqrt{n_x + \lambda} \end{aligned} \tag{5.22}$$

where λ is the composite scaling parameter and n_x is the dimension of the state vector. Parameters α and β determine the accuracy of the third- and higher-order moments.

for $k = 0, 1, \dots,$ **do**

Calculate the sigma points

$$\mathcal{X}_{k|k} = \left[\hat{\mathbf{x}}_{k|k} \quad \mathbf{1}_{n_x}^T \otimes \hat{\mathbf{x}}_{k|k} + \gamma \mathbf{P}_{k|k}^{\frac{1}{2}} \quad \mathbf{1}_{n_x}^T \otimes \hat{\mathbf{x}}_{k|k} - \gamma \mathbf{P}_{k|k}^{\frac{1}{2}}\right]$$
$$\mathbf{P}_{k|k} = \mathbf{P}_{k|k}^{\frac{1}{2}}\left(\mathbf{P}_{k|k}^{\frac{1}{2}}\right)^T$$

where $\mathbf{1}_{n_x}^T$ is an n_x-dimensional all-one row vector and $\otimes$ denotes the Kronecker product.

A priori state estimate

$$\mathcal{X}_{k+1|k}^i = \mathbf{f}(\mathcal{X}_{k|k}^i, \mathbf{u}_k)$$
$$\hat{\mathbf{x}}_{k+1|k} = \sum_{i=0}^{2n_x} w_i^x \mathcal{X}_{k+1|k}^i$$

where $\mathcal{X}_{k|k}^i$ denotes the ith column of matrix $\mathcal{X}_{k|k}$.

A priori error covariance

$$\mathbf{P}_{k+1|k} = \sum_{i=0}^{2n_x} w_i^P \left(\mathcal{X}_{k+1|k}^i - \hat{\mathbf{x}}_{k+1|k}\right)\left(\mathcal{X}_{k+1|k}^i - \hat{\mathbf{x}}_{k+1|k}\right)^T + \mathbf{Q}_k$$

A priori measurement estimate

$$\mathcal{Y}_{k+1|k}^i = \mathbf{g}\left(\mathcal{X}_{k+1|k}^i, \mathbf{u}_{k+1}\right)$$

$$\hat{\mathbf{y}}_{k+1|k} = \sum_{i=0}^{2n_x} w_i^x \mathcal{Y}_{k+1|k}^i$$

A priori output error estimate (innovation)

$$\mathbf{e}_{\mathbf{y}_{k+1|k}} = \mathbf{y}_{k+1} - \hat{\mathbf{y}}_{k+1|k}$$

Kalman gain

$$\mathbf{P}_{k+1|k}^{\mathbf{yy}} = \sum_{i=0}^{2n_x} w_i^P \left(\mathcal{Y}_{k+1|k}^i - \hat{\mathbf{y}}_{k+1|k}\right)\left(\mathcal{Y}_{k+1|k}^i - \hat{\mathbf{y}}_{k+1|k}\right)^T + \mathbf{R}_k$$

$$\mathbf{P}_{k+1|k}^{\mathbf{xy}} = \sum_{i=0}^{2n_x} w_i^P \left(\mathcal{X}_{k+1|k}^i - \hat{\mathbf{x}}_{k+1|k}\right)\left(\mathcal{Y}_{k+1|k}^i - \hat{\mathbf{y}}_{k+1|k}\right)^T$$

$$\mathbf{K}_{k+1} = \mathbf{P}_{k+1|k}^{\mathbf{xy}}\left(\mathbf{P}_{k+1|k}^{\mathbf{yy}}\right)^{-1}$$

A posteriori state estimate

$$\hat{\mathbf{x}}_{k+1|k+1} = \hat{\mathbf{x}}_{k+1|k} + \mathbf{K}_{k+1}\mathbf{e}_{\mathbf{y}_{k+1|k}}$$

A posteriori error covariance

$$\mathbf{P}_{k+1|k+1} = \mathbf{P}_{k+1|k} - \mathbf{K}_{k+1}\mathbf{P}_{k+1|k}^{\mathbf{yy}}\mathbf{K}_{k+1}^T$$

end

Algorithm 5.7: Cubature Kalman filter

State-space model

$$\mathbf{x}_{k+1} = \mathbf{f}(\mathbf{x}_k, \mathbf{u}_k) + \mathbf{v}_k$$
$$\mathbf{y}_k = \mathbf{g}(\mathbf{x}_k, \mathbf{u}_k) + \mathbf{w}_k$$

where $\mathbf{v}_k \sim \mathcal{N}(\mathbf{0}, \mathbf{Q}_k)$ and $\mathbf{w}_k \sim \mathcal{N}(\mathbf{0}, \mathbf{R}_k)$.

Initialization

$$\hat{\mathbf{x}}_0 = \mathbb{E}[\mathbf{x}_0]$$
$$\mathbf{P}_0 = \mathbb{E}\left[\left(\mathbf{x}_0 - \mathbb{E}[\mathbf{x}_0]\right)\left(\mathbf{x}_0 - \mathbb{E}[\mathbf{x}_0]\right)^T\right]$$

for $k = 0, 1, \ldots,$ **do**

Calculate the cubature points

$$\mathcal{X}_{k|k} = \left[\mathbf{1}_{n_x}^T \otimes \hat{\mathbf{x}}_{k|k} + \sqrt{n_x}\,\mathbf{P}_{k|k}^{\frac{1}{2}} \quad \mathbf{1}_{n_x}^T \otimes \hat{\mathbf{x}}_{k|k} - \sqrt{n_x}\,\mathbf{P}_{k|k}^{\frac{1}{2}}\right]$$
$$\mathbf{P}_{k|k} = \mathbf{P}_{k|k}^{\frac{1}{2}}\left(\mathbf{P}_{k|k}^{\frac{1}{2}}\right)^T$$

where $\mathbf{1}_{n_x}^T$ is an n_x-dimensional all-one row vector and $\otimes$ denotes the Kronecker product.

A priori state estimate

$$\mathcal{X}_{k+1|k}^i = \mathbf{f}(\mathcal{X}_{k|k}^i, \mathbf{u}_k)$$
$$\hat{\mathbf{x}}_{k+1|k} = \frac{1}{2n_x}\sum_{i=1}^{2n_x}\mathcal{X}_{k+1|k}^i$$

where $\mathcal{X}_{k|k}^i$ denotes the ith column of matrix $\mathcal{X}_{k|k}$.

A priori error covariance

$$\mathbf{P}_{k+1|k} = \frac{1}{2n_x}\sum_{i=1}^{2n_x}\left(\mathcal{X}_{k+1|k}^i - \hat{\mathbf{x}}_{k+1|k}\right)\left(\mathcal{X}_{k+1|k}^i - \hat{\mathbf{x}}_{k+1|k}\right)^T + \mathbf{Q}_k$$

A priori measurement estimate

$$\mathcal{Y}^i_{k+1|k} = \mathbf{g}\left(\mathcal{X}^i_{k+1|k}, \mathbf{u}_{k+1}\right)$$

$$\hat{\mathbf{y}}_{k+1|k} = \frac{1}{2n_x}\sum_{i=1}^{2n_x} \mathcal{Y}^i_{k+1|k}$$

A priori output error estimate (innovation)

$$\mathbf{e}_{\mathbf{y}_{k+1|k}} = \mathbf{y}_{k+1} - \hat{\mathbf{y}}_{k+1|k}$$

Kalman gain

$$\mathbf{P}^{\mathbf{yy}}_{k+1|k} = \frac{1}{2n_x}\sum_{i=1}^{2n_x}\left(\mathcal{Y}^i_{k+1|k} - \hat{\mathbf{y}}_{k+1|k}\right)\left(\mathcal{Y}^i_{k+1|k} - \hat{\mathbf{y}}_{k+1|k}\right)^T + \mathbf{R}_k$$

$$\mathbf{P}^{\mathbf{xy}}_{k+1|k} = \frac{1}{2n_x}\sum_{i=1}^{2n_x}\left(\mathcal{X}^i_{k+1|k} - \hat{\mathbf{x}}_{k+1|k}\right)\left(\mathcal{Y}^i_{k+1|k} - \hat{\mathbf{y}}_{k+1|k}\right)^T$$

$$\mathbf{K}_{k+1} = \mathbf{P}^{\mathbf{xy}}_{k+1|k}\left(\mathbf{P}^{\mathbf{yy}}_{k+1|k}\right)^{-1}$$

A posteriori state estimate

$$\hat{\mathbf{x}}_{k+1|k+1} = \hat{\mathbf{x}}_{k+1|k} + \mathbf{K}_{k+1}\mathbf{e}_{\mathbf{y}_{k+1|k}}$$

A posteriori error covariance

$$\mathbf{P}_{k+1|k+1} = \mathbf{P}_{k+1|k} - \mathbf{K}_{k+1}\mathbf{P}^{\mathbf{yy}}_{k+1|k}\mathbf{K}^T_{k+1}$$

end

5.10 Generalized PID Filter

In the Kalman filter algorithm, the a posteriori state estimate $\hat{\mathbf{x}}_{k+1|k+1}$ is obtained by adding a corrective term to the a priori state estimate $\hat{\mathbf{x}}_{k+1|k}$. The corrective term is proportional to the a priori output error estimate (innovation):

$$\mathbf{e}_{\mathbf{y}_{k+1|k}} = \mathbf{y}_{k+1} - \hat{\mathbf{y}}_{k+1|k}, \tag{5.24}$$

$$\hat{\mathbf{x}}_{k+1|k+1} = \hat{\mathbf{x}}_{k+1|k} + \mathbf{K}_{k+1}\mathbf{e}_{\mathbf{y}_{k+1|k}}, \tag{5.25}$$

where $\mathbf{K}_{k+1}$ denotes the Kalman gain. The corrective term $\mathbf{K}_{k+1}\mathbf{e}_{\mathbf{y}_{k+1|k}}$ is reminiscent of the proportional controller. Inspired by the proportional-integral-derivative (PID) controller, the *generalized PID filter* uses a more sophisticated corrective term [76]. Considering the a posteriori output error estimate:

$$\mathbf{e}_{\mathbf{y}_{k|k}} = \mathbf{y}_k - \hat{\mathbf{y}}_{k|k}, \tag{5.26}$$

the generalized PID filter uses the following three corrective terms:

$$\mathbf{x}^P_{k+1} = \mathbf{K}^P_{k+1}\mathbf{e}_{\mathbf{y}_{k+1|k}}, \tag{5.27}$$

$$\mathbf{x}^I_{k+1} = \mathbf{x}^I_k + \mathbf{K}^I_{k+1}\mathbf{e}_{\mathbf{y}_{k|k}}, \tag{5.28}$$

$$\mathbf{x}^D_{k+1} = \mathbf{K}^D_{k+1}\left(\mathbf{e}_{\mathbf{y}_{k|k}} - \mathbf{e}_{\mathbf{y}_{k-1|k-1}}\right), \tag{5.29}$$

with $\mathbf{x}^I_0 = \mathbf{0}$ and $\mathbf{e}_{\mathbf{y}_{0|0}} - \mathbf{e}_{\mathbf{y}_{-1|-1}} = \mathbf{0}$. Therefore, the a posteriori state estimate is computed as:

$$\hat{\mathbf{x}}_{k+1|k+1} = \hat{\mathbf{x}}_{k+1|k} + \mathbf{x}^P_{k+1} + \mathbf{x}^I_{k+1} + \mathbf{x}^D_{k+1}. \tag{5.30}$$

Regarding the a posteriori state estimate error and its covariance matrix:

$$\mathbf{e}_{\mathbf{x}_{k+1|k+1}} = \mathbf{x}_{k+1} - \hat{\mathbf{x}}_{k+1|k+1}, \tag{5.31}$$

$$\mathbf{P}_{k+1|k+1} = \mathbb{E}\left[\mathbf{e}_{\mathbf{x}_{k+1|k+1}}\mathbf{e}^T_{\mathbf{x}_{k+1|k+1}}\right], \tag{5.32}$$

filter gains $\mathbf{K}^P_{k+1}$, $\mathbf{K}^I_{k+1}$, and $\mathbf{K}^D_{k+1}$ are computed by solving the following equations:

$$\frac{\partial}{\partial \mathbf{K}^P_{k+1}} \operatorname{trace}\left(\mathbf{P}_{k+1|k+1}\right) = \mathbf{0}, \tag{5.33}$$

$$\frac{\partial}{\partial \mathbf{K}^I_{k+1}} \operatorname{trace}\left(\mathbf{P}_{k+1|k+1}\right) = \mathbf{0}, \tag{5.34}$$

$$\frac{\partial}{\partial \mathbf{K}^D_{k+1}} \operatorname{trace}\left(\mathbf{P}_{k+1|k+1}\right) = \mathbf{0}, \tag{5.35}$$

where trace$(\cdot)$ denotes the trace of the matrix (sum of the diagonal elements). From the generalized PID filter, the following filters can be retrieved as special cases [76]:

- Kalman filter by setting $\mathbf{K}^I = \mathbf{0}$ and $\mathbf{K}^D = \mathbf{0}$.
- Generalized proportional-integral (PI) filter by setting $\mathbf{K}^D = \mathbf{0}$.
- Generalized proportional-derivative (PD) filter by setting $\mathbf{K}^I = \mathbf{0}$.

The generalized PID filter can be used for both linear and nonlinear systems.

5.11 Gaussian-Sum Filter

The *Gaussian-sum filter* extends application of Kalman filtering algorithms to non-Gaussian systems via approximating the posterior at each time instant k by a mixture of n_G Gaussians. Therefore, the Gaussian-sum filter can be implemented by running n_G Kalman filters in parallel [77]. Let us consider the discrete latent variable $\lambda_k \in \{1, \dots, n_G\}$ as the index for the Gaussians. At time instant k, passing the prior:

$$p\left(\mathbf{x}_k, \lambda_k = i | \mathbf{u}_{0:k}, \mathbf{y}_{0:k}\right) = \pi_k^i \mathcal{N}\left(\mathbf{x}_k | \boldsymbol{\mu}_k^i, \boldsymbol{\Sigma}_k^i\right), \tag{5.36}$$

through n_G Kalman filters, leads to the following posteriors:

$$p\left(\mathbf{x}_{k+1}, \lambda_k = i, \lambda_{k+1} = j | \mathbf{u}_{0:k+1}, \mathbf{y}_{0:k+1}\right) = \pi_{k+1}^{ij} \mathcal{N}\left(\mathbf{x}_{k+1} | \boldsymbol{\mu}_{k+1}^{ij}, \boldsymbol{\Sigma}_{k+1}^{ij}\right), \tag{5.37}$$

where

$$\pi_{k+1}^{ij} = \pi_k^i \, p\left(\lambda_{k+1} = j | \lambda_k = i\right). \tag{5.38}$$

For each value of j, one Gaussian mixture is extracted from the n_G Gaussian mixtures as:

$$p\left(\mathbf{x}_{k+1}, \lambda_{k+1} = j | \mathbf{u}_{0:k+1}, \mathbf{y}_{0:k+1}\right) = \pi_{k+1}^{j} \mathcal{N}\left(\mathbf{x}_{k+1} | \boldsymbol{\mu}_{k+1}^{j}, \boldsymbol{\Sigma}_{k+1}^{j}\right), \tag{5.39}$$

A mixture of Gaussians, p, can be approximated by a single Gaussian, q, through minimizing the Kullback–Leibler divergence (KLD) between the two distributions:

$$q = \arg\min_{q} \ \text{KLD}\,(q||p), \tag{5.40}$$

which can be performed by *moment matching*. For Gaussian distributions, moment matching involves preserving the first two moments:

$$\pi_{k+1}^{j} = \sum_{i=1}^{n_G} \pi_{k+1}^{ij}, \tag{5.41}$$

$$\pi_{k+1}^{j|i} = \frac{\pi_{k+1}^{ij}}{\sum_{j'=1}^{n_G} \pi_{k+1}^{ij'}}, \tag{5.42}$$

$$\boldsymbol{\mu}_{k+1}^{j} = \sum_{i=1}^{n_G} \pi_{k+1}^{j|i} \boldsymbol{\mu}_{k+1}^{ij}, \tag{5.43}$$

$$\boldsymbol{\Sigma}_{k+1}^{j} = \sum_{i=1}^{n_G} \pi_{k+1}^{j|i} \left(\boldsymbol{\Sigma}_{k+1}^{ij} + \left(\boldsymbol{\mu}_{k+1}^{ij} - \boldsymbol{\mu}_{k+1}^{j}\right)\left(\boldsymbol{\mu}_{k+1}^{ij} - \boldsymbol{\mu}_{k+1}^{j}\right)^T\right). \tag{5.44}$$

This algorithm needs running n_G^2 Kalman filters. To reduce the computational burden, at each time instant k, the prior (5.36) can be marginalized over the discrete latent variable λ to be represented by a single Gaussian. Then, n_G Kalman filters are deployed to compute n_G posteriors (one for each value of λ). This is the idea behind the *interacting multiple model* (IMM) algorithm [47].

5.12 Applications

Six case studies are reviewed in this section regarding application of Kalman filter and its variants in information fusion, augmented reality, urban traffic network, cybersecurity in power systems, and healthcare.

5.12.1 Information Fusion

Information fusion refers to combining raw or processed data from different sources into a coherent structure by removing possible inconsistencies. Fusion can be performed at different levels including raw data, feature, and decision [78, 79].

5.12.2 Augmented Reality

Augmented reality heavily relies on fast and accurate estimation of the position and orientation of the camera. For this application, an EKF-based data fusion algorithm has been proposed to fuse measurements from inertial sensors and computer vision, which deploys a detailed dynamic model of the camera [80].

5.12.3 Urban Traffic Network

In urban networks, real-time traffic state estimation plays a key role in traffic control and management. Assuming that the connected-vehicle data is available, a two-layer state estimator has been proposed for urban traffic considering the link layer and the network layer [81]. This assumption allows for transforming the nonlinear model of the urban network with switching dynamics to a non-switching linear-in-parameter model. Therefore, Kalman filter can be used to estimate the traffic state. While in the link layer, state variables are the queue tail location and the number of vehicles in the queue, in the network layer, state variables include the total number of vehicles per link and the turning rates at the intersections. The number of sensor links and their locations are determined based on an observability analysis. The estimated link-layer states are used to estimate the network layer states. In the proposed method, detectors must be deployed only at the borders of the network. The effectiveness of the algorithm was evaluated based on realistic microscopic simulations.

5.12.4 Cybersecurity of Power Systems

Real-time monitoring of power systems rely on dynamic state estimation, which is vulnerable to false data injection cyberattacks. A secure hybrid estimation algorithm has been proposed that deploys a dynamic model for the attack vector [82].

In the proposed method, an unknown input observer (UIO) is used to obtain an initial estimate of the system state, which is used to obtain a dynamic model for the attack vector. Then, the main system model and the obtained attack model are combined. A Kalman filter is deployed for coestimation of the attack and the system states using the resulting augmented model. Occurrence of a cyberattack can be quickly detected based on an accurate estimate of the attack vector. The proposed method was evaluated via simulations on the IEEE 14-bus and 57-bus test systems. Furthermore, real-time feasibility of the proposed algorithm was investigated through software-in-the-loop experiments considering the typical scan rates of commercial phasor measurement units.

5.12.5 Incidence of Influenza

Centers for Disease Control and Prevention provide a measure of influenza incidence in the United States. Such a measure is defined as the weekly percentage of weighted influenza-like illness (wILI). Since data gathering and processing takes time, wILI values are released with one week of delay. However, wILI can be predicted in real time by paying attention to search volume for influenza-related terms, site traffic for influenza-related pages, and pharmaceutical sales for influenza-related products as side information [83]. While the number of visits to the influenza-related webpages of the Centers for Disease Control and Prevention are available for each state, the number of Wikipedia page visits may not be available at the same geographic scale. Hence, observations occur at varying temporal and geographic resolutions. Kalman filter can be used to estimate the wILI values at a geographic level and a temporal resolution of interest by fusing the available data. To predict the wILI, a multi-rate Kalman filter-based data fusion algorithm has been developed in [83], which uses a mean imputation approach to cope with missing data. In the corresponding state-space model, the measurement model reflects the geographic level to which the data belong.

5.12.6 COVID-19 Pandemic

According to the *susceptible-infected-recovered* model, the *effective reproduction number* of an infectious disease, $\mathscr{R}$, is related to the growth rate of the number of infected individuals. Kalman filter has been used to estimate this time-varying growth rate for COVID-19 in 124 countries from data on new cases [84]. The proposed algorithm was also used for evaluating the effectiveness of non-pharmaceutical coutermeasures in 14 European countries. The designed filter was effective even when the COVID-19 dynamics did not completely match the susceptible-infected-recovered model or the number of reported infected individuals was not accurate.

The daily evolution of susceptible, S_k, infected, I_k, and recovered, R_k, individuals can be expressed as [84]:

$$S_{k+1} = S_k - \beta_k I_k \frac{S_k}{N}, \tag{5.45}$$

$$I_{k+1} = I_k + \beta_k I_k \frac{S_k}{N} - \gamma I_k, \tag{5.46}$$

$$R_{k+1} = R_k + \gamma I_k, \tag{5.47}$$

where $N = S_k + I_k + R_k$ is the population size, β_k denotes the daily transmission rate, and γ denotes the daily transition rate from infected to recovered. The recovered group includes the fully recovered as well as the deceased individuals. Interactions between individuals may be reduced due to voluntarily adopted countermeasures or government policy restrictions, hence, β_k is time-varying.

The *basic reproduction number*, $\mathscr{R}_k^0$, refers to the average number of individuals, which are infected by a single infected individual, when everyone else is susceptible:

$$\mathscr{R}_{k+1}^0 = \frac{\beta_k}{\gamma}. \tag{5.48}$$

The *effective reproduction number*, $\mathscr{R}_k$, refers to the average number of individuals, which are infected by a single infected individual, when a fraction of individuals $\left(\frac{S_{k-1}}{N}\right)$ is susceptible:

$$\mathscr{R}_{k+1} = \mathscr{R}_{k+1}^0 \frac{S_k}{N}. \tag{5.49}$$

From (5.46), the daily growth rate of the number of infected individuals is computed as:

$$\begin{aligned} g_{k+1} &= \frac{I_{k+1} - I_k}{I_k} \\ &= \gamma \left(\mathscr{R}_{k+1} - 1 \right). \end{aligned} \tag{5.50}$$

Additive process noise will introduce randomness into deterministic equations (5.45)–(5.47) to take account of uncertainty. The observation sequence is constructed as a time series for I_k from data on new cases:

$$I_{k+1} = (1-\gamma) I_k + (\text{new cases})_{k+1}, \tag{5.51}$$

where additive measurement noise takes account of inaccurate reports on new cases. Kalman filter provides $\hat{I}_{k+1|k+1}$, which is used to estimate the daily growth rate, $\hat{g}_{k+1}$, from (5.50). Given $\hat{g}_{k+1}$, the effective reproduction number is estimated from (5.50) as [84]:

$$\hat{\mathscr{R}}_{k+1} = 1 + \frac{1}{\gamma}\, \hat{g}_{k+1}. \tag{5.52}$$

5.13 Concluding Remarks

Under the assumptions of linearity and Gaussianity, the celebrated Kalman filter algorithm provides a computationally tractable formulation of the optimal Bayesian filtering problem that achieves the optimal solution. Kalman filter is an optimal filter in the sense of minimizing the mean-square error. Furthermore, Kalman filter is an optimal filter regarding a variety of information-theoretic measures of optimality. For instance, the mutual information between the measurement and the estimation error [85] and the mutual information between the observation path and the signal [86] can be considered as such information-theoretic measures of optimality. Studying the Bayesian filtering problem from an information-theoretic perspective paves the way for designing a wide range of effective suboptimal filtering algorithms, which are able to handle nonlinearity and non-Gaussianity.

Another issue of interest is the effect of unreliable communication on the Kalman filter, which is critical in the context of *network control systems*. In such systems, information loss due to unreliable communication must be taken into account in control loops. To study the effect of unreliable communication on the Kalman filter, arrival of the observations can be modeled as a stochastic process. Then, the expected estimation error covariance depends on the system dynamics as well as the information loss probability. To be more precise, if the probability of arrival of an observation at a time instant is less than a critical value, the expectation of the estimation error covariance will become unbounded. Hence, in order to design an appropriate filter, attention must be paid to both the dynamics of the system whose state is estimated and the reliability of the communication channel through which measurements arrive [87].

6

Particle Filter

6.1 Introduction

The optimal nonlinear filtering problem relies on recursively calculating the posterior distribution. However, except for special cases, it is impossible to find closed-form solutions. To elaborate, when unlike Gaussian distributions, the corresponding probability distribution functions (PDFs) cannot be described by well-known mathematical functions, the whole PDF must be considered in the calculations. Regarding the fact that dealing with the whole PDF for a single variable would be similar to working with an infinite-dimensional vector, it is computationally intractable to compute the distributions of interest for general non-Gaussian cases. Therefore, function approximation techniques must be used to reasonably approximate the corresponding distributions in order to implement a suboptimal Bayesian filter, which is computationally tractable, for the general case of nonlinear non-Gaussian dynamic systems. We can look at the problem from another perspective as well. For systems with severe nonlinearity, function approximation based on local linearization such as Taylor series expansion used in the extended Kalman filter (EKF) is not adequate. An idea for handling such systems is that approximating a probability distribution may be easier than approximating a nonlinear function. Following this line of thinking, numerical methods are deployed to approximate the corresponding distributions. A popular numerical approximation scheme for this purpose is the *sequential Monte Carlo* (SMC) method. In the Monte Carlo method, a probability distribution of interest is approximated by a set of random samples, which are called *particles* [88]. Hence, the estimators that are implemented based on the Monte Carlo method are called *particle filters*. To be more precise, particle filters are simulation-based algorithms that recursively provide a set of samples, which are approximately distributed according to the desired posterior [89].

Nonlinear Filters: Theory and Applications, First Edition. Peyman Setoodeh, Saeid Habibi, and Simon Haykin.

6.2 Monte Carlo Method

Let us assume that it is desired to approximate a generic PDF, $\pi(\mathbf{x})$. If it is possible to draw $n_p \gg 1$ samples, $\{\mathbf{x}^i | i = 1, \ldots, n_p\}$, according to $\pi(\mathbf{x})$, then, the Monte Carlo method provides the following empirical approximation for $\pi(\mathbf{x})$:

$$\hat{\pi}(\mathbf{x}) = \frac{1}{n_p} \sum_{i=1}^{n_p} \delta(\mathbf{x} - \mathbf{x}^i), \tag{6.1}$$

where δ denotes the *Dirac delta function*. Similarly, the expected value of a function $\mathbf{f}(\mathbf{x})$ with respect to distribution $\pi(\mathbf{x})$:

$$\mathbf{I} = \int \pi(\mathbf{x})\mathbf{f}(\mathbf{x})d\mathbf{x} \tag{6.2}$$

can be approximated by the Monte Carlo method as:

$$\begin{aligned} \mathbf{I}_{n_p} &= \int \hat{\pi}(\mathbf{x})\mathbf{f}(\mathbf{x})d\mathbf{x} \\ &= \frac{1}{n_p} \sum_{i=1}^{n_p} \mathbf{f}(\mathbf{x}^i). \end{aligned} \tag{6.3}$$

If samples $\mathbf{x}^i$ are independent, then, $\mathbf{I}_{n_p}$ will be an unbiased estimate, which almost surely converges to $\mathbf{I}$ due to the law of large numbers. If the covariance matrix:

$$\mathbf{\Sigma} = \int \pi(\mathbf{x})(\mathbf{f}(\mathbf{x}) - \mathbf{I})(\mathbf{f}(\mathbf{x}) - \mathbf{I})^T d\mathbf{x}, \tag{6.4}$$

has finite entries, then, according to the central limit theorem, for the estimation error, $\mathbf{I} - \mathbf{I}_{n_p}$, we will have [46]:

$$\lim_{n_p \to \infty} \sqrt{n_p}(\mathbf{I} - \mathbf{I}_{n_p}) \sim \mathcal{N}(\mathbf{0}, \mathbf{\Sigma}), \tag{6.5}$$

where $\mathcal{N}(.,.)$ denotes the normal distribution. Therefore, the rate of convergence of the Monte Carlo method is independent of the dimension of $\mathbf{x}$ [46, 89]. However, we may not be able to directly draw samples from complex and high-dimensional distributions, $\pi(\mathbf{x})$. This issue is addressed using the *importance sampling* (IS) method.

6.3 Importance Sampling

When it is not possible to draw samples directly from the distribution of interest, $\pi(\mathbf{x})$, we proceed by drawing samples from a *proposal* or *importance* density, $q(\mathbf{x})$, which has the same support as the original PDF:

$$\pi(\mathbf{x}) > 0 \Rightarrow q(\mathbf{x}) > 0 \quad \forall \mathbf{x} \in \mathbb{R}^{n_x}, \tag{6.6}$$

hence, the name importance sampling. If the aforementioned condition is satisfied, and $\pi(\mathbf{x})/q(\mathbf{x})$ is upper bounded, then we have:

$$\mathbf{I} = \int \pi(\mathbf{x})\mathbf{f}(\mathbf{x})d\mathbf{x} \\ = \int q(\mathbf{x})\frac{\pi(\mathbf{x})}{q(\mathbf{x})}\mathbf{f}(\mathbf{x})d\mathbf{x}. \tag{6.7}$$

Drawing $n_p \gg 1$ independent samples, $\{\mathbf{x}^i | i = 1, \ldots, n_p\}$, according to $q(\mathbf{x})$, a Monte Carlo estimate of the integral $\mathbf{I}$ can be obtained from the following weighted sum:

$$\mathbf{I}_{n_p} = \frac{1}{n_p}\sum_{i=1}^{n_p} \tilde{w}(\mathbf{x}^i)\mathbf{f}(\mathbf{x}^i), \tag{6.8}$$

with the importance weights:

$$\tilde{w}(\mathbf{x}^i) = \frac{\pi(\mathbf{x}^i)}{q(\mathbf{x}^i)}. \tag{6.9}$$

Normalization of the importance weights is required in case that the normalizing factor of the density of interest $\pi(\mathbf{x})$ is not known. Then, we will have:

$$\mathbf{I}_{n_p} = \frac{\frac{1}{n_p}\sum_{i=1}^{n_p} \tilde{w}(\mathbf{x}^i)\mathbf{f}(\mathbf{x}^i)}{\frac{1}{n_p}\sum_{j=1}^{n_p} \tilde{w}(\mathbf{x}^j)} \\ = \sum_{i=1}^{n_p} w(\mathbf{x}^i)\mathbf{f}(\mathbf{x}^i), \tag{6.10}$$

with the following normalized importance weights:

$$w(\mathbf{x}^i) = \frac{\tilde{w}(\mathbf{x}^i)}{\sum_{j=1}^{n_p} \tilde{w}(\mathbf{x}^j)}. \tag{6.11}$$

The importance sampling method can be deployed to implement a recursive suboptimal Bayesian filter, in which the posterior distribution is iteratively approximated by a set of random samples and their corresponding importance weights. For very large number of samples, this *sequential importance sampling* approaches the optimal Bayesian filter [46, 89]. The algorithm is developed in Section 6.4.

6.4 Sequential Importance Sampling

Let us consider the set of samples (support points) and their corresponding normalized weights, $\{\mathbf{x}^i_{0:k+1}, w^i_{k+1}\}$ for $i = 1, \ldots, n_p$, where $\sum_{i=1}^{n_p} w^i_{k+1} = 1$, and samples are drawn according to an importance posterior density, $q(\mathbf{x}_{0:k+1}|\mathbf{u}_{0:k+1},\mathbf{y}_{0:k+1})$. The posterior density can be approximated as:

$$p(\mathbf{x}_{0:k+1}|\mathbf{u}_{0:k+1},\mathbf{y}_{0:k+1}) \approx \sum_{i=1}^{n_p} w_{k+1}^i \delta(\mathbf{x}_{0:k+1} - \mathbf{x}_{0:k+1}^i). \tag{6.12}$$

From equation (6.9), we know that:

$$w_{k+1}^i \propto \frac{p(\mathbf{x}_{0:k+1}^i|\mathbf{u}_{0:k+1},\mathbf{y}_{0:k+1})}{q(\mathbf{x}_{0:k+1}^i|\mathbf{u}_{0:k+1},\mathbf{y}_{0:k+1})}. \tag{6.13}$$

If the selected importance density has the following property:

$$q(\mathbf{x}_{0:k+1}^i|\mathbf{u}_{0:k+1},\mathbf{y}_{0:k+1}) = q(\mathbf{x}_{k+1}|\mathbf{x}_{0:k},\mathbf{u}_{0:k+1},\mathbf{y}_{0:k+1})q(\mathbf{x}_{0:k}|\mathbf{u}_{0:k},\mathbf{y}_{0:k}), \tag{6.14}$$

then, the existing samples

$$\mathbf{x}_{0:k}^i \sim q(\mathbf{x}_{0:k}|\mathbf{u}_{0:k},\mathbf{y}_{0:k}) \tag{6.15}$$

can be augmented with the updated state

$$\mathbf{x}_{k+1}^i \sim q(\mathbf{x}_{k+1}|\mathbf{x}_{0:k},\mathbf{u}_{0:k+1},\mathbf{y}_{0:k+1}) \tag{6.16}$$

to obtain

$$\mathbf{x}_{0:k+1}^i \sim q(\mathbf{x}_{0:k+1}|\mathbf{u}_{0:k+1},\mathbf{y}_{0:k+1}). \tag{6.17}$$

In the Bayesian framework, the posterior distribution provides the inference about the state, $\mathbf{x}_{0:k+1}$, based on a realization of the measurements, $\mathbf{y}_{0:k+1}$:

$$\begin{aligned}
p(\mathbf{x}_{0:k+1}|\mathbf{u}_{0:k+1},\mathbf{y}_{0:k+1}) &= p(\mathbf{x}_{0:k+1}|\mathbf{u}_{0:k+1},\mathbf{y}_{k+1},\mathbf{y}_{0:k}) \\
&= \frac{p(\mathbf{y}_{k+1}|\mathbf{x}_{0:k+1},\mathbf{u}_{0:k+1},\mathbf{y}_{0:k})p(\mathbf{x}_{0:k+1}|\mathbf{u}_{0:k},\mathbf{y}_{0:k})}{p(\mathbf{y}_{k+1}|\mathbf{u}_{0:k+1},\mathbf{y}_{0:k})} \\
&= \frac{p(\mathbf{y}_{k+1}|\mathbf{x}_{0:k+1},\mathbf{u}_{0:k+1},\mathbf{y}_{0:k})p(\mathbf{x}_{k+1},\mathbf{x}_{0:k}|\mathbf{u}_{0:k},\mathbf{y}_{0:k})}{p(\mathbf{y}_{k+1}|\mathbf{u}_{0:k+1},\mathbf{y}_{0:k})} \\
&= \frac{p(\mathbf{y}_{k+1}|\mathbf{x}_{0:k+1},\mathbf{u}_{0:k+1},\mathbf{y}_{0:k})p(\mathbf{x}_{k+1}|\mathbf{x}_{0:k},\mathbf{u}_{0:k},\mathbf{y}_{0:k})}{p(\mathbf{y}_{k+1}|\mathbf{u}_{0:k+1},\mathbf{y}_{0:k})} \\
&\quad \times p(\mathbf{x}_{0:k}|\mathbf{u}_{0:k},\mathbf{y}_{0:k}) \\
&= \frac{p(\mathbf{y}_{k+1}|\mathbf{x}_{k+1},\mathbf{u}_{k+1})p(\mathbf{x}_{k+1}|\mathbf{x}_k,\mathbf{u}_k)}{p(\mathbf{y}_{k+1}|\mathbf{u}_{0:k+1},\mathbf{y}_{0:k})} p(\mathbf{x}_{0:k}|\mathbf{u}_{0:k},\mathbf{y}_{0:k}).
\end{aligned} \tag{6.18}$$

The property of the first-order Markov process was used to reach the last line of the aforementioned equation. Equation (6.18) provides a recursive equation for computing the posterior. It can be rewritten as:

$$\begin{aligned}
p(\mathbf{x}_{0:k+1}|\mathbf{u}_{0:k+1},\mathbf{y}_{0:k+1}) &\propto p(\mathbf{y}_{k+1}|\mathbf{x}_{k+1},\mathbf{u}_{k+1})p(\mathbf{x}_{k+1}|\mathbf{x}_k,\mathbf{u}_k) \\
&\quad \times p(\mathbf{x}_{0:k}|\mathbf{u}_{0:k},\mathbf{y}_{0:k}).
\end{aligned} \tag{6.19}$$

A recursive formula for updating the importance weights can be obtained by substituting equations (6.19) and (6.14) into the equation (6.13):

$$
\begin{aligned}
w^i_{k+1} &\propto \frac{p(\mathbf{x}^i_{0:k+1}|\mathbf{u}_{0:k+1}, \mathbf{y}_{0:k+1})}{q(\mathbf{x}^i_{0:k+1}|\mathbf{u}_{0:k+1}, \mathbf{y}_{0:k+1})} \\
&\propto \frac{p(\mathbf{y}_{k+1}|\mathbf{x}^i_{k+1}, \mathbf{u}_{k+1})p(\mathbf{x}^i_{k+1}|\mathbf{x}^i_k, \mathbf{u}_k)p(\mathbf{x}^i_{0:k}|\mathbf{u}_{0:k}, \mathbf{y}_{0:k})}{q(\mathbf{x}^i_{k+1}|\mathbf{x}^i_{0:k}, \mathbf{u}_{0:k+1}, \mathbf{y}_{0:k+1})q(\mathbf{x}^i_{0:k}|\mathbf{u}_{0:k}, \mathbf{y}_{0:k})} \\
&= \frac{p(\mathbf{y}_{k+1}|\mathbf{x}^i_{k+1}, \mathbf{u}_{k+1})p(\mathbf{x}^i_{k+1}|\mathbf{x}^i_k, \mathbf{u}_k)}{q(\mathbf{x}^i_{k+1}|\mathbf{x}^i_{0:k}, \mathbf{u}_{0:k+1}, \mathbf{y}_{0:k+1})} \frac{p(\mathbf{x}^i_{0:k}|\mathbf{u}_{0:k}, \mathbf{y}_{0:k})}{q(\mathbf{x}^i_{0:k}|\mathbf{u}_{0:k}, \mathbf{y}_{0:k})} \\
&\propto \frac{p(\mathbf{y}_{k+1}|\mathbf{x}^i_{k+1}, \mathbf{u}_{k+1})p(\mathbf{x}^i_{k+1}|\mathbf{x}^i_k, \mathbf{u}_k)}{q(\mathbf{x}^i_{k+1}|\mathbf{x}^i_{0:k}, \mathbf{u}_{0:k+1}, \mathbf{y}_{0:k+1})} w^i_k.
\end{aligned}
\tag{6.20}
$$

If the importance density satisfies the following property:

$$q(\mathbf{x}^i_{k+1}|\mathbf{x}^i_{0:k}, \mathbf{u}_{0:k+1}, \mathbf{y}_{0:k+1}) = q(\mathbf{x}^i_{k+1}|\mathbf{x}^i_k, \mathbf{u}_{k:k+1}, \mathbf{y}_{k+1}), \tag{6.21}$$

then, we have:

$$w^i_{k+1} \propto \frac{p(\mathbf{y}_{k+1}|\mathbf{x}^i_{k+1}, \mathbf{u}_{k+1})p(\mathbf{x}^i_{k+1}|\mathbf{x}^i_k, \mathbf{u}_k)}{q(\mathbf{x}^i_{k+1}|\mathbf{x}^i_k, \mathbf{u}_{k:k+1}, \mathbf{y}_{k+1})} w^i_k. \tag{6.22}$$

This means that at each iteration of the filter, storing $\mathbf{x}^i_{k+1}$ will suffice, and there is no need to keep the entire history of the state trajectory and the observations. Having the updated weights, the approximate posterior density is obtained from the following equation:

$$p(\mathbf{x}_{k+1}|\mathbf{u}_{0:k+1}, \mathbf{y}_{0:k+1}) \approx \sum_{i=1}^{n_p} w^i_{k+1}\delta(\mathbf{x}_{k+1} - \mathbf{x}^i_{k+1}). \tag{6.23}$$

As $n_p \to \infty$, this approximation converges to the true posterior [46].

6.5 Resampling

The sequential importance sampling method as formulated so far, is prone to *degeneracy problem*, which means that after a few iterations, most particles will have negligible weights. This reduces the number of effective particles, which in turn, deteriorates the performance. The effective sample size, N_{eff}, provides a measure of degeneracy of an algorithm, which can be estimated as follows:

$$N_{eff} = \frac{1}{\sum_{i=1}^{n_p} (w^i_k)^2}. \tag{6.24}$$

Appropriate choice of the importance density is a way to address this issue, which will be discussed in more detail later. Using *resampling* is another way. This idea leads to a generic particle filter algorithm that is referred to as *sampling importance resampling* (SIR) particle filter [46, 89].

In resampling, particles with small weights are removed and particles with large weights are replicated. This leads to an unchanged number of particles with weights, which are equal to $1/n_p$. A popular method for resampling is the systematic resampling strategy introduced in [90] that results in a minimum variance among children of a particle. In this method, a set of n_p points is chosen over the interval $[0, 1]$ in a way that the distance between every two points will be $1/n_p$. The first point is drawn from $[0, 1/n_p]$, and the number of children of the ith sample, n_i, is taken to be the number of points that lie between $\sum_{j=1}^{i-1} w_k^j$ and $\sum_{j=1}^{i} w_k^j$.

6.6 Sample Impoverishment

In practice, resampling, may introduce the *sample impoverishment* problem. Since resampling is performed using a discretized version of the state-space, particles with large weights are selected many times and the resampled set of particles will have many repeated points, which results in loss of diversity. This sample-impoverishment problem, which arises especially in cases with small process noise, can be handled in different ways:

- *Regularization*: Since the sample-impoverishment problem occurs due to choosing the samples from a limited number of discrete points, using a continuous function can help to overcome this problem. Thus, in the *regularized particle filter*, resampling is performed using a continuous approximation of the posterior distribution:

$$p(\mathbf{x}_{k+1}|\mathbf{u}_{0:k+1}, \mathbf{y}_{0:k+1}) \approx \sum_{i=1}^{n_p} w_{k+1}^i \mathscr{K}_h(\mathbf{x}_{k+1} - \mathbf{x}_{k+1}^i), \tag{6.25}$$

  where

$$\mathscr{K}_h(\mathbf{x}) = \frac{1}{h^{n_x}} \mathscr{K}\left(\frac{\mathbf{x}}{h}\right) \tag{6.26}$$

  is a rescaled version of the kernel function $\mathscr{K}(.)$, which is a zero-mean PDF with limited energy, and h is the kernel bandwidth. $\mathscr{K}(.)$ and h are chosen to minimize some error criteria [91].
- *The resample-move algorithm*: Another technique to avoid sample impoverishment, which is more rigorous, is the *resample-move* algorithm described

in [92]. This method guarantees that the particles approximate samples from the posterior based on a *Markov chain Monte Carlo* (MCMC) step after resampling.

As mentioned previously, while the resampling step creates multiple copies of the samples with greater weights, particles with small weights may end up with no children. In the extreme case, all resampled particles might be created from one prior particle, which results in a severe loss of diversity. The idea behind the resample-move strategy is to introduce diversity after the resampling step by moving new particles in a way that preserves the distribution.

The move step performs one or more iterations of the MCMC algorithm on each of the particles selected in the resampling step. A Markov chain transition kernel and its corresponding invariant distribution are used for moving each resampled particle. Any of the standard MCMC methods such as the *Gibbs* sampler or the *Metropolis–Hastings* algorithm may be used for this purpose.

6.7 Choosing the Proposal Distribution

Equation (6.22) provides the update procedure for the particle weights at each step. The updating and sampling, and thus, the estimation performance highly depends on the proper choice of the proposal distribution q. This is where our prior knowledge about the underlying system can be extremely valuable. In fact, in the successful application of particle filtering, the proposal distribution may require even a more detailed attention than the formulation of the predictive model. The proposal distribution can be selected in different ways:

- *The prior proposal distribution*: A simple choice for the proposal distribution used in the generic SIR filter is the prior:

$$q(\mathbf{x}^i_{k+1}|\mathbf{x}^i_k, \mathbf{u}_{k:k+1}, \mathbf{y}_{k+1}) \propto p(\mathbf{x}_{k+1}|\mathbf{x}_k, \mathbf{u}_k), \tag{6.27}$$

  which results in an easy to implement weight update procedure. This proposal distribution, however, does not incorporate the latest measurement $\mathbf{y}_{k+1}$, as is obvious from equation (6.27).
- *The likelihood proposal distribution*: The prior contains only our knowledge of the past state without incorporating the latest measurement. An alternative is to use only our knowledge about the current measurement. This can be done by choosing the likelihood function as the importance distribution. A likelihood-based importance density samples $\mathbf{x}^i_{k+1}$ based on the measurement

$\mathbf{y}_{k+1}$, and independent from $\mathbf{x}_k^i$:

$$q(\mathbf{x}_{k+1}^i | \mathbf{x}_k^i, \mathbf{u}_{k:k+1}, \mathbf{y}_{k+1}) \propto p(\mathbf{y}_{k+1} | \mathbf{x}_{k+1}^i, \mathbf{u}_{k+1}). \tag{6.28}$$

- *The optimal proposal distribution and suboptimal approximates*: Another method that may not always be practical is to incorporate all of our knowledge by choosing an *optimal importance density*:

$$q(\mathbf{x}_{k+1}^i | \mathbf{x}_k^i, \mathbf{u}_{k:k+1}, \mathbf{y}_{k+1}) \propto p(\mathbf{x}_{0:k+1} | \mathbf{x}_{0:k}^i, \mathbf{u}_{0:k+1}, \mathbf{y}_{0:k+1}). \tag{6.29}$$

This distribution minimizes the variance of importance weights, and therefore, to some extent, prevents the degeneracy problem. The optimal importance function is theoretically the best choice because it contains all of our knowledge about the previous states and the inputs as well as the current and past measurements. However, in practice, it is often a difficult method to implement, and therefore, suboptimal methods have been developed to estimate such a solution. Among the proposed methods, the *extended Kalman particle filter* uses the EKF to estimate the posterior mean and variance for each particle and offers a Gaussian approximation of the posterior. Another proposed algorithm is the *unscented particle filter* that uses the unscented Kalman filter (UKF) for the same purpose [75].

6.8 Generic Particle Filter

Algorithm 6.1 shows the procedure for the sequential importance sampling particle filter with systematic resampling [88, 93]. At each time instant, the mean state vector and the estimation error covariance matrix can be computed as:

$$\hat{\mathbf{x}}_{k+1} = \sum_{i=1}^{n_p} w_{k+1}^i \mathbf{x}_{k+1}^i, \tag{6.30}$$

$$\mathbf{P}_{k+1|k+1} = \sum_{i=1}^{n_p} w_{k+1}^i \left(\mathbf{x}_{k+1}^i - \hat{\mathbf{x}}_{k+1}\right)\left(\mathbf{x}_{k+1}^i - \hat{\mathbf{x}}_{k+1}\right)^T. \tag{6.31}$$

To improve the particle filter efficiency, size of the particle set can be adjusted during the estimation process. At each iteration of the particle filter, the number of particles can be determined in such a way to bound the *Kullback–Leibler divergence* (KLD) between the true posterior and its sample-based approximation. To be more precise, for cases with low state uncertainty, a small number of particles are chosen, and for cases with high state uncertainty, a large number of particles are selected [94].

Algorithm 6.1: Sequential importance sampling particle filter with systematic resampling

State-space model

$$p(\mathbf{x}_{k+1}|\mathbf{x}_k, \mathbf{u}_k)$$
$$p(\mathbf{y}_k|\mathbf{x}_k, \mathbf{u}_k)$$

Initialization

$i = 1, \ldots, n_p$

$$\mathbf{x}_0^i \sim q(\mathbf{x}_0), \quad w_0^i = \frac{1}{n_p}$$

for $k = 0, 1, \ldots,$ **do**

Sequential importance sampling

$$\mathbf{x}_{k+1}^i \sim q(\mathbf{x}_{k+1}^i|\mathbf{x}_k^i, \mathbf{u}_{k:k+1}, \mathbf{y}_{k+1})$$
$$\widetilde{w}_{k+1}^i = \frac{p(\mathbf{y}_{k+1}|\mathbf{x}_{k+1}^i, \mathbf{u}_{k+1})p(\mathbf{x}_{k+1}^i|\mathbf{x}_k^i, \mathbf{u}_k)}{q(\mathbf{x}_{k+1}^i|\mathbf{x}_k^i, \mathbf{u}_{k:k+1}, \mathbf{y}_{k+1})} w_k^i$$

Weight normalization

$$w_{k+1}^i = \frac{\widetilde{w}_{k+1}^i}{\sum_{i=1}^{n_p} \widetilde{w}_{k+1}^i}$$

Effective sample size and weight reset

$$N_{eff} = \frac{1}{\sum_{i=1}^{n_p} (w_{k+1}^i)^2}$$

if $N_{eff} \ll n_p$ **then**

Resampling and weight reset

Generate n_p ordered numbers from a uniform distribution $\mathcal{U}$

$$\tilde{a} \sim \mathcal{U}[0, 1), \quad a^\ell = \frac{\tilde{a} + (\ell - 1)}{n_p}$$

Use them to select the new set of particles $\mathbf{x}_{k+1}^{\ell *}$

$$\mathbf{x}_{k+1}^{\ell *} = \mathbf{x}_{k+1}^i, \quad \text{for } a^\ell \in \left[\sum_{j=1}^{i-1} w_k^j, \sum_{j=1}^{i} w_k^j\right)$$
$$w_{k+1}^{\ell *} = \frac{1}{n_p}$$

end

end

As an illustrative example, let us consider a scalar nonlinear system with the following state-space model [88]:

$$x_{k+1} = 0.5x_k + \frac{25x_k}{1+x_k^2} + 8\cos(1.2k) + v_k, \tag{6.32}$$

$$y_k = \frac{x_k^2}{20} + w_k, \tag{6.33}$$

where $x_0 = 0.1$, $v_k \sim \mathcal{N}(0, 10)$, and $w_k \sim \mathcal{N}(0, 1)$. The likelihood, $p(y_k|x_k)$, is bimodal for positive measurements with modes at $\pm\sqrt{20y_k}$, and unimodal at zero for negative measurements. In either case, the likelihood is symmetric about zero. Figure 6.1 shows typical posterior estimate trajectories for this system, which are obtained by the SIR particle filter, the SIR particle filter with MCMC step, the likelihood particle filter, and the extended Kalman particle filter. Comparing Figure 6.1a–d shows that the bimodal structure of the estimated posterior distribution is

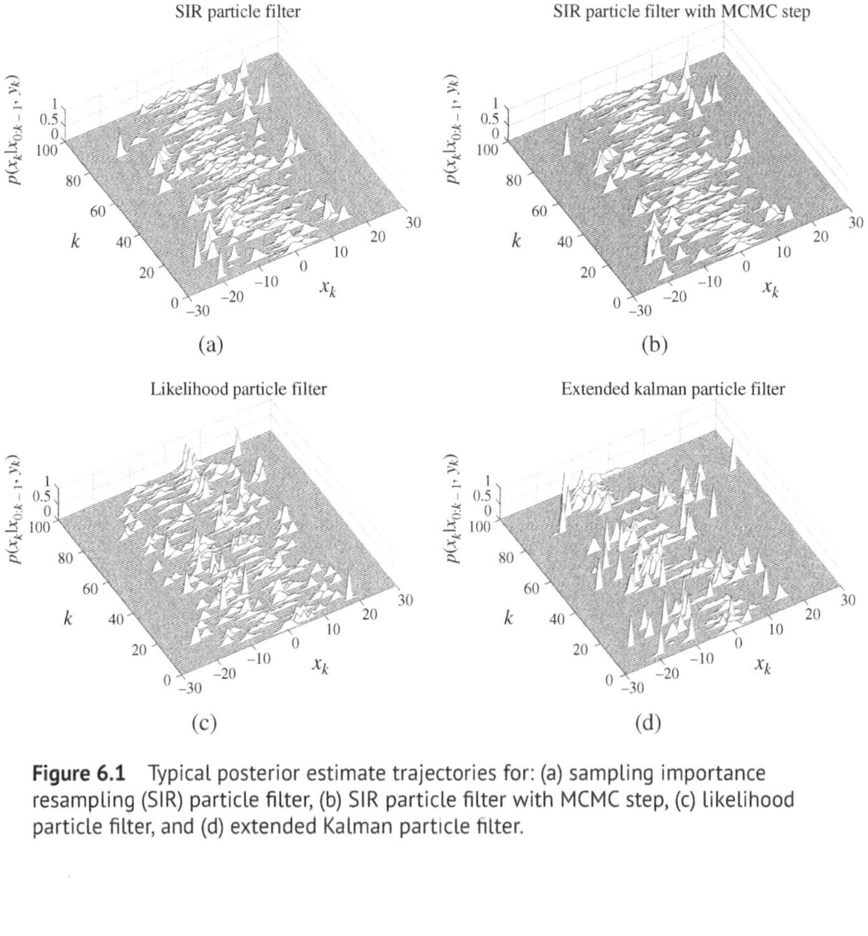

Figure 6.1 Typical posterior estimate trajectories for: (a) sampling importance resampling (SIR) particle filter, (b) SIR particle filter with MCMC step, (c) likelihood particle filter, and (d) extended Kalman particle filter.

more emphasized in Figure 6.1c. Hence, in this case, the likelihood particle filter provides a better estimate of the true posterior compared to the other three filters.

6.9 Applications

Particle filtering algorithms have been successfully used for underwater positioning using a topographic map, surface positioning using a sea chart, vehicle positioning using a road map, and aircraft positioning using a topographic map [95]. This section reviews application of the particle filter for *simultaneous localization and mapping* (SLAM), which can be viewed as an extension of the positioning problem in an unknown environment. In SLAM, a map of the unknown environment is constructed and continually updated, while the location of an agent is simultaneously determined in the evolving map.

6.9.1 Simultaneous Localization and Mapping

FastSLAM is a promising algorithm that takes advantage of inherent structure of the SLAM problem to reduce the burden on the deployed particle filter. To formulate the SLAM problem, an augmented state vector is considered that consists of two subvectors; $\mathbf{x}_k$ that contains information about the platform under study (a mobile robot), and $\mathbf{m}_k$ that contains information about positions of the landmarks [95]:

$$\mathbf{m}_k = \begin{bmatrix} \mathbf{m}_k^1 \\ \vdots \\ \mathbf{m}_k^{M_k} \end{bmatrix}, \tag{6.34}$$

where $\mathbf{m}_k^i$ denotes position of the ith landmark at time instant k. At this time instant, it is assumed that the map consists of M_k landmarks, which is time-dependent and implies that the number of landmarks changes as the map evolves over time. The idea behind FastSLAM is to model the landmarks as linear Gaussian subsystems and use Kalman filter to estimate their states. In this way, particle filter can be used for estimating only the platform states, $\mathbf{x}_k$. Similarly, for a complex platform with a high-dimensional state space, the state vector, $\mathbf{x}_k$, may be decomposed into two subvectors; $\mathbf{x}_k^{\mathscr{P}}$ that contains the state variables with highly nonlinear dynamics, and $\mathbf{x}_k^{\mathscr{K}}$ that contains the state variables with almost linear dynamics. Then, $\mathbf{x}_k^{\mathscr{P}}$ and $\mathbf{x}_k^{\mathscr{K}}$ will be estimated by the particle filter and the (extended) Kalman filter, respectively. The state-space model includes the following state-transition model [96]:

$$\mathbf{x}_{k+1}^{\mathcal{P}} = \mathbf{f}_k^{\mathcal{P}}\left(\mathbf{x}_k^{\mathcal{P}}, \mathbf{u}_k\right) + \mathbf{A}_k^{\mathcal{P}}\left(\mathbf{x}_k^{\mathcal{P}}\right)\mathbf{x}_k^{\mathcal{K}} + \mathbf{B}_k^{\mathcal{P}}\left(\mathbf{x}_k^{\mathcal{P}}\right)\mathbf{u}_k + \mathbf{G}_k^{\mathcal{P}}\left(\mathbf{x}_k^{\mathcal{P}}\right)\mathbf{v}_k^{\mathcal{P}}, \tag{6.35}$$

$$\mathbf{x}_{k+1}^{\mathcal{K}} = \mathbf{f}_k^{\mathcal{K}}\left(\mathbf{x}_k^{\mathcal{P}}, \mathbf{u}_k\right) + \mathbf{A}_k^{\mathcal{K}}\left(\mathbf{x}_k^{\mathcal{P}}\right)\mathbf{x}_k^{\mathcal{K}} + \mathbf{B}_k^{\mathcal{K}}\left(\mathbf{x}_k^{\mathcal{P}}\right)\mathbf{u}_k + \mathbf{G}_k^{\mathcal{K}}\left(\mathbf{x}_k^{\mathcal{P}}\right)\mathbf{v}_k^{\mathcal{K}}, \tag{6.36}$$

$$\mathbf{m}_{k+1} = \mathbf{m}_k, \tag{6.37}$$

and the following measurement model:

$$\mathbf{y}_k^{a} = \mathbf{g}_k^{a}\left(\mathbf{x}_k^{\mathcal{P}}, \mathbf{u}_k\right) + \mathbf{C}_k^{a}\left(\mathbf{x}_k^{\mathcal{P}}\right)\mathbf{x}_k^{\mathcal{K}} + \mathbf{D}_k^{a}\left(\mathbf{x}_k^{\mathcal{P}}\right)\mathbf{u}_k + \mathbf{w}_k^{a}, \tag{6.38}$$

$$\mathbf{y}_k^{b} = \mathbf{g}_k^{b}\left(\mathbf{x}_k^{\mathcal{P}}, \mathbf{u}_k\right) + \mathbf{C}_k^{b}\left(\mathbf{x}_k^{\mathcal{P}}\right)\mathbf{m}_k + \mathbf{D}_k^{b}\left(\mathbf{x}_k^{\mathcal{P}}\right)\mathbf{u}_k + \mathbf{w}_k^{b}, \tag{6.39}$$

where

$$\begin{bmatrix}\mathbf{v}_k^{\mathcal{P}} \\ \mathbf{v}_k^{\mathcal{K}}\end{bmatrix} \sim \mathcal{N}\left(\mathbf{0}, \begin{bmatrix}\mathbf{Q}_k^{\mathcal{P}} & \mathbf{Q}_k^{\mathcal{P}\mathcal{K}} \\ \left(\mathbf{Q}_k^{\mathcal{P}\mathcal{K}}\right)^T & \mathbf{Q}_k^{\mathcal{K}}\end{bmatrix}\right), \tag{6.40}$$

$$\mathbf{w}_k^{a} \sim \mathcal{N}\left(\mathbf{0}, \mathbf{R}_k^{a}\right), \tag{6.41}$$

$$\mathbf{w}_k^{b} \sim \mathcal{N}\left(\mathbf{0}, \mathbf{R}_k^{b}\right). \tag{6.42}$$

The posterior can be factorized as [96]:

$$\begin{aligned} & p(\mathbf{x}_{0:k}^{\mathcal{P}}, \mathbf{x}_{0:k}^{\mathcal{K}}, \mathbf{m}_k | \mathbf{u}_{0:k}, \mathbf{y}_{0:k}) \\ & \quad = p\left(\mathbf{m}_k | \mathbf{x}_{0:k}^{\mathcal{P}}, \mathbf{x}_{0:k}^{\mathcal{K}}, \mathbf{u}_{0:k}, \mathbf{y}_{0:k}\right) \, p\left(\mathbf{x}_k^{\mathcal{K}} | \mathbf{x}_{0:k}^{\mathcal{P}}, \mathbf{u}_{0:k}, \mathbf{y}_{0:k}\right) \, p\left(\mathbf{x}_k^{\mathcal{P}} | \mathbf{u}_{0:k}, \mathbf{y}_{0:k}\right). \end{aligned} \tag{6.43}$$

Assuming that landmarks are independent, the aforementioned posterior can be rewritten as:

$$\begin{aligned} & p(\mathbf{x}_{0:k}^{\mathcal{P}}, \mathbf{x}_{0:k}^{\mathcal{K}}, \mathbf{m}_k | \mathbf{u}_{0:k}, \mathbf{y}_{0:k}) \\ & = \underbrace{\prod_{i=1}^{M_k} p\left(\mathbf{m}_k^i | \mathbf{x}_{0:k}^{\mathcal{P}}, \mathbf{x}_{0:k}^{\mathcal{K}}, \mathbf{u}_{0:k}, \mathbf{y}_{0:k}\right) \, p\left(\mathbf{x}_k^{\mathcal{K}} | \mathbf{x}_{0:k}^{\mathcal{P}}, \mathbf{u}_{0:k}, \mathbf{y}_{0:k}\right)}_{\text{(extended) Kalman filter}} \, \underbrace{p\left(\mathbf{x}_k^{\mathcal{P}} | \mathbf{u}_{0:k}, \mathbf{y}_{0:k}\right)}_{\text{particle filter}}. \end{aligned} \tag{6.44}$$

6.10 Concluding Remarks

Particle filter is a powerful tool for nonlinear/non-Gaussian estimation, which easily outperforms other classic Bayesian filters in cases of severe nonlinearities and multimodal distributions. In order to design an effective particle filtering algorithm, several points must be taken into consideration:

- Choosing a suitable proposal distribution is critical for the success or failure of the algorithm.
- A sufficient number of particles must be determined in order to minimize the estimation error without unnecessarily increasing the computational burden.
- Countermeasures must be taken against sample impoverishment.

The *Rao–Blackwellized particle filter* (RBPF), which is also known as the *marginalized particle filter* (MPF), is the technical enabler for applying particle filtering algorithms to problems with high-dimensional state-space models when only a small subset of the state variables are affected by severe nonlinearities. For such systems, Kalman filter is used for linear Gaussian subsystems, EKF or UKF is used for slightly nonlinear/non-Gaussian subsystems, and particle filter is used for highly nonlinear/non-Gaussian subsystems. In this way, the burden on the particle filter is reduced [95]. FastSLAM has been developed based on this idea.

7

Smooth Variable-Structure Filter

7.1 Introduction

The *smooth variable-structure filter* (SVSF) algorithm has been derived based on a stability theorem [97]. Inspired by the variable-structure control (VSC) theory, the SVSF uses an inherent switching action to guarantee convergence of the estimated states to within a neighborhood of their true values. Similar to the mentioned Bayesian filters in the previous chapters, the SVSF has been formulated in a predictor–corrector form [56, 58, 62, 98]. Robustness against bounded uncertainties is an inherent characteristic of the VSC, which has been inherited by the SVSF [99]. The distinguishing features of the SVSF from other filters can be summarized as follows [97]:

- The SVSF takes advantage of the inherent robustness of the VSC against bounded uncertainties. Hence, its convergence can be guaranteed for bounded uncertainty and noise. Moreover, a fairly precise estimate of the upper bound on the uncertainties will enhance the performance of the SVSF.
- Unlike other filtering strategies that implicitly consider uncertainty and rely on trial and error for tuning, the SVSF formulation allows for explicit identification of the source of uncertainty and assigning a bound to it. Taking account of this information in the design, alleviates tuning by trial and error to a large extent.
- In order to quantify the degree of uncertainty and modeling mismatch associated with each estimated state or parameter, the SVSF uses a secondary set of performance indicators. This is in addition to the filter innovation vector, which is normally used as a performance measure by other filters including Kalman filter, cubature Kalman filter (CKF), unscented Kalman filter (UKF), and particle filter.

The distinguishing features of the SVSF such as robustness, existence of multiple performance indicators, and the ability to identify the source of uncertainty,

Nonlinear Filters: Theory and Applications, First Edition. Peyman Setoodeh, Saeid Habibi, and Simon Haykin.

enable this filter to dynamically refine its internal model, and therefore, gradually improve its performance. The SVSF owes this capability to its formulation as a variable-structure estimation algorithm in the predictor–corrector form. The mentioned distinguishing features make the SVSF a valid candidate for applications such as fault detection and health monitoring [97].

7.2 The Switching Gain

Since the SVSF algorithm is built on the VSC, its convergence can be proved using the Lyapunov stability theory. Variable-structure systems were originally proposed in the Soviet literature in the 1950s, and VSC was initially derived for continuous-time systems [100–102]. The state equations of variable-structure systems contain discontinuities, and the state space of these systems is divided into different regions by *discontinuity hyperplanes*. In each one of these regions, a different continuous state equation describes the system dynamics. In such systems, the state space is segmented, and as the state trajectory crosses a discontinuity hyperplane, the continuous characterization of the governing dynamics changes [101, 102]. In VSC, the control input is a discontinuous function of the state vector, and artificially introduces discontinuity hyperplanes. A simple discontinuous input $\mathbf{u}_k^i(\mathbf{x})$ subject to a discontinuity surface $s^i(\mathbf{x})$ is expressed as:

$$\mathbf{u}_k^i(\mathbf{x}) = \begin{cases} \mathbf{u}_k^{i+}(\mathbf{x}), & s^i(\mathbf{x}) > 0, \\ \mathbf{u}_k^{i-}(\mathbf{x}), & s^i(\mathbf{x}) < 0, \end{cases} \tag{7.1}$$

where $\mathbf{u}_k^{i+}(\mathbf{x})$ and $\mathbf{u}_k^{i-}(\mathbf{x})$ are continuous.

Sliding-mode control is a special class of VSC in which the discontinuous control input is designed in a way to force the states toward the switching hyperplane. Then, the control signal forces the states to remain on the switching hyperplane and slide along it [99, 102]. While constrained to the sliding surface, the system shows robustness against uncertainties and external disturbances. The variable-structure and sliding-mode control methods were restricted to continuous-time systems until the mid-1980s. Regarding discrete implementations of the sliding-mode control, the real sliding-mode condition was studied prior to 1985 [101], and then, discrete sliding mode was presented under quasi-sliding mode in 1985 [103]. Subsequently, articulation of a stability condition for discrete-time sliding-mode control [104] paved the way for discrete-time controller designs [105, 106].

For continuous-time systems, a number of observers with a switching error nature have been proposed in the literature that rely on VSC in order to guarantee convergence [102, 107, 108]. The notion of equivalent control has played a key role in the design of sliding mode observers [101, 102, 109], and has led to optimal strategies [110, 111]. Lyapunov-based designs were proposed to guarantee asymptotic estimation error decay in the presence of bounded uncertainties in the input channel [112–115]. The proposed design method in [114], which was extended in [116], aimed at obtaining an explicit design procedure rather than relying on symbolic manipulation. The Lyapunov-based approach has been further developed in [117, 118]. Variable-structure and sliding-mode observers have been successfully used for fault detection [119, 120]. Moreover, such observers have been combined with Ackermann's formula [121] and contraction analysis [122].

The mentioned design methods for sliding-mode observers do not use a predictor–corrector form. The *variable-structure filter* (VSF), which is designed based on the variable-structure theory in a predictor–corrector form, has been proposed as a state/parameter estimation algorithm [123]. Although the SVSF can be viewed as an extension of the VSF, these two filters are different regarding their characteristics, derivations, and filter corrective terms. The novelty of the SVSF compared with other state estimation algorithms is due to [97]:

- Its predictor–corrector form in a variable-structure context
- The use of a stability theorem
- The subsequent formulation of its corrective term

The basic concept behind the SVSF is depicted in Figure 7.1. Let us consider the state trajectory of a system and its estimate, which is generated based on an uncertain model of the system. The estimated state is forced toward its true value until it reaches the *existence subspace*, which is time-varying and encloses the true state trajectory. The estimated state is forced to switch back and forth across the true state trajectory, but it is not allowed to leave the existence subspace. The width of the existence subspace depends on uncertainties and external disturbances, hence,

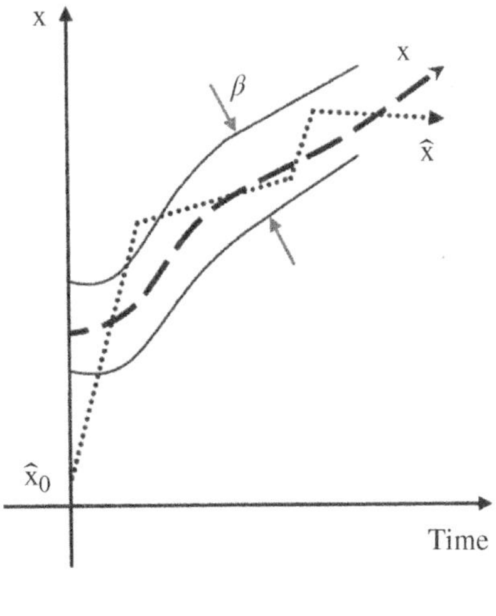

Figure 7.1 The SVSF state estimation concept.

it is unknown. However, for bounded uncertainties and disturbances, a bound can be found for the width of the existence subspace. After the *reachability phase* [124, 125], which refers to the convergence to the existence subspace, the estimated state will remain within this bound.

Let $\mathbf{e}_k$ denote the estimation error at time instant k. Assume that the system is stochastic due to the presence of process noise in the state equation and the measurement noise in the measurement equation. Then, the estimation error, $\mathbf{e}_k$, will have two components:

$$\mathbf{e}_k = \mathbf{s}_k + \mathbf{n}_k, \tag{7.2}$$

where $\mathbf{s}_k$ denotes the estimation error for the corresponding deterministic system, and $\mathbf{n}_k$ denotes the error due to the noise contents. The width of the existence subspace cannot be reduced below a limit, which is determined by the random element of $\mathbf{e}_k$. Let us assume that the random component of $\mathbf{e}_k$ is norm bounded, $\| \mathbf{n}_k \| \leq \boldsymbol{\beta}$, or amplitude bounded, $|\mathbf{n}_k| \leq \boldsymbol{\beta}$. Then, in the absence of modeling uncertainties, an upper bound for the width of the existence subspace can be found as a function of $\boldsymbol{\beta}$. In the discrete-time sliding-mode context, $\boldsymbol{\beta}$ is significantly affected by the sampling interval and the switching delay regarding the real sliding-mode condition [102]. In addition to this conceptual description, outside the boundary of the existence subspace, the estimation error decreases according to the following lemma [97].

Lemma 7.1 *The estimation process is stable and convergent, if* $|\mathbf{e}_k| < |\mathbf{e}_{k-1}|$.

Proof: Let us assume that the random component $\mathbf{n}_k$ is amplitude bounded, $|\mathbf{n}_k| \leq \boldsymbol{\beta}$. During the reachability phase, where $|\mathbf{e}_k| > \boldsymbol{\beta}$, the Lyapunov stability theory can be used to prove that estimation error decreases over time. Considering the Lyapunov function $\mathbf{V}_k = \mathbf{e}_k^2$, the estimation process is stable, if $\Delta\mathbf{V}_k = \mathbf{e}_k^2 - \mathbf{e}_{k-1}^2 < 0$. This stability condition is satisfied by $|\mathbf{e}_k| < |\mathbf{e}_{k-1}|$ [126].

The stability condition of Lemma 7.1 leads to $|\mathbf{s}_k + \mathbf{n}_k| < |\mathbf{s}_{k-1} + \mathbf{n}_{k-1}|$. For $|\mathbf{n}_k| \leq \boldsymbol{\beta}$ and $|\mathbf{e}_k| > \boldsymbol{\beta}$, the worst and the best cases are defined as $|\mathbf{s}_k| + \boldsymbol{\beta} < |\mathbf{s}_{k-1}| - \boldsymbol{\beta}$ and $|\mathbf{s}_k| - \boldsymbol{\beta} < |\mathbf{s}_{k-1}| + \boldsymbol{\beta}$, respectively. These two conditions are satisfied, if:

$$|\mathbf{s}_k| < |\mathbf{s}_{k-1}|, \tag{7.3}$$

$$\lim_{k\to\infty} \big| |\mathbf{s}_{k-1}| - |\mathbf{s}_k| \big| < 2\boldsymbol{\beta}. \tag{7.4}$$

Equation (7.4) satisfies the *Cauchy's convergence principle* [127], and implies that for a bounded $\mathbf{n}_k$, the width of the existence subspace will be less than $2\boldsymbol{\beta}$. The condition of Lemma 7.1 has been used to define the switching hyperplane for discrete-time sliding-mode control [104]. Stability of the SVSF algorithm is proved based on this condition. If it holds, the error magnitude reduces at each iteration, which implies stability of the estimation algorithm as well as confinement of the estimated state trajectory within the existence subspace. The existence subspace provides an upper bound on the accuracy of the estimation algorithm. Hence, performance of the filter is strongly influenced by the nature of $\mathbf{n}_k$ and its corresponding bound $\boldsymbol{\beta}$, because the width of the existence subspace is a function of $\boldsymbol{\beta}$. Since unmodeled drift in an observed signal and structural uncertainties in the model contribute to $\mathbf{n}_k$, they have negative impacts on the filter performance.

In the SVSF, the true state trajectory plays the role of a discontinuity surface or hyperplane, in the sense that in the vicinity of this surface, the estimated state trajectory is directed toward it. In order to direct the estimated state toward the surface and force it to go back and forth across it, a discontinuous corrective term is applied to the estimated state. Anytime that the estimated state crosses the hyperplane, direction of the corrective term is changed in order to force the estimated state to move back toward the true state. Hence, the hyperplane is called the *switching surface*. The SVSF corrective term is determined in a way that the magnitude of the estimation error is reduced during the switching action. In general, the ground truth for the state trajectory is not available, and therefore, cannot be used as a switching hyperplane. To address this issue, the switching hyperplane is defined as a function of the measured output, $\mathbf{y}$:

$$\mathbf{k}_k^i(\mathbf{y}) = \begin{cases} \mathbf{k}_k^{i+}(\mathbf{y}), & s^i(\mathbf{y}) > 0, \\ \mathbf{k}_k^{i-}(\mathbf{y}), & s^i(\mathbf{y}) < 0. \end{cases} \tag{7.5}$$

To be more precise, the output prediction error (the innovation vector) is used as an indicator of the state estimation error.

In control systems, sensors are designed to be linear over their operating range and are accurately calibrated. Therefore, it would be reasonable to assume a linear measurement equation in which the relationship between measurements and states is described by a constant, pseudo-diagonal, and positive matrix, $\mathbf{C}$ [97]. For a constant and positive matrix $\mathbf{C}$, a reduction in the magnitude of the output prediction error reflects a reduction in the expectation of the state estimation error. Thus, for completely observable and completely controllable systems, accurate output prediction reflects accurate state estimation [16]. Regarding the linear relationship between the system states and measurements, the measured output trajectory can be used as the switching surface.

Building on the intuitive description presented in this section, in what follows, the SVSF algorithm will be derived in mathematical terms. In this chapter, unless otherwise stated, $|\cdot|$ denotes the absolute value.

7.3 Stability Analysis

The SVSF can be applied to systems described by the following state-space model:

$$\mathbf{x}_{k+1} = \mathbf{f}(\mathbf{x}_k, \mathbf{u}_k, \mathbf{v}_k), \tag{7.6}$$

$$\mathbf{y}_k = \mathbf{C}\mathbf{x}_k + \mathbf{w}_k, \tag{7.7}$$

where $\mathbf{x} \in \mathbb{R}^{n_x}$, $\mathbf{u} \in \mathbb{R}^{n_u}$, and $\mathbf{y} \in \mathbb{R}^{n_y}$ denote the state, the input, and the output vectors, respectively. Process and measurement noise are denoted by $\mathbf{v}_k$ and $\mathbf{w}_k$, respectively. Furthermore, the state-space model (7.6) and (7.7) is assumed to be *consecutive bijective* based on the following definition [97].

Definition 7.1 ***(Consecutive bijective model)*** *The state-space model in (7.6) and (7.7) is consecutive bijective, if and only if in the absence of process and measurement noise, for the corresponding deterministic system:*

$$\mathbf{x}_{k+1} = \mathbf{f}(\mathbf{x}_k, \mathbf{u}_k), \tag{7.8}$$

$$\mathbf{y}_k = \mathbf{C}\mathbf{x}_k, \tag{7.9}$$

there exists a mapping, $\mathbf{h}$*, that uniquely determines the state,* $\mathbf{x}_k$*, by consecutive time iterations of the output vector:*

$$\mathbf{x}_k = \mathbf{h}\left(\mathbf{C}^{\dagger}\mathbf{y}_{k+1}, \mathbf{C}^{\dagger}\mathbf{y}_k, \mathbf{u}_k\right), \tag{7.10}$$

where $\mathbf{C}^{\dagger}$ *denotes the pseudo-inverse of* $\mathbf{C}$.

According to this definition, completely observable and completely controllable linear systems are consecutive bijective. A nonlinear system with state-space model in (7.6) and (7.7) is consecutive bijective, if and only if the nonlinear function $\mathbf{f}$ is smooth with continuous partial derivatives of any order with respect to its arguments. The word "smooth" in the SVSF reflects this condition [97].

The SVSF is a predictor–corrector algorithm that predicts an a priori estimate of the state using a state-space model. Then, the a posteriori state estimate is obtained from this a priori estimate through using a corrective term, $\mathbf{k}_k$. According to Lemma 7.1, the estimated state converges to its true value through using this corrective term, which is discontinuous. Algorithm 7.1 summarizes the SVSF estimation process. In the algorithm, $\hat{\mathbf{f}}(\hat{\mathbf{x}}_{k|k}, \mathbf{u}_k)$ and $\hat{\mathbf{C}}\hat{\mathbf{x}}_{k+1|k}$ denote the uncertain estimations of the exact forms $\mathbf{f}(\mathbf{x}_k, \mathbf{u}_k, \mathbf{v}_k)$ and $\mathbf{C}\mathbf{x}_k + \mathbf{w}_k$ presented in (7.6) and (7.7). To be more precise, the a priori estimates of the state $\hat{\mathbf{x}}_{k+1|k}$ and output $\hat{\mathbf{y}}_{k+1|k}$ vectors are computed using these uncertain estimated functions rather than the exact functions in (7.6) and (7.7). The other corrective term, $\hat{\mathbf{d}}^{bias}_{k+1}$, which is used to compensate for the effect of unmodeled disturbances, does not propagate through the a priori estimate. Therefore, adding this corrective term will not adversely affect the stability of $\hat{\mathbf{x}}_{k+1|k+1}$ [97]. In the SVSF algorithm, the a posteriori and the a priori output error estimates:

$$\mathbf{e}_{\mathbf{y}_{k|k}} = \mathbf{y}_k - \hat{\mathbf{y}}_{k|k}, \tag{7.11}$$

$$\mathbf{e}_{\mathbf{y}_{k+1|k}} = \mathbf{y}_{k+1} - \hat{\mathbf{y}}_{k+1|k}, \tag{7.12}$$

are used as indicators of the a priori and the a posteriori state estimation errors, respectively. Regarding the intuitive description, the SVSF algorithm converges, if the magnitude of the a posteriori estimation error is reduced over time until the estimated state trajectory reaches the existence subspace. The estimated state trajectory will remain confined to this subspace afterwards. The following theorem provides a basis for derivation of the SVSF corrective term, $\mathbf{k}_k$, which supports the conceptual description. The theorem presents conditions on the range and the sign of $\mathbf{k}_k$ for which the stability condition of Lemma 7.1 holds [97].

Theorem 7.1 *For a system, which is stable and consecutive bijective (or completely observable and completely controllable in the case of linear systems), the SVSF corrective term, $\mathbf{k}_k$, that satisfies the stability condition of Lemma 7.1 is subject to the following conditions:*

$$\left|\mathbf{e}_{\mathbf{y}_{k|k-1}}\right| \leq \left|\hat{\mathbf{C}}\mathbf{k}_k\right| < \left|\mathbf{e}_{\mathbf{y}_{k|k-1}}\right| + \left|\mathbf{e}_{\mathbf{y}_{k-1|k-1}}\right|, \tag{7.13}$$

$$\operatorname{sign}\left(\hat{\mathbf{C}}\mathbf{k}_k\right) = \operatorname{sign}\left(\mathbf{e}_{\mathbf{y}_{k|k-1}}\right). \tag{7.14}$$

Proof: From Lemma 7.1, the estimation process is stable, if:

$$\left|\mathbf{e}_{\mathbf{y}_{k|k}}\right| < \left|\mathbf{e}_{\mathbf{y}_{k-1|k-1}}\right|. \tag{7.15}$$

From (7.13), if $\left|\hat{\mathbf{C}}\mathbf{k}_k\right| < \left|\mathbf{e}_{\mathbf{y}_{k|k-1}}\right| + \left|\mathbf{e}_{\mathbf{y}_{k-1|k-1}}\right|$, then

$$\left|\hat{\mathbf{C}}\mathbf{k}_k\right| - \left|\mathbf{e}_{\mathbf{y}_{k|k-1}}\right| < \left|\mathbf{e}_{\mathbf{y}_{k-1|k-1}}\right|. \tag{7.16}$$

From (7.13) and (7.14), if $\left|\mathbf{e}_{\mathbf{y}_{k|k-1}}\right| \leq \left|\hat{\mathbf{C}}\mathbf{k}_k\right|$ and $\operatorname{sign}\left(\hat{\mathbf{C}}\mathbf{k}_k\right) = \operatorname{sign}\left(\mathbf{e}_{\mathbf{y}_{k|k-1}}\right)$, then from (7.16), we have:

$$\left|\mathbf{e}_{\mathbf{y}_{k|k-1}} - \hat{\mathbf{C}}\mathbf{k}_k\right| < \left|\mathbf{e}_{\mathbf{y}_{k-1|k-1}}\right|. \tag{7.17}$$

Regarding Algorithm 7.1, from

$$\hat{\mathbf{y}}_{k+1|k} = \hat{\mathbf{C}}\hat{\mathbf{x}}_{k+1|k}, \tag{7.18}$$

$$\hat{\mathbf{x}}_{k+1|k+1} = \hat{\mathbf{x}}_{k+1|k} + \mathbf{k}_{k+1}, \tag{7.19}$$

(7.11), and (7.12), we obtain:

$$\mathbf{e}_{\mathbf{y}_{k|k}} = \mathbf{e}_{\mathbf{y}_{k|k-1}} - \hat{\mathbf{C}}\mathbf{k}_k. \tag{7.20}$$

Algorithm 7.1: Smooth variable-structure filter

State-space model:

$$\mathbf{x}_{k+1} = \mathbf{f}(\mathbf{x}_k, \mathbf{u}_k, \mathbf{v}_k)$$
$$\mathbf{y}_k = \mathbf{C}\mathbf{x}_k + \mathbf{w}_k$$

Initialization:

$$\hat{\mathbf{x}}_0 = \mathbb{E}[\mathbf{x}_0]$$

for $k = 0, 1, \ldots,$ **do**

A priori state estimate

$$\hat{\mathbf{x}}_{k+1|k} = \hat{\mathbf{f}}(\hat{\mathbf{x}}_{k|k}, \mathbf{u}_k)$$

A priori measurement estimate

$$\hat{\mathbf{y}}_{k+1|k} = \hat{\mathbf{C}}\hat{\mathbf{x}}_{k+1|k}$$

A posteriori and a priori output error estimates

$$\mathbf{e}_{\mathbf{y}_{k|k}} = \mathbf{y}_k - \hat{\mathbf{y}}_{k|k}$$
$$\mathbf{e}_{\mathbf{y}_{k+1|k}} = \mathbf{y}_{k+1} - \hat{\mathbf{y}}_{k+1|k}$$

Filter corrective term

$$\mathbf{k}_{k+1} = \hat{\mathbf{C}}^{\dagger}\left(\left|\mathbf{e}_{\mathbf{y}_{k+1|k}}\right| + \mathbf{\Gamma}\left|\mathbf{e}_{\mathbf{y}_{k|k}}\right|\right) \odot \operatorname{sign}\left(\mathbf{e}_{\mathbf{y}_{k+1|k}}\right)$$
$$\mathbf{\Gamma}_{ij} = \begin{cases} \in [0, 1), & i = j \\ 0, & i \neq j \end{cases}$$

where $\odot$ denotes the element-wise product.

A posteriori state estimate

$$\hat{\mathbf{x}}_{k+1|k+1} = \hat{\mathbf{x}}_{k+1|k} + \mathbf{k}_{k+1}$$

Correction for unmodeled disturbance

$$\hat{\mathbf{x}}^{corrected}_{k+1|k+1} = \hat{\mathbf{x}}_{k+1|k+1} + \hat{\mathbf{d}}^{bias}_{k+1}$$

end

Substituting (7.20) into (7.17) results in $\left|\mathbf{e}_{\mathbf{y}_{k|k}}\right| < \left|\mathbf{e}_{\mathbf{y}_{k-1|k-1}}\right|$, and the stability condition of Lemma 7.1 is satisfied for the output estimate.

From (7.7), (7.18), (7.11), and (7.12), for a known matrix **C**, the stability condition in (7.15) can be expressed in terms of the state estimation error as:

$$\left|\mathbf{C}\mathbf{e}_{\mathbf{x}_{k|k}} - \mathbf{w}_k\right| < \left|\mathbf{C}\mathbf{e}_{\mathbf{x}_{k-1|k-1}} - \mathbf{w}_{k-1}\right|. \tag{7.21}$$

For a consecutive bijective system with a positive diagonal output matrix **C** and white measurement noise, $\mathbf{w}_k$, during the reachability phase and outside of the existence subspace, where $|\mathbf{e}_{\mathbf{x}_{k|k}}| > \boldsymbol{\beta}$, the condition (7.21) leads to:

$$\mathbb{E}\left[\mathbf{e}_{\mathbf{x}_{k|k}}\right] < \mathbb{E}\left[\mathbf{e}_{\mathbf{x}_{k-1|k-1}}\right]. \tag{7.22}$$

According to (7.22), the expected value of the state estimation error iteratively reduces in line with the reduction of the output estimation error. In case that **C** is not a square matrix, the correspondence of (7.15)–(7.22) may not be immediately obvious. However, for a consecutive bijective system or equivalently for a completely observable and completely controllable linear system, condition (7.15), which signifies a stable and convergent output estimate, is a sufficient condition for internal stability as well as convergence of the state estimation as discussed in [16] for linear systems.

7.4 Smoothing Subspace

Regarding the SVSF algorithm, according to Theorem 7.1, the magnitude of the output estimation error and the expected value of the state estimation error are respectively confined to a neighborhood of the measured output and the system state trajectory. As shown in Figure 7.1, this neighborhood is referred to as the existence subspace. Its time-varying width reflects the inaccuracy of the state-space model used by the filter due to uncertainties. Although the width of the existence subspace is unknown, an upper bound, $\boldsymbol{\beta}$, can be found for it. Once the estimated state trajectory enters the existence subspace, it will switch back and forth within its extremities. This high-frequency switching phenomenon, which is known as *chattering*, occurs due to the discontinuity of the filter corrective term $\mathbf{k}_k$. It can be viewed as an artificial noise, which limits the applicability of the SVSF.

In order to filter out the chattering, a smoothing subspace can be considered, which has a known boundary, $\boldsymbol{\psi}$, around the switching surface [99]. In algorithm 7.1, the filter corrective term is computed as:

$$\mathbf{k}_{k+1} = \hat{\mathbf{C}}^{\dagger}\left(\left|\mathbf{e}_{\mathbf{y}_{k+1|k}}\right| + \boldsymbol{\Gamma}\left|\mathbf{e}_{\mathbf{y}_{k|k}}\right|\right) \odot \operatorname{sign}\left(\mathbf{e}_{\mathbf{y}_{k+1|k}}\right), \tag{7.23}$$

where $\odot$ denotes the element-wise product. To ensure stability, in (7.23), the sign function is maintained when the estimated state trajectory is outside of the smoothing boundary, $\boldsymbol{\psi}$. However, when the estimated state trajectory is inside the smoothing boundary, $\mathbf{k}_k$ is interpolated to obtain a smooth function. This is achieved by replacing the function sign($\mathbf{e}$) for any error vector $\mathbf{e}$ by the function sat($\mathbf{e}, \boldsymbol{\psi}$) with the following elements [97]:

$$\text{sat}(\mathbf{e}, \boldsymbol{\psi}) = \begin{cases} \frac{\mathbf{e}_i}{\psi_i}, & \left|\frac{\mathbf{e}_i}{\psi_i}\right| \leq 1, \\ \text{sign}\left(\frac{\mathbf{e}_i}{\psi_i}\right), & \left|\frac{\mathbf{e}_i}{\psi_i}\right| > 1. \end{cases} \tag{7.24}$$

A conservative choice for the width of the smoothing subspace is the estimated upper bound on the width of the existence subspace:

$$\boldsymbol{\psi} = \hat{\boldsymbol{\beta}}. \tag{7.25}$$

Hence, the width of the smoothing subspace depends on the corresponding upper bounds on disturbances and uncertainties. If the width of the smoothing subspace is greater than the width of the existence subspace, $\boldsymbol{\psi} > \boldsymbol{\beta}$, then chattering will be removed as shown in Figure 7.2a. If due to unexpected large disturbances, the width of the existence subspace increases beyond its assumed upper bound, which was used for $\boldsymbol{\psi}$, then we will have $\boldsymbol{\psi} < \boldsymbol{\beta}$. In this case, the estimated state trajectory will exit the existence subspace, and in effect therefore, the smoothing subspace. Then, according to Theorem 7.1, the estimated state trajectory will reconverge. However, if the width of the existence subspace persistently exceeds its assumed upper bound due to disturbances, then the smoothing subspace will be ineffective and chattering will occur as shown in Figure 7.2b. As suggested in [128], a less conservative design for removing chattering can be obtained through adaptively varying the width of the smoothing subspace, $\boldsymbol{\psi}$.

The SVSF concept intuitively highlights the importance of the chattering effect. If the estimated state trajectory and the existence subspace are within the smoothing subspace, then the filter is pessimistic in the sense that the assumed level of uncertainty in the model is either reasonable or over-estimated. In this case, the corrective action applied by the filter will be continuous and smooth. On the other hand, if the estimated state trajectory exits the smoothing subspace, the corrective action will be discontinuous and switching will occur beyond the outer limits of the smoothing subspace. In this case, chattering occurs as the estimated state switches back and forth across the switching hyperplane. The presence of chattering implies that the smoothing subspace is inside the existence subspace. Hence, the filter is optimistic in the sense that the assumed upper bounds on the level of uncertainties in the model, which are used in the calculation of the width of the smoothing subspace, are incorrect. The severity of chattering can be viewed as a measure of the distance between the widths of the existence and smoothing

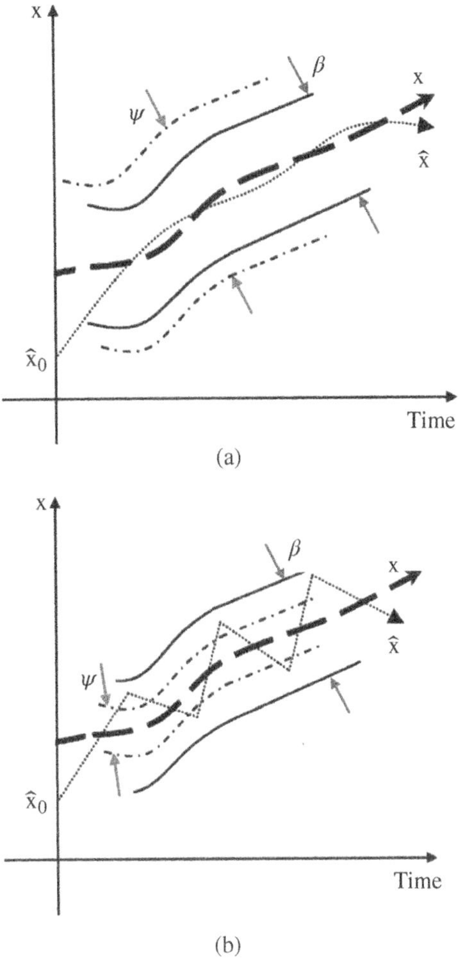

Figure 7.2 Effect of the smoothing subspace on chattering: (a) $\psi > \beta$ and (b) $\psi < \beta$.

subspaces. Since the variance of chattering in the estimated states, which do not have direct measurements associated with them, is a measure of model uncertainty, it can be used to adaptively modify the system model. Hence, the SVSF can benefit from two performance indicators [97]:

- *The output estimation error (the innovation vector)*: This performance indicator is conventionally used in other filtering algorithms as well.
- *The variance or energy of chattering*: It applies separately to every estimated parameter or state including the states without any associated explicit measurements. Therefore, it would be possible to track the source of model uncertainty and correct it based on the magnitude of chattering.

The first indicator implies that a more accurate model leads to a smaller *mean-squared error* (MSE). However, dynamic model refinement cannot be performed by paying attention only to this indicator. This is due to the fact that the states with associated direct measurements can be accurately estimated. This may lead to a small output estimation error, although the estimation error for other states or parameters may be considerable due to measurement noise. Therefore, complementary information provided by the second indicator would be needed to evaluate the assumed range of uncertainty in the model. As a result, the combination of these two performance indicators provides a powerful mean for dynamic model refinement. However, dynamic model refinement can benefit from other problem-specific a priori information as well.

7.5 Filter Corrective Term for Linear Systems

In this section, using Theorem 7.1, the SVSF gain is derived for linear systems in its full and reduced order forms [129]. Let us consider the following discrete-time state-space model:

$$\mathbf{x}_{k+1} = \mathbf{A}\mathbf{x}_k + \mathbf{B}\mathbf{u}_k + \mathbf{v}_k, \tag{7.26}$$

$$\mathbf{y}_k = \mathbf{C}\mathbf{x}_k + \mathbf{w}_k. \tag{7.27}$$

The a priori state estimate and the a priori and a posteriori measurement estimates are computed as:

$$\hat{\mathbf{x}}_{k+1|k} = \hat{\mathbf{A}}\hat{\mathbf{x}}_{k|k} + \hat{\mathbf{B}}\mathbf{u}_k, \tag{7.28}$$

$$\hat{\mathbf{y}}_{k+1|k} = \hat{\mathbf{C}}\hat{\mathbf{x}}_{k+1|k}, \tag{7.29}$$

$$\hat{\mathbf{y}}_{k|k} = \hat{\mathbf{C}}\hat{\mathbf{x}}_{k|k}. \tag{7.30}$$

If the output matrix $\mathbf{C}$ is diagonal, positive, and square ($\mathbf{C} \in \mathbb{R}^{n_x \times n_x}$), then these conditions also imply a direct correspondence between measurements and states. Theorem 7.1 is directly applicable to the system in (7.26) and (7.27). According to Theorem 7.1, given (7.13) and (7.14), a stable SVSF corrective gain can be computed as:

$$\mathbf{k}_k = \hat{\mathbf{C}}^{-1}\left(\left|\mathbf{e}_{\mathbf{y}_{k|k-1}}\right| + \boldsymbol{\Gamma}\left|\mathbf{e}_{\mathbf{y}_{k-1|k-1}}\right|\right) \odot \operatorname{sign}\left(\mathbf{e}_{\mathbf{y}_{k|k-1}}\right), \tag{7.31}$$

where $\boldsymbol{\Gamma} \in \mathbb{R}^{n_x \times n_x}$ is a diagonal matrix with elements $\boldsymbol{\Gamma}_{ii} \in [0,1)$.

The rate of convergence is obtained from the estimation error dynamics. Substituting (7.31) into the error equation (7.20), results in:

$$\begin{aligned}\mathbf{e}_{\mathbf{y}_{k|k}} &= \mathbf{e}_{\mathbf{y}_{k|k-1}} - \hat{\mathbf{C}}\hat{\mathbf{C}}^{-1}\left(\left|\mathbf{e}_{\mathbf{y}_{k|k-1}}\right| + \boldsymbol{\Gamma}\left|\mathbf{e}_{\mathbf{y}_{k-1|k-1}}\right|\right) \odot \operatorname{sign}\left(\mathbf{e}_{\mathbf{y}_{k|k-1}}\right) \\ &= -\boldsymbol{\Gamma}\left|\mathbf{e}_{\mathbf{y}_{k-1|k-1}}\right| \odot \operatorname{sign}\left(\mathbf{e}_{\mathbf{y}_{k|k-1}}\right).\end{aligned} \tag{7.32}$$

Since $\boldsymbol{\Gamma}$ is diagonal with elements $\boldsymbol{\Gamma}_{ii} < 1$, (7.32) leads to a time series of a posteriori output estimation errors with decreasing magnitudes:

$$\left|\mathbf{e}_{\mathbf{y}_{k|k}}\right| = \boldsymbol{\Gamma}\left|\mathbf{e}_{\mathbf{y}_{k-1|k-1}}\right|. \tag{7.33}$$

According to (7.33), the rate of convergence is determined by the choice of $\boldsymbol{\Gamma}$. Since eigenvalues of $\boldsymbol{\Gamma}$, which are associated with the error dynamics, are within the unit circle, the estimation process is stable.

Regarding the filter corrective term of (7.31), $\mathbf{e}_{\mathbf{y}_{k|k-1}} = 0$ is considered as the switching hyperplane. In the absence of a smoothing subspace, this corrective term leads to chattering. The uncertainty associated with the a priori state estimate can be quantified as:

$$\mathbf{d}_k = \tilde{\mathbf{A}}\left(\hat{\mathbf{C}}^{-1}(\mathbf{y}_k - \mathbf{w}_k)\right) + \tilde{\mathbf{B}}\mathbf{u}_k + \mathbf{v}_k, \tag{7.34}$$

where

$$\tilde{\mathbf{A}} = \mathbf{A} - \hat{\mathbf{A}}, \tag{7.35}$$

$$\tilde{\mathbf{B}} = \mathbf{B} - \hat{\mathbf{B}}. \tag{7.36}$$

Hence, a conservative upper bound on the width of the existence subspace can be obtained as:

$$\hat{\beta} = \sup\left(\mathbf{d}_k\right). \tag{7.37}$$

To remove the chattering effect, a smoothing subspace is introduced. The estimate of the upper bound on $\boldsymbol{\beta}$ in (7.37) is used as the width of the smoothing subspace, $\boldsymbol{\psi} = \hat{\boldsymbol{\beta}}$. The filter corrective term of (7.31) is modified by replacing the sign($\cdot$) function with the sat($\cdot$, $\boldsymbol{\psi}$) function defined in (7.24) as follows:

$$\mathbf{k}_k = \hat{\mathbf{C}}^{-1}\left(\left|\mathbf{e}_{\mathbf{y}_{k|k-1}}\right| + \boldsymbol{\Gamma}\left|\mathbf{e}_{\mathbf{y}_{k-1|k-1}}\right|\right) \odot \operatorname{sat}\left(\mathbf{e}_{\mathbf{y}_{k|k-1}}, \boldsymbol{\psi}\right). \tag{7.38}$$

Chattering will be removed if $\boldsymbol{\psi} = \hat{\boldsymbol{\beta}} > \boldsymbol{\beta}$, otherwise chattering occurs with a magnitude, which is proportional to $\boldsymbol{\psi} - \mathbf{d}_k$.

In many applications, it is unnecessary and prohibitively expensive to have a measurement corresponding to each state variable. Therefore, in real systems, the number of measurements (outputs) is usually less than the number of states, $n_y < n_x$. In such systems, if the output matrix $\mathbf{C} \in \mathbb{R}^{n_y \times n_x}$ is full rank, its rank will be n_y. In this case, by properly arranging the state variables in the state vector or transforming the state vector, if it is necessary, the output matrix can be restructured as:

$$\mathbf{C} = \begin{bmatrix}\mathbf{C}_1 & \mathbf{C}_2\end{bmatrix}, \tag{7.39}$$

where $\mathbf{C}_1 \in \mathbb{R}^{n_y \times n_y}$ is a square matrix of rank n_y, and $\mathbf{C}_2 \in \mathbb{R}^{n_y \times (n_x - n_y)}$ is either a null matrix or its columns are linear combinations of the columns of $\mathbf{C}_1$.

The SVSF is a filter that estimates the state vector in the presence of process and measurement noise, hence, it is not simply an observer. In this regard, there is benefit in using the SVSF to estimate the states associated with $\mathbf{C}_1$. Hence, the state vector is partitioned or if necessary transformed into two subvectors as:

$$\mathbf{x}_k = \begin{bmatrix} \mathbf{x}_k^u \\ \mathbf{x}_k^l \end{bmatrix}. \tag{7.40}$$

While $\mathbf{x}_k^u \in \mathbb{R}^{n_y}$ is directly linked to measurements, $\mathbf{x}_k^l \in \mathbb{R}^{n_x - n_y}$ is not. The filter corrective term $\mathbf{k}_k$ is partitioned accordingly:

$$\mathbf{k}_k = \begin{bmatrix} \mathbf{k}_k^u \\ \mathbf{k}_k^l \end{bmatrix}, \tag{7.41}$$

where

$$\mathbf{k}_k^u = \hat{\mathbf{C}}^\dagger \left(\left| \mathbf{e}_{\mathbf{y}_{k|k-1}} \right| + \mathbf{\Gamma} \left| \mathbf{e}_{\mathbf{y}_{k-1|k-1}} \right| \right) \odot \operatorname{sign}\left(\mathbf{e}_{\mathbf{y}_{k|k-1}} \right), \tag{7.42}$$

where $\mathbf{\Gamma} \in \mathbb{R}^{n_y \times n_y}$ is a diagonal matrix with elements $\mathbf{\Gamma}_{ii} \in [0,1)$.

For completely observable and completely controllable systems, stability of the output estimate in the sense of (7.15) implies stability of the state estimate [16]. According to Theorem 7.1, the SVSF gain of (7.42) guarantees stability of the state-estimation process. For completely observable and completely controllable systems, a reduced order estimator can be designed for the lower subvector of the state vector, $\mathbf{x}_k^l$. Following a similar strategy to the Luenberger's reduced order observer [129], the state vector is transformed into a partitioned form in which the upper subvector has the identity relationship with the measurement vector [97]:

$$\mathbf{z}_k = \begin{bmatrix} \mathbf{z}_k^u \\ \mathbf{z}_k^l \end{bmatrix} = \mathbf{T}\mathbf{x}_k, \tag{7.43}$$

where $\mathbf{T}$ is a transformation matrix, and

$$\mathbf{y}_k = \mathbf{z}_k^u + \mathbf{w}_k. \tag{7.44}$$

According to Luenberger's approach, the transformed state vector can be expressed in terms of measurements, albeit a noisy one in this case:

$$\mathbf{z}_k = \begin{bmatrix} \mathbf{y}_k \\ \mathbf{z}_k^l \end{bmatrix}. \tag{7.45}$$

Now, the goal is to estimate the lower subvector, $\mathbf{z}_k^l$. Regarding the transformed state in (7.45), the state-space model in (7.26) and (7.27) can be restated in its partitioned form as:

$$\begin{bmatrix} \mathbf{y}_{k+1} \\ \mathbf{z}_{k+1}^l \end{bmatrix} = \begin{bmatrix} \mathbf{\Phi}_{11} & \mathbf{\Phi}_{12} \\ \mathbf{\Phi}_{21} & \mathbf{\Phi}_{22} \end{bmatrix} \begin{bmatrix} \mathbf{y}_k \\ \mathbf{z}_k^l \end{bmatrix} + \begin{bmatrix} \mathbf{G}_1 \\ \mathbf{G}_2 \end{bmatrix} \mathbf{u}_k + \begin{bmatrix} \bar{\mathbf{v}}_{1_k} \\ \bar{\mathbf{v}}_{2_k} \end{bmatrix}, \tag{7.46}$$

where

$$\boldsymbol{\Phi} = \begin{bmatrix} \boldsymbol{\Phi}_{11} & \boldsymbol{\Phi}_{12} \\ \boldsymbol{\Phi}_{21} & \boldsymbol{\Phi}_{22} \end{bmatrix} = \mathbf{TAT}^{-1}, \tag{7.47}$$

$$\mathbf{G} = \begin{bmatrix} \mathbf{G}_1 \\ \mathbf{G}_2 \end{bmatrix} = \mathbf{TB}, \tag{7.48}$$

$$\bar{\mathbf{v}}_k = \begin{bmatrix} \bar{\mathbf{v}}_{1_k} \\ \bar{\mathbf{v}}_{2_k} \end{bmatrix} = \mathbf{Tv}_k - \begin{bmatrix} \boldsymbol{\Phi}_{11} \\ \boldsymbol{\Phi}_{21} \end{bmatrix} \mathbf{w}_k. \tag{7.49}$$

The corresponding output matrix is now $\begin{bmatrix} \mathbf{I} & \mathbf{0} \end{bmatrix}$. Regarding (7.46), the a priori state estimate is computed as:

$$\begin{bmatrix} \hat{\mathbf{y}}_{k+1|k} \\ \hat{\mathbf{z}}^l_{k+1|k} \end{bmatrix} = \begin{bmatrix} \hat{\boldsymbol{\Phi}}_{11} & \hat{\boldsymbol{\Phi}}_{12} \\ \hat{\boldsymbol{\Phi}}_{21} & \hat{\boldsymbol{\Phi}}_{22} \end{bmatrix} \begin{bmatrix} \mathbf{y}_k \\ \hat{\mathbf{z}}^l_{k|k} \end{bmatrix} + \begin{bmatrix} \hat{\mathbf{G}}_1 \\ \hat{\mathbf{G}}_2 \end{bmatrix} \mathbf{u}_k, \tag{7.50}$$

Subtracting (7.50) from (7.46), the following error equation is obtained:

$$\begin{bmatrix} \mathbf{e}_{\mathbf{y}_{k+1|k}} \\ \mathbf{e}_{\mathbf{z}^l_{k+1|k}} \end{bmatrix} = \begin{bmatrix} \hat{\boldsymbol{\Phi}}_{11} & \hat{\boldsymbol{\Phi}}_{12} \\ \hat{\boldsymbol{\Phi}}_{21} & \hat{\boldsymbol{\Phi}}_{22} \end{bmatrix} \begin{bmatrix} \mathbf{0} \\ \mathbf{e}_{\mathbf{z}^l_{k|k}} \end{bmatrix} + \mathbf{d}_k, \tag{7.51}$$

where

$$\mathbf{d}_k = \begin{bmatrix} \mathbf{d}_{1_k} \\ \mathbf{d}_{2_k} \end{bmatrix} = \tilde{\boldsymbol{\Phi}} \begin{bmatrix} \mathbf{y}_k \\ \mathbf{z}_k \end{bmatrix} + \tilde{\mathbf{G}}\mathbf{u}_k + \begin{bmatrix} \bar{\mathbf{v}}_{1_k} \\ \bar{\mathbf{v}}_{2_k} \end{bmatrix}, \tag{7.52}$$

$$\tilde{\boldsymbol{\Phi}} = \boldsymbol{\Phi} - \hat{\boldsymbol{\Phi}}, \tag{7.53}$$

$$\tilde{\mathbf{G}} = \mathbf{G} - \hat{\mathbf{G}}. \tag{7.54}$$

From (7.51), we have:

$$\mathbf{e}_{\mathbf{z}^l_{k|k}} = \hat{\boldsymbol{\Phi}}_{12}^{-1} \mathbf{e}_{\mathbf{y}_{k+1|k}} - \hat{\boldsymbol{\Phi}}_{12}^{-1} \mathbf{d}_{1_k}, \tag{7.55}$$

$$\mathbf{e}_{\mathbf{z}^l_{k+1|k}} = \hat{\boldsymbol{\Phi}}_{22} \hat{\boldsymbol{\Phi}}_{12}^{-1} \mathbf{e}_{\mathbf{y}_{k+1|k}} - \hat{\boldsymbol{\Phi}}_{22} \hat{\boldsymbol{\Phi}}_{12}^{-1} \mathbf{d}_{1_k} + \mathbf{d}_{2_k}. \tag{7.56}$$

Here, for a linear system, equation (7.55) provides the mapping mentioned in Definition 7.1 for a consecutive bijective system. According to Theorem 7.1, the process of estimating the lower subvector of the state vector, $\mathbf{z}^l_k$, by the SVSF will be stable, if the following two conditions hold:

$$\left|\mathbf{e}_{\mathbf{z}^l_{k|k-1}}\right| \le |\mathbf{k}_k| < \left|\mathbf{e}_{\mathbf{z}^l_{k|k-1}}\right| + \left|\mathbf{e}_{\mathbf{z}^l_{k-1|k-1}}\right|, \tag{7.57}$$

$$\text{sign}\,(\mathbf{k}_k) = \text{sign}\left(\mathbf{e}_{\mathbf{z}^l_{k|k-1}}\right). \tag{7.58}$$

Regarding (7.56), the only known component of $\mathbf{e}_{\mathbf{z}^l_{k|k-1}}$ in the aforementioned conditions, which can be measured, is $\hat{\boldsymbol{\Phi}}_{22} \hat{\boldsymbol{\Phi}}_{12}^{-1} \mathbf{e}_{\mathbf{y}_{k|k-1}}$. This quantity can be viewed as a mapping of the output estimation error to the unmeasured lower subvector of the transformed state vector, $\mathbf{z}_k^l$. Therefore, it is used to define the switching hyperplane for the lower subvector $\mathbf{z}_k^l$ as $\hat{\mathbf{\Phi}}_{22}\hat{\mathbf{\Phi}}_{12}^{-1}\mathbf{e}_{\mathbf{y}_{k+1|k}} = 0$. The corresponding SVSF corrective term for estimating $\mathbf{z}_k^l$ is obtained as:

$$\mathbf{k}_k^l = \left(\left|\hat{\mathbf{\Phi}}_{22}\hat{\mathbf{\Phi}}_{12}^{-1}\mathbf{e}_{\mathbf{y}_{k|k-1}}\right| + \mathbf{\Gamma}\left|\hat{\mathbf{\Phi}}_{12}^{-1}\mathbf{e}_{\mathbf{y}_{k|k-1}}\right|\right) \odot \operatorname{sign}\left(\hat{\mathbf{\Phi}}_{22}\hat{\mathbf{\Phi}}_{12}^{-1}\mathbf{e}_{\mathbf{y}_{k|k-1}}\right). \tag{7.59}$$

The upper bound on the width of the existence subspace, $\boldsymbol{\beta}$, around the switching hyperplane $\hat{\mathbf{\Phi}}_{22}\hat{\mathbf{\Phi}}_{12}^{-1}\mathbf{e}_{\mathbf{y}_{k+1|k}} = 0$ is computed in what follows.

Subtracting $\left|\mathbf{e}_{\mathbf{z}_{k|k-1}^l}\right|$ from all elements in the inequality of (7.57), this condition is restated as:

$$0 \le |\mathbf{k}_k| - \left|\mathbf{e}_{\mathbf{z}_{k|k-1}^l}\right| < \left|\mathbf{e}_{\mathbf{z}_{k-1|k-1}^l}\right|. \tag{7.60}$$

From (7.56), we have:

$$\begin{aligned}|\mathbf{k}_k| - \left|\mathbf{e}_{\mathbf{z}_{k|k-1}^l}\right| &= \left|\hat{\mathbf{\Phi}}_{22}\hat{\mathbf{\Phi}}_{12}^{-1}\mathbf{e}_{\mathbf{y}_{k|k-1}}\right| + \mathbf{\Gamma}\left|\hat{\mathbf{\Phi}}_{12}^{-1}\mathbf{e}_{\mathbf{y}_{k|k-1}}\right| \\ &\quad - \left|\hat{\mathbf{\Phi}}_{22}\hat{\mathbf{\Phi}}_{12}^{-1}\mathbf{e}_{\mathbf{y}_{k|k-1}} - \hat{\mathbf{\Phi}}_{22}\hat{\mathbf{\Phi}}_{12}^{-1}\mathbf{d}_{1_{k-1}} + \mathbf{d}_{2_{k-1}}\right| \\ &\le \mathbf{\Gamma}\left|\hat{\mathbf{\Phi}}_{12}^{-1}\mathbf{e}_{\mathbf{y}_{k|k-1}}\right| + \left|\mathbf{d}_{2_{k-1}} - \hat{\mathbf{\Phi}}_{22}\hat{\mathbf{\Phi}}_{12}^{-1}\mathbf{d}_{1_{k-1}}\right|.\end{aligned} \tag{7.61}$$

From (7.55) and (7.61), we obtain:

$$\min\left|\mathbf{e}_{\mathbf{z}_{k-1|k-1}^l}\right| = \left|\hat{\mathbf{\Phi}}_{12}^{-1}\mathbf{e}_{\mathbf{y}_{k|k-1}}\right| - \left|\hat{\mathbf{\Phi}}_{12}^{-1}\mathbf{d}_{1_{k-1}}\right|, \tag{7.62}$$

$$\max\left(|\mathbf{k}_k| - \left|\mathbf{e}_{\mathbf{z}_{k|k-1}^l}\right|\right) = \mathbf{\Gamma}\left|\hat{\mathbf{\Phi}}_{12}^{-1}\mathbf{e}_{\mathbf{y}_{k|k-1}}\right| + \left|\mathbf{d}_{2_{k-1}} - \hat{\mathbf{\Phi}}_{22}\hat{\mathbf{\Phi}}_{12}^{-1}\mathbf{d}_{1_{k-1}}\right|, \tag{7.63}$$

which leads to

$$\min\left|\mathbf{e}_{\mathbf{z}_{k-1|k-1}^l}\right| > \max\left(|\mathbf{k}_k| - \left|\mathbf{e}_{\mathbf{z}_{k|k-1}^l}\right|\right), \tag{7.64}$$

if:

$$\left|\hat{\mathbf{\Phi}}_{12}^{-1}\mathbf{e}_{\mathbf{y}_{k|k-1}}\right| > (\mathbf{I} - \mathbf{\Gamma})^{-1}\left(\left|\mathbf{d}_{2_{k-1}} - \hat{\mathbf{\Phi}}_{22}\hat{\mathbf{\Phi}}_{12}^{-1}\mathbf{d}_{1_{k-1}}\right| + \left|\hat{\mathbf{\Phi}}_{12}^{-1}\mathbf{d}_{1_{k-1}}\right|\right). \tag{7.65}$$

Regarding the argument of the sign$(\cdot)$ function in (7.59), the width of the existence subspace is considered with respect to $\hat{\mathbf{\Phi}}_{22}\hat{\mathbf{\Phi}}_{12}^{-1}\mathbf{e}_{\mathbf{y}_{k|k-1}}$, hence, condition (7.65) can be rewritten as:

$$\left|\hat{\mathbf{\Phi}}_{22}\hat{\mathbf{\Phi}}_{12}^{-1}\mathbf{e}_{\mathbf{y}_{k|k-1}}\right| > (\mathbf{I} - \mathbf{\Gamma})^{-1}\left|\hat{\mathbf{\Phi}}_{22}\right|\left(\left|\mathbf{d}_{2_{k-1}} - \hat{\mathbf{\Phi}}_{22}\hat{\mathbf{\Phi}}_{12}^{-1}\mathbf{d}_{1_{k-1}}\right| + \left|\hat{\mathbf{\Phi}}_{12}^{-1}\mathbf{d}_{1_{k-1}}\right|\right). \tag{7.66}$$

A condition that would satisfy the right hand side of inequality (7.57) is:

$$\left|\hat{\mathbf{\Phi}}_{22}\hat{\mathbf{\Phi}}_{12}^{-1}\mathbf{e}_{\mathbf{y}_{k|k-1}}\right| > \boldsymbol{\beta}, \tag{7.67}$$

where $\boldsymbol{\beta}$ is the width of the existence subspace for $\hat{\boldsymbol{\Phi}}_{22}\hat{\boldsymbol{\Phi}}_{12}^{-1}\mathbf{e}_{\mathbf{y}_{k|k-1}}$ as:

$$\boldsymbol{\beta} > \sup\left((\mathbf{I}-\boldsymbol{\Gamma})^{-1}\left|\hat{\boldsymbol{\Phi}}_{22}\right|\left(\left|\mathbf{d}_{2_{k-1}} - \hat{\boldsymbol{\Phi}}_{22}\hat{\boldsymbol{\Phi}}_{12}^{-1}\mathbf{d}_{1_{k-1}}\right| + \left|\hat{\boldsymbol{\Phi}}_{12}^{-1}\mathbf{d}_{1_{k-1}}\right|\right)\right). \tag{7.68}$$

Let us consider the left hand side of inequality (7.57), to satisfy Theorem 7.1, from (7.60) and (7.61), we obtain:

$$0 \leq \min\left(|\mathbf{k}_k| - \left|\mathbf{e}_{\mathbf{z}^l_{k|k-1}}\right|\right) = \boldsymbol{\Gamma}\left|\hat{\boldsymbol{\Phi}}_{12}^{-1}\mathbf{e}_{\mathbf{y}_{k|k-1}}\right| - \left|\mathbf{d}_{2_{k-1}} - \hat{\boldsymbol{\Phi}}_{22}\hat{\boldsymbol{\Phi}}_{12}^{-1}\mathbf{d}_{1_{k-1}}\right|, \tag{7.69}$$

Since $\boldsymbol{\Gamma}$ is a matrix with non-negative elements, and we have:

$$\left|\hat{\boldsymbol{\Phi}}_{22}\right| \cdot \left|\hat{\boldsymbol{\Phi}}_{12}^{-1}\mathbf{e}_{\mathbf{y}_{k|k-1}}\right| \geq \left|\hat{\boldsymbol{\Phi}}_{22}\hat{\boldsymbol{\Phi}}_{12}^{-1}\mathbf{e}_{\mathbf{y}_{k|k-1}}\right|, \tag{7.70}$$

condition (7.69) is satisfied, if the following inequality holds:

$$\left|\hat{\boldsymbol{\Phi}}_{22}\hat{\boldsymbol{\Phi}}_{12}^{-1}\mathbf{e}_{\mathbf{y}_{k|k-1}}\right| > \boldsymbol{\Gamma}^{-1}\left|\hat{\boldsymbol{\Phi}}_{22}\right| \cdot \left|\mathbf{d}_{2_{k-1}} - \hat{\boldsymbol{\Phi}}_{22}\hat{\boldsymbol{\Phi}}_{12}^{-1}\mathbf{d}_{1_{k-1}}\right|. \tag{7.71}$$

According to (7.67), (7.68), and (7.71), for $\hat{\boldsymbol{\Phi}}_{22}\hat{\boldsymbol{\Phi}}_{12}^{-1}\mathbf{e}_{\mathbf{y}_{k|k-1}}$, the width of the existence subspace can be obtained as follows in order to satisfy both sides of inequality (7.57):

$$\boldsymbol{\beta} = \max\left((\mathbf{I}-\boldsymbol{\Gamma})^{-1}, \boldsymbol{\Gamma}^{-1}\right) \cdot \left|\hat{\boldsymbol{\Phi}}_{22}\right| \cdot \sup\left(\left|\mathbf{d}_{2_{k-1}} - \hat{\boldsymbol{\Phi}}_{22}\hat{\boldsymbol{\Phi}}_{12}^{-1}\mathbf{d}_{1_{k-1}}\right| + \left|\hat{\boldsymbol{\Phi}}_{12}^{-1}\mathbf{d}_{1_{k-1}}\right|\right). \tag{7.72}$$

If the state trajectory is outside the existence subspace, Theorem 7.1 applies, such that:

$$\left|\hat{\boldsymbol{\Phi}}_{22}\hat{\boldsymbol{\Phi}}_{12}^{-1}\mathbf{e}_{\mathbf{y}_{k|k-1}}\right| > \boldsymbol{\beta}. \tag{7.73}$$

If:

$$\boldsymbol{\beta} > \sup\left(\hat{\boldsymbol{\Phi}}_{22}\hat{\boldsymbol{\Phi}}_{12}^{-1}\mathbf{d}_{1_{k-1}} - \mathbf{d}_{2_{k-1}}\right), \tag{7.74}$$

then, from (7.56), outside the existence subspace, we have:

$$\operatorname{sign}\left(\mathbf{z}^l_{k|k-1}\right) = \operatorname{sign}\left(\hat{\boldsymbol{\Phi}}_{22}\hat{\boldsymbol{\Phi}}_{12}^{-1}\mathbf{y}_{k|k-1}\right). \tag{7.75}$$

Regarding (7.56), this is achieved by modifying the expression for the width of the existence subspace in (7.72) as follows:

$$\hat{\boldsymbol{\beta}} = \max\left((\mathbf{I}-\boldsymbol{\Gamma})^{-1}\left|\hat{\boldsymbol{\Phi}}_{22}\right|, \boldsymbol{\Gamma}^{-1}\left|\hat{\boldsymbol{\Phi}}_{22}\right|, (\mathbf{I}-\boldsymbol{\Gamma})^{-1}, \boldsymbol{\Gamma}^{-1}, \mathbf{I}\right) \times \sup\left(\left|\mathbf{d}_{2_{k-1}} - \hat{\boldsymbol{\Phi}}_{22}\hat{\boldsymbol{\Phi}}_{12}^{-1}\mathbf{d}_{1_{k-1}}\right| + \left|\hat{\boldsymbol{\Phi}}_{12}^{-1}\mathbf{d}_{1_{k-1}}\right|\right). \tag{7.76}$$

In the reachability phase, when $\left|\hat{\boldsymbol{\Phi}}_{22}\hat{\boldsymbol{\Phi}}_{12}^{-1}\mathbf{e}_{\mathbf{y}_{k|k-1}}\right| > \boldsymbol{\beta}$, regarding (7.69) and (7.71), the sign condition of (7.58) will hold by letting:

$$\operatorname{sign}(\mathbf{k}_k) = \operatorname{sign}\left(\hat{\boldsymbol{\Phi}}_{22}\hat{\boldsymbol{\Phi}}_{12}^{-1}\mathbf{y}_{k|k-1}\right). \tag{7.77}$$

Hence, the reduced order SVSF corrective term of (7.59) satisfies the magnitude and sign conditions of (7.57) and (7.58) for the width of the existence subspace given in (7.76). Now, the width the of the smoothing subspace can be specified as [97]:

$$\psi = \hat{\beta}. \tag{7.78}$$

Accordingly, the SVSF corrective term of (7.59) is modified as follows to take account of the smoothing subspace:

$$\mathbf{k}_k^l = \left(\left| \hat{\mathbf{\Phi}}_{22} \hat{\mathbf{\Phi}}_{12}^{-1} \mathbf{e}_{\mathbf{y}_{k|k-1}} \right| + \mathbf{\Gamma} \left| \hat{\mathbf{\Phi}}_{12}^{-1} \mathbf{e}_{\mathbf{y}_{k|k-1}} \right| \right) \odot \operatorname{sat} \left(\hat{\mathbf{\Phi}}_{22} \hat{\mathbf{\Phi}}_{12}^{-1} \mathbf{e}_{\mathbf{y}_{k|k-1}}, \psi \right). \tag{7.79}$$

The rate of convergence of the estimated trajectory to the existence subspace is obtained by considering the error equation in the reduced-order form:

$$\mathbf{e}_{\mathbf{z}^l_{k|k}} = \mathbf{e}_{\mathbf{z}^l_{k|k-1}} - \mathbf{k}_k^l. \tag{7.80}$$

Let us substitute the error terms from (7.55) and (7.56) in (7.80), and multiply both sides of equation (7.80) by $\hat{\mathbf{\Phi}}_{22}$. After some algebraic manipulation, the error equation is obtained as:

$$\begin{aligned} \left| \hat{\mathbf{\Phi}}_{22} \hat{\mathbf{\Phi}}_{12}^{-1} \mathbf{e}_{\mathbf{y}_{k+1|k}} \right| &= \mathbf{\Gamma} \left| \hat{\mathbf{\Phi}}_{22} \hat{\mathbf{\Phi}}_{12}^{-1} \mathbf{e}_{\mathbf{y}_{k|k-1}} \right| \\ &\pm \left| \hat{\mathbf{\Phi}}_{22} \left(\hat{\mathbf{\Phi}}_{12}^{-1} \mathbf{d}_{1_k} - \hat{\mathbf{\Phi}}_{22} \hat{\mathbf{\Phi}}_{12}^{-1} \mathbf{d}_{1_{k-1}} + \mathbf{d}_{2_{k-1}} \right) \right|. \end{aligned} \tag{7.81}$$

According to (7.81), the convergence rate of the filter is determined by the choice of the non-negative matrix $\mathbf{\Gamma}$.

7.6 Filter Corrective Term for Nonlinear Systems

Many real-world systems are inherently nonlinear, and can be described by the state-space model (7.6) and (7.7). In order to apply the SVSF to such systems with smooth nonlinear time-invariant dynamics and measurements that are directly related to some states, a linearized strategy is deployed, which is similar to the one used in the extended Kalman filter (EKF). However, if the nonlinear function $\mathbf{f}$ in (7.6) is consecutive bijective, linearization of the state equation would not be necessary for using the SVSF. The adopted strategy for applying the SVSF to nonlinear systems is quite similar to the one presented for linear systems [97]. Therefore, the state vector is partitioned according to (7.40) as:

$$\mathbf{x}_k = \begin{bmatrix} \mathbf{x}_k^u \\ \mathbf{x}_k^l \end{bmatrix}. \tag{7.82}$$

The state vector is partitioned in a way to have one measurement corresponding to each state variable in the upper subvector, $\mathbf{x}_k^u$. For estimating $\mathbf{x}_k^u$, the following SVSF corrective term satisfies the conditions of Theorem 7.1:

$$\mathbf{k}_k^u = \hat{\mathbf{C}}^{\dagger} \left(\left| \mathbf{e}_{\mathbf{y}_{k|k-1}} \right| + \boldsymbol{\Gamma} \left| \mathbf{e}_{\mathbf{y}_{k-1|k-1}} \right| \right) \odot \operatorname{sign} \left(\mathbf{e}_{\mathbf{y}_{k|k-1}} \right). \tag{7.83}$$

Regarding the existence subspace, the corresponding uncertain dynamics can be obtained as:

$$\mathbf{d}_{\mathbf{x}_{k+1}^u} = \tilde{\mathbf{f}}^u \left((\hat{\mathbf{C}} + \tilde{\mathbf{C}})^{\dagger} \mathbf{y}_k, (\hat{\mathbf{C}} + \tilde{\mathbf{C}})^{\dagger} \mathbf{w}_k, \mathbf{u}_k, \mathbf{v}_k \right), \tag{7.84}$$

with the following associated estimate for the upper bound on the width of the existence subspace:

$$\hat{\boldsymbol{\beta}}_{\mathbf{x}^u} = \sup \left(\mathbf{d}_{\mathbf{x}^u} \right). \tag{7.85}$$

This estimated upper bound on the width of the existence subspace is used as the width of the smoothing subspace:

$$\boldsymbol{\psi}_{\mathbf{x}^u} = \hat{\boldsymbol{\beta}}_{\mathbf{x}^u}. \tag{7.86}$$

Hence, for the upper subvector of the state vector, $\mathbf{x}_k^u$, the SVSF corrective term is modified as:

$$\mathbf{k}_k^u = \hat{\mathbf{C}}^{\dagger} \left(\left| \mathbf{e}_{\mathbf{y}_{k|k-1}} \right| + \boldsymbol{\Gamma} \left| \mathbf{e}_{\mathbf{y}_{k-1|k-1}} \right| \right) \odot \operatorname{sat} \left(\mathbf{e}_{\mathbf{y}_{k|k-1}}, \boldsymbol{\psi}_{\mathbf{x}^u} \right). \tag{7.87}$$

For the state variables in the lower subvector, $\mathbf{x}_k^l$, a reduced-order filter can be designed following the same strategy, which was presented for linear systems. Let us consider the following transformed state vector:

$$\mathbf{z}_k = \mathbf{T}\mathbf{x}_k, \tag{7.88}$$

which can be represented with elements that are directly obtained from measurements:

$$\mathbf{z}_k = \begin{bmatrix} \mathbf{z}_k^u \\ \mathbf{z}_k^l \end{bmatrix} = \mathbf{T} \begin{bmatrix} \mathbf{y}_k - \mathbf{w}_k \\ \mathbf{x}_k^l \end{bmatrix}. \tag{7.89}$$

Regarding the state equation (7.6) and the aforementioned transformation, we have:

$$\mathbf{z}_{k+1} = \mathbf{h} \left(\mathbf{z}_k, \mathbf{u}_k, \mathbf{v}_k, \mathbf{w}_k \right), \tag{7.90}$$

where

$$\mathbf{h} \left(\mathbf{z}_k, \mathbf{u}_k, \mathbf{v}_k, \mathbf{w}_k \right) = \mathbf{T}\mathbf{f} \left(\mathbf{x}_k, \mathbf{u}_k, \mathbf{v}_k \right). \tag{7.91}$$

For instance, if $\mathbf{C}$ has the following form:

$$\mathbf{C} = \begin{bmatrix} \mathbf{I} & \mathbf{0} \end{bmatrix}, \tag{7.92}$$

where $\mathbf{I}$ is an identity matrix of dimension $n_y \times n_y$, then $\mathbf{T} = \mathbf{I}$ of dimension $n_x \times n_x$, and

$$\mathbf{h}\left(\mathbf{z}_k, \mathbf{u}_k, \mathbf{v}_k, \mathbf{w}_k\right) = \mathbf{f}\left(\begin{bmatrix} \mathbf{y}_k - \mathbf{w}_k \\ \mathbf{x}_k^l \end{bmatrix}, \mathbf{u}_k, \mathbf{v}_k\right). \tag{7.93}$$

For a reduced-order filter, equation (7.90) is partitioned as:

$$\mathbf{z}_{k+1} = \begin{bmatrix} \mathbf{y}_{k+1} - \mathbf{w}_{k+1} \\ \mathbf{x}_{k+1}^l \end{bmatrix} = \begin{bmatrix} \mathbf{h}^u\left(\mathbf{z}_k^l, \mathbf{u}_k, \mathbf{y}_k, \mathbf{v}_k, \mathbf{w}_k\right) \\ \mathbf{h}^l\left(\mathbf{z}_k^l, \mathbf{u}_k, \mathbf{y}_k, \mathbf{v}_k, \mathbf{w}_k\right) \end{bmatrix}. \tag{7.94}$$

If functions $\mathbf{h}^u$ and $\mathbf{h}^l$ are consecutive bijective over the operating range of the input variables, then the mapping between $\mathbf{z}_{k+1}$ and $\left(\mathbf{z}_k^l, \mathbf{u}_k, \mathbf{y}_k, \mathbf{v}_k, \mathbf{w}_k\right)$ will be unique, and therefore, the following inverse functions exist:

$$\mathbf{z}_k^l = \left(\mathbf{h}^u\right)^{-1}\left(\mathbf{u}_k, \mathbf{y}_{k:k+1}, \mathbf{v}_k, \mathbf{w}_{k:k+1}\right), \tag{7.95}$$

$$\mathbf{z}_k^l = \left(\mathbf{h}^l\right)^{-1}\left(\mathbf{z}_{k+1}^l, \mathbf{u}_k, \mathbf{y}_k, \mathbf{v}_k, \mathbf{w}_k\right). \tag{7.96}$$

For the class of consecutive bijective nonlinear systems, reduced-order observers can be designed based on equations (7.95) and (7.96). These two equations can be expressed in terms of the estimated functions and uncertainty terms as:

$$\mathbf{z}_k^l = \left(\hat{\mathbf{h}}^u\right)^{-1}\left(\mathbf{u}_k, \mathbf{y}_{k:k+1}\right) + \left(\tilde{\mathbf{h}}_k^u\right)^{-1}\left(\mathbf{u}_k, \mathbf{y}_{k:k+1}, \mathbf{v}_k, \mathbf{w}_{k:k+1}\right), \tag{7.97}$$

$$\mathbf{z}_{k+1}^l = \hat{\mathbf{h}}^l\left(\mathbf{z}_k^l, \mathbf{u}_k, \mathbf{y}_k\right) + \tilde{\mathbf{h}}_k^l\left(\mathbf{z}_k^l, \mathbf{u}_k, \mathbf{y}_k, \mathbf{v}_k, \mathbf{w}_k\right). \tag{7.98}$$

Then, we obtain:

$$\hat{\mathbf{z}}_{k+1|k}^l = \hat{\mathbf{h}}^l\left(\left(\hat{\mathbf{h}}^u\right)^{-1}\left(\mathbf{u}_k, \mathbf{y}_{k:k+1}\right), \mathbf{u}_k, \mathbf{y}_k\right), \tag{7.99}$$

$$\hat{\mathbf{z}}_{k+1|k+1}^l = \hat{\mathbf{z}}_{k+1|k}^l + \mathbf{k}_{k+1}. \tag{7.100}$$

The a priori estimation of the transformed state is computed in (7.99) based on the known parts of (7.97) and (7.98). The corresponding uncertain dynamics can be obtained from (7.95)–(7.100) as:

$$\begin{aligned} \mathbf{d}_{\mathbf{z}_k^l} = {} & \hat{\mathbf{h}}^l\left(\left(\hat{\mathbf{h}}^u\right)^{-1}\left(\mathbf{u}_{k-1}, \mathbf{y}_{k-1:k}\right)\right. \\ & \left. + \left(\tilde{\mathbf{h}}_k^u\right)^{-1}\left(\mathbf{u}_{k-1}, \mathbf{y}_{k-1:k}, \mathbf{v}_{k-1}, \mathbf{w}_{k-1:k}\right), \mathbf{u}_{k-1}, \mathbf{y}_{k-1}\right) \\ & - \hat{\mathbf{h}}^l\left(\left(\hat{\mathbf{h}}^u\right)^{-1}\left(\mathbf{u}_{k-1}, \mathbf{y}_{k-1:k}\right), \mathbf{u}_{k-1}, \mathbf{y}_{k-1}\right) \\ & + \tilde{\mathbf{h}}_k^l(\mathbf{z}_k^l, \mathbf{u}_k, \mathbf{y}_k, \mathbf{v}_k, \mathbf{w}_k). \end{aligned} \tag{7.101}$$

For bounded $\left(\tilde{\mathbf{h}}_k^u\right)^{-1}\left(\mathbf{u}_{k-1}, \mathbf{y}_{k-1:k}, \mathbf{v}_{k-1}, \mathbf{w}_{k-1:k}\right)$ and $\tilde{\mathbf{h}}_k^l\left(\mathbf{z}_k^l, \mathbf{u}_k, \mathbf{y}_k, \mathbf{v}_k, \mathbf{w}_k\right)$, an upper bound on the width of the existence subspace can be estimated for $\hat{\mathbf{z}}_{k|k}^l$ as:

$$\hat{\beta}_{\mathbf{z}^l} = \sup\left(\mathbf{d}_{\mathbf{z}_k^l}\right), \tag{7.102}$$

with the following corresponding upper bound for $\hat{\mathbf{x}}^{l}_{k|k}$:

$$\hat{\beta}_{\mathbf{x}^{l}} = \sup\left(\mathbf{T}\mathbf{d}_{\mathbf{z}^{l}_{k}}\right). \tag{7.103}$$

This estimated upper bound on the width of the existence subspace is considered as the width of the smoothing subspace:

$$\psi_{\mathbf{x}^{l}} = \hat{\beta}_{\mathbf{x}^{l}}. \tag{7.104}$$

Subtracting (7.99) and (7.100) from $\left(\hat{\mathbf{h}}^{u}\right)^{-1}\left(\mathbf{u}_{k}, \mathbf{y}_{k:k+1}\right)$ provides the a priori and the a posteriori estimation errors as:

$$\mathbf{e}_{\mathbf{z}^{l}_{k|k-1}} = \left(\hat{\mathbf{h}}^{u}\right)^{-1}\left(\mathbf{u}_{k}, \mathbf{y}_{k:k+1}\right) - \hat{\mathbf{z}}^{l}_{k|k-1}, \tag{7.105}$$

$$\begin{aligned}\mathbf{e}_{\mathbf{z}^{l}_{k|k}} &= \left(\hat{\mathbf{h}}^{u}\right)^{-1}\left(\mathbf{u}_{k}, \mathbf{y}_{k:k+1}\right) - \hat{\mathbf{z}}^{l}_{k|k} \\ &= \mathbf{e}_{\mathbf{z}^{l}_{k|k-1}} - \mathbf{k}^{l}_{k}.\end{aligned} \tag{7.106}$$

Using the following SVSF corrective term, the estimated transformed state $\hat{\mathbf{z}}^{l}_{k|k}$ converges toward the existence subspace according to Theorem 7.1 and the aforementioned error terms:

$$\mathbf{k}^{l}_{k} = \left(\left|\mathbf{e}_{\mathbf{z}^{l}_{k|k-1}}\right| + \boldsymbol{\Gamma}\left|\mathbf{e}_{\mathbf{z}^{l}_{k-1|k-1}}\right|\right) \odot \operatorname{sign}\left(\mathbf{e}_{\mathbf{z}^{l}_{k|k-1}}\right), \tag{7.107}$$

where $\boldsymbol{\Gamma}$ is a diagonal matrix with $\boldsymbol{\Gamma}_{ii} \in (0,1)$. Regarding the chosen width for the smoothing subspace in (7.104), equations (7.87) and (7.107) can be used to combine the results for the two subvectors of the state vector in order to obtain the following SVSF corrective term [97]:

$$\mathbf{k}_{k} = \begin{bmatrix} \hat{\mathbf{C}}^{\dagger}\left(\left|\mathbf{e}_{\mathbf{y}_{k|k-1}}\right| + \boldsymbol{\Gamma}\left|\mathbf{e}_{\mathbf{y}_{k-1|k-1}}\right|\right) \odot \operatorname{sat}\left(\mathbf{e}_{\mathbf{y}_{k|k-1}}, \psi_{\mathbf{x}^{u}}\right) \\ \left(\left|\mathbf{e}_{\mathbf{z}^{l}_{k|k-1}}\right| + \boldsymbol{\Gamma}\left|\mathbf{e}_{\mathbf{z}^{l}_{k-1|k-1}}\right|\right) \odot \operatorname{sat}\left(\mathbf{e}_{\mathbf{z}^{l}_{k|k-1}}, \psi_{\mathbf{x}^{l}}\right) \end{bmatrix}. \tag{7.108}$$

The corresponding switching hyperplane is defined as:

$$\begin{bmatrix} \mathbf{e}_{\mathbf{y}_{k|k-1}} \\ \mathbf{e}_{\mathbf{z}^{l}_{k|k-1}} \end{bmatrix} = \mathbf{0}. \tag{7.109}$$

7.7 Bias Compensation

Uncertain dynamics and low-frequency drift significantly affect the performance of the state estimation algorithms such as Kalman filter, sliding-mode observer, and SVSF. In the presence of drift in the system dynamics, performance of such estimation algorithms can be greatly improved by deploying bias compensation strategies. In the absence of modeling uncertainty, if the system is subjected to white noise, the expected value of the state estimation error will converge to zero according to Theorem 7.1. Otherwise, convergence of the estimated state trajectory to the true state trajectory is guaranteed within the existence subspace. Accuracy of the SVSF within this subspace can be improved by obtaining a fairly accurate estimate of the worst-case disturbance. For a system with directly related states and measurements [129], a bias in the measurement noise or an uncertainty in the output matrix, $\tilde{\mathbf{C}}$, would lead to an error term $\mathbf{d}_k^{signal}$, which affects the estimated state trajectory as:

$$\mathbb{E}\left[\mathbf{e}_{\mathbf{x}_{k|k}}\right] = \tilde{\mathbf{C}}^{\dagger}\mathbb{E}\left[\mathbf{e}_{\mathbf{y}_{k|k}}\right] - \mathbb{E}\left[\mathbf{d}_k^{signal}\right]. \tag{7.110}$$

The governing dynamics of $\mathbf{d}_k^{signal}$ can be divided into two parts regarding its medium to high-frequency contents and its DC bias and low-frequency contents, $\mathbf{d}_k^{bias}$. The low-frequency drift can be partially identified and removed through using a proper low-pass filter. In applications, where states are derivatives of outputs, the noise amplification due to numerical differentiation may lead to a noisy state estimate. For a system with a single measurement and a unity measurement gain, we have:

$$\hat{\mathbf{x}}_k^{filtered} = \begin{bmatrix} y_k^{filtered} \\ \left(\frac{y_k - y_{k-1}}{\tau_s}\right)^{filtered} \\ \vdots \end{bmatrix}, \tag{7.111}$$

where τ_s is the sampling time.

Over a relatively broad range of frequencies, the estimated state vector in (7.111) may be too noisy. Hence, a low-pass filter with a relatively low cut-off frequency may be required to remove the noise. However, deploying such a filter will lead to information loss. Therefore, minimizing the mean value of the estimation error would be a better option for estimates of the form of (7.111), but this method leads to a higher error variance for higher derivatives compared with the Kalman filter and the SVSF. For systems with states that are derivatives of outputs and $\mathbf{C} = \mathbf{I}$, an estimate $\mathbf{d}_k^{bias}$ of the low-frequency contents of $\mathbf{d}_k^{signal}$ can be obtained from (7.111) as:

$$\hat{\mathbf{d}}_k^{bias} = \left(\hat{\mathbf{x}}_k^{filtered} - \hat{\mathbf{x}}_{k|k}\right)^{filtered}. \tag{7.112}$$

Now, compensating for the estimated bias in the existence subspace will further improve the accuracy of the a posteriori state estimate:

$$\hat{\mathbf{x}}_k^{corrected} = \hat{\mathbf{x}}_{k|k} + \hat{\mathbf{d}}_k^{bias}. \tag{7.113}$$

As shown in Algorithm 7.1, this correction is performed in the last step of the SVSF procedure. Since the bias compensation of (7.113) does not propagate through the a priori estimate in the next iteration, stability of the estimation process will not be adversely affected by this step [97].

7.8 The Secondary Performance Indicator

In addition to the estimation error, the SVSF can use the magnitude of chattering, which occurs due to the switching gain, as a secondary performance indicator [97]. The secondary performance indicator for the states with directly related measurements is defined as:

$$\boldsymbol{\Xi}_{\mathbf{y}_i} = \begin{cases} \mathbf{0}, & \mathbf{e}_{\mathbf{y}_{i_{k|k}}} < \psi_i^u, \\ \frac{1}{c_n^2}\left(\mathbf{e}_{\mathbf{y}_{i_{k|k}}} - \psi_i^u\right)^2, & \mathbf{e}_{\mathbf{y}_{i_{k|k}}} > \psi_i^u. \end{cases} \tag{7.114}$$

Similarly, the secondary performance indicator for the states without associated measurements is defined as:

$$\boldsymbol{\Xi}_{\mathbf{z}_i} = \begin{cases} \mathbf{0}, & \mathbf{e}_{\mathbf{z}_{i_{k|k}}} < \psi_i^l, \\ \frac{1}{c_n^2}\left(\mathbf{e}_{\mathbf{z}_{i_{k|k}}} - \psi_i^l\right)^2, & \mathbf{e}_{\mathbf{z}_{i_{k|k}}} > \psi_i^l, \end{cases} \tag{7.115}$$

where c_n is used for normalization and scaling. These indicators allow for quantification of the uncertainty level, change detection, and fault diagnosis. Presence of uncertainty will have a more significant impact on the estimation accuracy for states without directly related measurements. Therefore, for fault detection, the SVSF will outperform other methods that rely only on filter innovation (or the output estimation error) as the indicator of performance. Taking advantage of the secondary indicators of performance, the SVSF is capable of change detection in the system despite a seemingly normal output. This is a unique characteristic of the SVSF compared to other filters.

The dynamic model-refinement strategy can be implemented using a fast inner loop and a slow outer loop. The inner loop deploys the SVSF for parameter and state estimation. It can tolerate model uncertainties and detect changes. The outer loop can deploy a neural network, which can modify the model used by the filter after fault occurrence. The modeling error can be reduced by monitoring the performance indicators. At every sampling interval of the slower outer loop, the model parameters, which are not estimated by the SVSF, are adjusted by the neural network in a way to enhance the overall performance of the estimation algorithm through improving the model accuracy. The SVSF in the fast inner loop uses the updated model parameters provided by the neural network to retune the estimated states. Such a dynamic model-refinement strategy allows for applying the SVSF to systems that may not be fully observable. This demonstrates the benefit of using the secondary indicators of performance. The additional information provided by these indicators makes the SVSF a suitable estimation algorithm for high-order dynamic systems, which may not be fully observable.

7.9 Second-Order Smooth Variable Structure Filter

In the second-order SVSF, the corrective term is computed based on the measurement errors from the two previous time steps. The extra information used in the update stage leads to smoother state estimates as well as improved estimation performance in terms of accuracy and robustness against uncertainties. However, accuracy is improved at the expense of increasing the computational complexity. Similar to (7.108), the corrective term for the second-order SVSF is calculated as [130]:

$$\mathbf{k}_k^u = \hat{\mathbf{C}}^\dagger \left(\mathbf{e}_{\mathbf{y}_{k|k-1}} - \frac{1}{2}\mathbf{e}_{\mathbf{y}_{k-1|k-1}} \right) - \hat{\mathbf{C}}^\dagger \boldsymbol{\Gamma} \sqrt{\frac{1}{4}\mathbf{e}^{\odot 2}_{\mathbf{y}_{k-1|k-1}} + \frac{1}{2}\left(\mathbf{e}_{\mathbf{y}_{k|k}} - \mathbf{e}_{\mathbf{y}_{k-1|k-1}} \right)^{\odot 2}}, \tag{7.116}$$

$$\mathbf{k}_k^l = \hat{\boldsymbol{\Phi}}_{22}\hat{\boldsymbol{\Phi}}_{12}^{-1}\mathbf{e}_{\mathbf{y}_{k|k-1}} - \frac{1}{2}\hat{\boldsymbol{\Phi}}_{12}^{-1}\mathbf{e}_{\mathbf{y}_{k-1|k-1}} - \boldsymbol{\Gamma}\hat{\boldsymbol{\Phi}}_{12}^{-1} \sqrt{\frac{1}{4}\mathbf{e}^{\odot 2}_{\mathbf{y}_{k-1|k-1}} + \frac{1}{2}\left(\mathbf{e}_{\mathbf{y}_{k|k}} - \mathbf{e}_{\mathbf{y}_{k-1|k-1}} \right)^{\odot 2}}, \tag{7.117}$$

where $(.)^{\odot 2}$ denotes element-wise raising to the power of 2.

7.10 Optimal Smoothing Boundary Design

Algorithm 7.2 shows a revised form of the SVSF with estimation error covariance, which is beneficial for comparative purposes [131]. The covariance derivation paves the way for optimally choosing the width of the smoothing subspace. To remove chattering, SVSF uses a smoothing subspace. The width of this subspace is specified by an upper bound on the corresponding uncertainties in the estimation process. This leads to a conservative design in the sense that the SVSF corrective term is computed regarding a larger smoothing subspace than is necessary. More accurate estimates can be obtained through optimally choosing the width of the smoothing subspace, which is referred to as *optimal smoothing boundary design*. Minimizing the trace of the a posteriori estimation error covariance matrix leads to the following optimal width for the smoothing subspace [132]:

$$\boldsymbol{\psi}_{k+1} = \left(\mathbf{P}_{k+1|k} + \mathbf{R}_{k+1} \right) \mathbf{P}^{-1}_{k+1|k} \left(\left| \mathbf{e}_{\mathbf{y}_{k+1|k}} \right| + \boldsymbol{\Gamma} \left| \mathbf{e}_{\mathbf{y}_{k|k}} \right| \right), \tag{7.118}$$

which is time-varying.

Algorithm 7.2: Smooth variable-structure filter with covariance derivation

State-space model:

$$\mathbf{x}_{k+1} = \mathbf{f}(\mathbf{x}_k, \mathbf{u}_k, \mathbf{v}_k)$$
$$\mathbf{y}_k = \mathbf{C}\mathbf{x}_k + \mathbf{w}_k$$

where $\mathbf{v}_k \sim \mathcal{N}(\mathbf{0}, \mathbf{Q}_k)$ and $\mathbf{w}_k \sim \mathcal{N}(\mathbf{0}, \mathbf{R}_k)$.

Initialization:

$$\hat{\mathbf{x}}_0 = \mathbb{E}[\mathbf{x}_0]$$
$$\mathbf{P}_0 = \mathbb{E}\left[\left(\mathbf{x}_0 - \mathbb{E}[\mathbf{x}_0]\right)\left(\mathbf{x}_0 - \mathbb{E}[\mathbf{x}_0]\right)^T\right]$$

for $k = 0, 1, \dots,$ **do**

Linearization

$$\hat{\mathbf{A}}_k = \frac{\partial \hat{\mathbf{f}}(\mathbf{x}_k, \mathbf{u}_k)}{\partial \mathbf{x}_k}\Big|_{\mathbf{x}_k = \hat{\mathbf{x}}_{k|k}}$$

A priori state estimate

$$\hat{\mathbf{x}}_{k+1|k} = \hat{\mathbf{f}}(\hat{\mathbf{x}}_{k|k}, \mathbf{u}_k)$$

A priori error covariance

$$\mathbf{P}_{k+1|k} = \hat{\mathbf{A}}_k \mathbf{P}_{k|k} \hat{\mathbf{A}}_k^T + \mathbf{Q}_k$$

A priori measurement estimate

$$\hat{\mathbf{y}}_{k+1|k} = \hat{\mathbf{C}}\hat{\mathbf{x}}_{k+1|k}$$

A posteriori and a priori output error estimates

$$\mathbf{e}_{\mathbf{y}_{k|k}} = \mathbf{y}_k - \hat{\mathbf{y}}_{k|k}$$
$$\mathbf{e}_{\mathbf{y}_{k+1|k}} = \mathbf{y}_{k+1} - \hat{\mathbf{y}}_{k+1|k}$$

Filter gain

$$\mathbf{K}_{k+1} = \text{diag}\left(\left(|\mathbf{e}_{\mathbf{y}_{k+1|k}}| + \mathbf{\Gamma}|\mathbf{e}_{\mathbf{y}_{k|k}}|\right) \odot \text{sat}\left(\mathbf{e}_{\mathbf{y}_{k+1|k}}, \boldsymbol{\psi}\right)\right) \times \left(\text{diag}\left(\mathbf{e}_{\mathbf{y}_{k+1|k}}\right)\right)^{-1}$$

A posteriori state estimate

$$\hat{\mathbf{x}}_{k+1|k+1} = \hat{\mathbf{x}}_{k+1|k} + \mathbf{K}_{k+1}\mathbf{e}_{\mathbf{y}_{k+1|k}}$$

A posteriori error covariance

$$\mathbf{P}_{k+1|k+1} = \left(\mathbf{I} - \mathbf{K}_{k+1}\hat{\mathbf{C}}\right)\mathbf{P}_{k+1|k}\left(\mathbf{I} - \mathbf{K}_{k+1}\hat{\mathbf{C}}\right)^T + \mathbf{K}_{k+1}\mathbf{R}_{k+1}\mathbf{K}_{k+1}^T$$

end

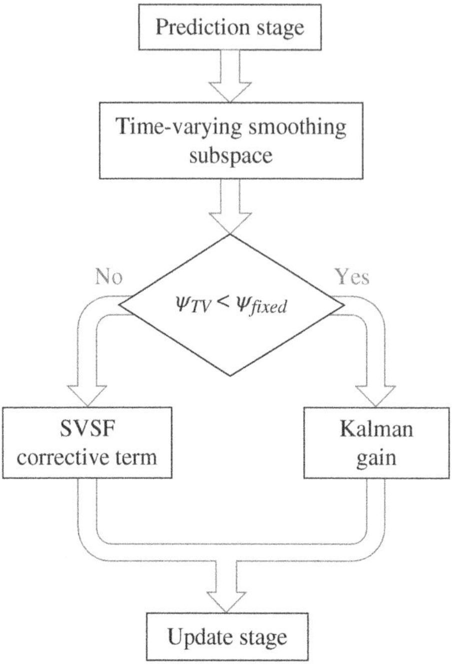

Figure 7.3 Combining the SVSF with Bayesian filters.

7.11 Combination of SVSF with Other Filters

The SVSF can be combined with Bayesian filters such as EKF, UKF, and CKF to achieve a trade-off between robustness and optimality. Such hybrid algorithms benefit from the robustness of the SVSF as well as the accuracy of other filters [133]. In the update stage, the time-varying width of the smoothing subspace in (7.118), ψ_{TV}, is compared with the fixed width of the smoothing subspace in (7.86) and (7.104), ψ_{fixed}. If $\psi_{TV} \geq \psi_{fixed}$, then the SVSF corrective term will be used in the update stage, otherwise the multiplication of the other filter's gain, which is the counterpart of the Kalman gain, and the innovation term will be used as the corrective term. This procedure is depicted in Figure 7.3.

7.12 Applications

The SVSF and its variants have been extensively used in a wide range of applications including robotics, navigation, tracking, and fault detection. In this section, three applications are reviewed, which are focused on multiple target tracking in the presence of clutter, state-of-charge (SoC) estimation for lithium-ion batteries, and a four degree-of-freedom robotic manipulator.

7.12.1 Multiple Target Tracking

The SVSF has been used in combination with the *joint probabilistic data association* for multiple target tracking in the presence of clutter [134]. The proposed algorithm benefits from deriving a generalized form of the estimation error covariance matrix for optimal smoothing boundary design. Driving through the highway between Hamilton and Toronto, Ontario, Canada, several LiDAR datasets were collected, which were used to evaluate the proposed algorithm in terms of the true and false positive rates as well as the rate of track breakups. According to the results, the proposed algorithm provides a reliable multiple target tracking method for real road car tracking scenarios.

7.12.2 Battery State-of-Charge Estimation

The SVSF and its combination with the *interacting multiple model* (IMM) estimator have been used for battery SoC estimation in electric vehicles [135]. In this application, accuracy and robustness of the estimation method is critical for maximizing battery utilization and ensuring safe operation regarding a wide range of operating circumstances. For SoC estimation, model-based filters rely on *equivalent circuit models* for batteries. In order to cover low, medium, and high drive cycle discharge current magnitudes (C-rates) for lithium-ion batteries, three different parameterized equivalent circuit models were considered. The proposed method takes advantage of these three models, the IMM estimator, and the SVSF to significantly improve the SoC estimation accuracy, especially at low temperatures.

7.12.3 Robotics

The SVSF and its combinations with UKF and CKF have been applied to a robotic manipulator with four degrees of freedom [133]. The considered manipulator, which has one prismatic and three revolute joints, was used as a case study for a thorough analysis and comparison of the SVSF and its variants against nonlinear Kalman filters such as UKF and CKF.

7.13 Concluding Remarks

The SVSF is a predictor–corrector method that is based on a stability theorem with well-defined performance measures. If the system under consideration is completely observable and completely controllable, then a realization of the filter may be attained to confine the estimated states to within a neighborhood of their true values. In the presence of severe and changing modeling uncertainties and noise, the SVSF provides a performance improvement over simple designs of Kalman and infinite impulse response (IIR) filters. An important advantage of the SVSF is the availability of a set of secondary performance indicators that pertain to each estimate. This allows for dynamic refinement of the filter model. The combination of the SVSF's robust stability and its secondary indicators of performance makes it a powerful estimation tool capable of compensating for uncertainties that are abruptly introduced in the system as a result of events such as fault conditions.

8

Deep Learning

8.1 Introduction

The *Dartmouth conference* in 1956 is regarded as the birth of *artificial intelligence* [136, 137]. A carefully selected group of scientists gathered together to work on the following topics:

- Automatic computers
- How can a computer be programmed to use a language
- Neural nets
- Theory of the size of a calculation
- Self-improvement
- Abstraction
- Randomness and creativity

Since then, artificial intelligence has been increasingly deployed in different branches of science and engineering. Nowadays, learning algorithms have become part and parcel of daily life. Learning algorithms are categorized as [138]:

- *Supervised learning* refers to learning with a teacher and relies on labeled datasets. Teacher's knowledge is represented as input-output examples.
- *Unsupervised learning* refers to learning without a teacher and relies on unlabeled datasets consisting of only input signals (stimuli) without the desired or target outputs (responses). The learning algorithm is tuned to the statistical regularities of the input data based on a task-independent measure of the quality of the learned representation.
- *Semi-supervised learning* relies on datasets that consist of labeled as well as unlabeled samples.

Nonlinear Filters: Theory and Applications, First Edition. Peyman Setoodeh, Saeid Habibi, and Simon Haykin.

- *Reinforcement learning* refers to learning through continued interaction with the environment. The learning agent tries to optimize a scalar index of performance, which is the expected collected reward received from the environment.

Most of the *machine-learning* algorithms are built on the following four pillars [139]:

- A *dataset*, which may consist of labeled and/or unlabeled samples.
- A *model*, which may be linear or nonlinear.
- A *cost function* or *loss function*, which is based on an error measure.
- An *optimization procedure* to search for the point, where the gradient of the cost function is zero.

Different options for each one of these pillars have led to a wide range of learning algorithms. For nonlinear models, it is generally impossible to find a closed-form optimal solution, hence, iterative numerical optimization procedures such as different variations of the *gradient descent* method are deployed to approximately minimize the cost function. Such iterative methods are effective even in dealing with computationally intractable problems, as long as the corresponding gradient can be estimated [140]. Therefore, statistical estimation plays a key role in machine learning.

8.2 Gradient Descent

The gradient descent method is considered as an essential tool in machine-learning algorithms [141]. Let $\mathcal{L}(\boldsymbol{\theta})$ denote the scalar cost function, which depends on the parameter vector $\boldsymbol{\theta}$. It is desired to solve the following optimization problem:

$$\min_{\boldsymbol{\theta}} \ \mathcal{L}(\boldsymbol{\theta}). \tag{8.1}$$

At a specific point in the parameter space, $\boldsymbol{\theta}$, the gradient vector, $\nabla_{\boldsymbol{\theta}}\mathcal{L}(\boldsymbol{\theta})$, shows the direction at which the cost function increases. Hence, $-\nabla_{\boldsymbol{\theta}}\mathcal{L}(\boldsymbol{\theta})$, which is known as the *steepest descent* direction, shows the direction at which the cost function decreases. Starting from an initial point $\boldsymbol{\theta}_0$, the basic gradient descent method iteratively takes steps along the steepest descent direction until convergence. Algorithm 8.1 shows the basic gradient descent method. While gradient $\nabla_{\boldsymbol{\theta}}\mathcal{L}(\boldsymbol{\theta})$ provides the right direction, its magnitude may be misleading. Therefore, it is preferred to rescale the stepsize according to the local properties of $\mathcal{L}(\boldsymbol{\theta})$. Two simple heuristic rules can be used for this purpose [141]:

- If after an iteration, the value of the cost function increases, that iteration is ignored and the stepsize is decreased.
- If after an iteration, the value of the cost function decreases, the stepsize may be increased.

Algorithm 8.1: Gradient descent

Hyper-parameters

η : stepsize

ϵ : tolerance

Initialization

θ_0

for $k = 0, 1, \ldots,$ *until convergence* **do**

$\theta_{k+1} = \theta_k - \eta \nabla_\theta \mathcal{L}(\theta_k)$

end

8.3 Stochastic Gradient Descent

If a *batch* of data is available, machine learning can be performed in an *offline* manner. However, for *streaming data*, *online* learning is needed, and the gradient estimate is updated as a new data point arrives. In offline learning, for a batch of n_B data points, the following loss function is minimized with respect to θ [77]:

$$\mathcal{L}(\theta, \mathbf{z}_{0:n_B-1}) = \frac{1}{n_B} \sum_{i=0}^{n_B-1} \mathcal{L}(\theta, \mathbf{z}_i), \tag{8.2}$$

where $\mathbf{z}_i$ consists of a set of input and target output data points, $\mathbf{z}_i = (\mathbf{u}_i, \mathbf{y}_i)$, in supervised learning, and a set of input data points, $\mathbf{z}_i = \mathbf{u}_i$ in unsupervised learning. The gradient of loss function (8.2) is computed as:

$$\nabla_\theta \mathcal{L}(\theta, \mathbf{z}_{0:n_B-1}) = \frac{1}{n_B} \nabla_\theta \sum_{i=0}^{n_B-1} \mathcal{L}(\theta, \mathbf{z}_i). \tag{8.3}$$

In online learning, at iteration k, parameters are updated by *online gradient descent* as [77]:

$$\theta_{k+1} = \theta_k - \eta_k \nabla_\theta \mathcal{L}(\theta_k, \mathbf{z}_k). \tag{8.4}$$

In case that the parameters must be constrained to a certain subspace of the parameter space, Θ, a projection operator is used in the right hand side of equation (8.4):

$$\theta_{k+1} = \Pi_\Theta \left(\theta_k - \eta_k \nabla_\theta \mathcal{L}(\theta_k, \mathbf{z}_k)\right), \tag{8.5}$$

where Π_Θ denotes the projection into subspace Θ:

$$\Pi_\Theta \left(\theta_k - \eta_k \nabla_\theta \mathcal{L}(\theta_k, \mathbf{z}_k)\right) = \arg\min_{\theta \in \Theta} \left\| \theta - \left(\theta_k - \eta_k \nabla_\theta \mathcal{L}(\theta_k, \mathbf{z}_k)\right) \right\|. \tag{8.6}$$

Stochastic optimization refers to problems, where the objective function includes random variables such as:

$$\mathcal{L}(\boldsymbol{\theta}) = \mathbb{E}\left[\mathcal{L}(\boldsymbol{\theta}, \mathbf{z})\right]. \tag{8.7}$$

Stochastic optimization problems can be solved using the update rule in (8.4) or (8.5). This leads to the *stochastic gradient descent* method. The *Robbins–Monro* conditions are sufficient conditions on the learning rate, η_k, for convergence of the stochastic gradient descent method:

$$\sum_{k=1}^{\infty} \eta_k = \infty, \tag{8.8}$$

$$\sum_{k=1}^{\infty} \eta_k^2 < \infty. \tag{8.9}$$

The set of values of η_k over time, which is referred to as the *learning rate schedule*, is obtained as $\eta_k = \frac{1}{k}$ or:

$$\eta_k = (\tau_0 + k)^{\kappa}, \tag{8.10}$$

where during the early iterations, $\tau_0 \geq 0$ slows down the algorithm, and $\kappa \in (0.5, 1]$ controls the rate at which old values are forgotten. Instead of using the same step size for all parameters, the adaptive gradient (*adagrad*) method uses a different step size for each parameter that adapts to the curvature of the loss function. Algorithm 8.2 shows the stochastic gradient descent procedure [77].

The training dataset may be divided into a number of *minibatches*. *Minibatch gradient descent* combines the advantages of both offline batch learning and online stochastic gradient descent by updating the parameters for every minibatch of n_b training data points [139]:

$$\boldsymbol{\theta}_{k+1} = \boldsymbol{\theta}_k - \eta_k \nabla_{\boldsymbol{\theta}} \mathcal{L}(\boldsymbol{\theta}_k, \mathbf{z}_{k:k+n_b-1}), \tag{8.11}$$

where

$$\nabla_{\boldsymbol{\theta}} \mathcal{L}(\boldsymbol{\theta}_k, \mathbf{z}_{k:k+n_b-1}) = \frac{1}{n_b} \nabla_{\boldsymbol{\theta}} \sum_{i=0}^{n_b-1} \mathcal{L}(\boldsymbol{\theta}_k, \mathbf{z}_i). \tag{8.12}$$

In order to train deep neural networks with the stochastic gradient descent method, the learning rate and the batch size must be carefully chosen. Smaller batch sizes require fewer training epochs to converge, but larger batch sizes improve parallelism and computational efficiency. Hence, it has been suggested to adaptively increase the batch size during the training process [142]. As mentioned previously, it is common practice to decay the learning rate during the training process. However, it has been shown that similar learning curves can be obtained by increasing the batch size during training, instead of decreasing the learning rate. The same number of training epochs with a smaller number

Algorithm 8.2: Stochastic gradient descent

Training data

$$\mathbf{z}_k, \quad k = 0, 1, \dots, n_B$$

Hyper-parameters

ϵ : tolerance

Initialization

θ_0

η_0

while *not Converged* **do**

  Randomly permute data

  for $k = 0, 1, \dots, n_B$ **do**

    Update θ_{k+1}

$$\theta_{k+1} = \theta_k - \eta_k \nabla_\theta \mathcal{L}(\theta_k, \mathbf{z}_k)$$

    Update η_{k+1}

  end

end

of parameter updates would provide equivalent test accuracies. The number of parameter updates can be further reduced by increasing the learning rate. Fewer number of parameter updates leads to greater parallelism and shorter training times [143].

Another adaptive learning-rate optimization algorithm is *Adam*, which is a popular optimizer for training deep neural networks. Adam is a first-order gradient-based optimization method for stochastic objective functions. Since Adam is based on adaptive estimates of the first two moments, its name was derived from *adaptive moments*. Algorithm 8.3 shows the Adam procedure, where the following default values has been suggested for the hyper-parameters [144]:

$$\alpha = 0.001,$$
$$\beta_1 = 0.9,$$
$$\beta_2 = 0.999,$$
$$\varepsilon = 10^{-8}.$$

Algorithm 8.3: Adam

Hyper-parameters

$$\begin{aligned} \epsilon &: \text{tolerance} \\ \eta &: \text{learning rate} \\ \beta_1, \beta_2 \in [0,1) &: \text{exponential decay rates for the moment estimates} \\ \varepsilon & \end{aligned}$$

Initialization

$$\theta_0$$

$$\mathbf{m}_0 = \mathbf{0}$$

$$\mathbf{v}_0 = \mathbf{0}$$

where **m** and **v** denote first and second moment vectors, respectively.

for $k = 0, 1, \ldots$ *until convergence* **do**

Compute gradient

$$\mathbf{g}_k = \nabla_\theta \mathcal{L}(\theta_k)$$

Update biased moment estimates

$$\mathbf{m}_{k+1} = \beta_1 \mathbf{m}_k + (1 - \beta_2)\, \mathbf{g}_k$$

$$\mathbf{v}_{k+1} = \beta_1 \mathbf{v}_k + (1 - \beta_2)\, \mathbf{g}_k \odot \mathbf{g}_k$$

Compute bias-corrected moment estimates

$$\widehat{\mathbf{m}}_{k+1} = \frac{1}{1 - \beta_1^k} \mathbf{m}_{k+1}$$

$$\widehat{\mathbf{v}}_{k+1} = \frac{1}{1 - \beta_2^k} \mathbf{v}_{k+1}$$

Update parameters

$$(\theta_{k+1})_i = (\theta_k)_i - \eta \frac{(\widehat{\mathbf{m}}_{k+1})_i}{\sqrt{(\widehat{\mathbf{v}}_{k+1})_i} + \varepsilon}$$

where $(.)_i$ denotes the ith element the corresponding vector.

end

8.4 Natural Gradient Descent

Around a point $\boldsymbol{\theta}$ in the parameter space, the Fisher information matrix, $\mathbf{F}_{\boldsymbol{\theta}}$, plays the role of the metric that defines the *inner product* between two vectors $\mathbf{v}_1$ and $\mathbf{v}_2$ as:

$$\langle \mathbf{v}_1, \mathbf{v}_2 \rangle_{\boldsymbol{\theta}} = \mathbf{v}_1^T \, \mathbf{F}_{\boldsymbol{\theta}} \, \mathbf{v}_2, \tag{8.13}$$

which provides a local measure of distance. The *natural gradient descent* method modifies the gradient of the loss function according to the local curvature of the Kullback–Leibler divergence (KLD) surface as:

$$\mathbf{F}_{\boldsymbol{\theta}}^{-1} \, \nabla_{\boldsymbol{\theta}} \mathcal{L}(\boldsymbol{\theta}), \tag{8.14}$$

which is called the *natural gradient* [145].

It is desired to find a small $\Delta\boldsymbol{\theta}$ that minimizes the first-order Taylor series expansion of $\mathcal{L}(\boldsymbol{\theta} + \Delta\boldsymbol{\theta})$:

$$\mathcal{L}(\boldsymbol{\theta} + \Delta\boldsymbol{\theta}) \approx \mathcal{L}(\boldsymbol{\theta}) + \nabla_{\boldsymbol{\theta}} \mathcal{L}(\boldsymbol{\theta}) \, \Delta\boldsymbol{\theta}, \tag{8.15}$$

while KLD $\left(p_{\boldsymbol{\theta}}(\mathbf{z}) || p_{\boldsymbol{\theta}(\mathbf{z}) + \Delta\boldsymbol{\theta}}(\mathbf{z})\right)$ remains constant [146]:

$$\begin{aligned} \min_{\Delta\boldsymbol{\theta}} \quad & \mathcal{L}(\boldsymbol{\theta} + \Delta\boldsymbol{\theta}) \\ \text{s.t.} \quad & \text{KLD}\left(p_{\boldsymbol{\theta}}(\mathbf{z}) || p_{\boldsymbol{\theta}(\mathbf{z}) + \Delta\boldsymbol{\theta}}(\mathbf{z})\right) = \text{constant} \end{aligned} \tag{8.16}$$

This constraint guarantees a constant speed for moving along the steepest descent direction. Furthermore, it provides robustness against model reparametrization. Assuming that $\Delta\boldsymbol{\theta} \to \mathbf{0}$, the KLD can be approximated by the following second order Taylor series [146]:

$$\begin{aligned} \text{KLD}\left(p_{\boldsymbol{\theta}}(\mathbf{z}) || p_{\boldsymbol{\theta}(\mathbf{z}) + \Delta\boldsymbol{\theta}}(\mathbf{z})\right) \approx \; & \left(\mathbb{E}[\log p_{\boldsymbol{\theta}}(\mathbf{z})] - \mathbb{E}[\log p_{\boldsymbol{\theta}}(\mathbf{z})]\right) \\ & - \mathbb{E}[\nabla_{\boldsymbol{\theta}} \log p_{\boldsymbol{\theta}}(\mathbf{z})] \Delta\boldsymbol{\theta} \\ & - \frac{1}{2} (\Delta\boldsymbol{\theta})^T \mathbb{E}\left[\nabla_{\boldsymbol{\theta}}^2 \log p_{\boldsymbol{\theta}}(\mathbf{z})\right] \Delta\boldsymbol{\theta}, \end{aligned} \tag{8.17}$$

where

$$\begin{aligned} \mathbb{E}[\nabla_{\boldsymbol{\theta}} \log p_{\boldsymbol{\theta}}(\mathbf{z})] &= \sum_{\mathbf{z}} p_{\boldsymbol{\theta}}(\mathbf{z}) \frac{1}{p_{\boldsymbol{\theta}}(\mathbf{z})} \frac{\partial p_{\boldsymbol{\theta}}(\mathbf{z})}{\partial \boldsymbol{\theta}} \\ &= \frac{\partial}{\partial \boldsymbol{\theta}} \sum_{\mathbf{z}} p_{\boldsymbol{\theta}}(\mathbf{z}) \\ &= \frac{\partial}{\partial \boldsymbol{\theta}} 1. \\ &= 0. \end{aligned} \tag{8.18}$$

Therefore, we have:

$$\begin{aligned} \text{KLD}\left(p_{\boldsymbol{\theta}}(\mathbf{z}) || p_{\boldsymbol{\theta}(\mathbf{z}) + \Delta\boldsymbol{\theta}}(\mathbf{z})\right) &\approx -\frac{1}{2} (\Delta\boldsymbol{\theta})^T \mathbb{E}\left[\nabla_{\boldsymbol{\theta}}^2 \log p_{\boldsymbol{\theta}}(\mathbf{z})\right] \Delta\boldsymbol{\theta} \\ &= \frac{1}{2} (\Delta\boldsymbol{\theta})^T \, \mathbf{F}_{\boldsymbol{\theta}} \, \Delta\boldsymbol{\theta}. \end{aligned} \tag{8.19}$$

Let us substitute (8.15) and (8.19) into (8.16). Then, the constrained optimization problem (8.16) can be rewritten as the following unconstrained optimization problem [146]:

$$\min_{\Delta\theta} \quad \mathcal{L}(\theta) + \nabla_\theta \mathcal{L}(\theta)\ \Delta\theta + \frac{1}{2}(\Delta\theta)^T\ \mathbf{F}_\theta\ \Delta\theta. \tag{8.20}$$

This optimization problem can be solved as follows:

$$\frac{\partial}{\partial \Delta\theta}\left(\mathcal{L}(\theta) + \nabla_\theta \mathcal{L}(\theta)\ \Delta\theta + \frac{1}{2}(\Delta\theta)^T\ \mathbf{F}_\theta\ \Delta\theta\right) = 0, \tag{8.21}$$

which leads to:

$$\nabla_\theta \mathcal{L}(\theta) + \mathbf{F}_\theta\ \Delta\theta = 0. \tag{8.22}$$

The solution is obtained as [146]:

$$\Delta\theta = -\mathbf{F}_\theta^{-1}\ \nabla_\theta \mathcal{L}(\theta). \tag{8.23}$$

Hence, the parameter update rule of the natural gradient descent is expressed as:

$$\theta_{k+1} = \theta_k - \mathbf{F}_\theta^{-1}\ \nabla_\theta \mathcal{L}(\theta_k). \tag{8.24}$$

8.5 Neural Networks

Inspired by biological neural networks in the *central nervous system*, artificial *neural networks* have been proposed as computational models aimed at mimicking the brain. As an adaptive machine, a neural network has been defined as follows [138]:

Definition 8.1 ***(Neural network)*** *A neural network is a massively parallel distributed processor made up of simple processing units that has a natural propensity for storing experiential knowledge and making it available for use. It resembles the brain in two respects:*

- *Knowledge is acquired by the network from its environment through a learning process.*
- *Interneuron connection strengths, known as synaptic weights, are used to store the acquired knowledge.*

Neural networks can be used to implement both *regression* and *classification* algorithms. While regression algorithms are used to predict continuous values, classification algorithms are used to predict or classify discrete values. In a neural network, information-processing units are referred to as *neurons*. In a neural network, neurons are arranged in layers, and neurons of a layer are connected to neurons in other layers. Therefore, a neural network can be viewed as a directed graph. Neural networks are categorized as *feedforward neural networks* and *recurrent neural networks*. While there is no feedback path in feedforward networks, recurrent networks deploy feedback as a facilitator of intelligence.

Perceptron as a simple model of a neuron is the structural building block of many neural networks. A *multilayer perceptron* is a fully-connected network of perceptrons in the sense that in each layer, every neuron is connected to all of the neurons in the previous layer as well as the next layer. In mathematical terms, a perceptron is described as:

$$\hat{y} = f\left(\mathbf{w}^T\mathbf{u} + b\right), \tag{8.25}$$

where $\hat{y}$ denotes the neuron's output, $\mathbf{u}$ denotes the input vector, $\mathbf{w}$ denotes the weights, b denotes the bias, and f denotes the activation function. The set of weights and biases form the adjustable set of parameters in a neural network, which are tuned during the training process. Common activation functions are:

- *Sigmoid or logistic function*

$$\sigma(x) = \frac{1}{1 + e^{-x}}, \tag{8.26}$$

$$\sigma'(x) = \sigma(x)\left(1 - \sigma(x)\right), \tag{8.27}$$

- *Hyperbolic tangent*

$$\tanh(x) = \frac{e^x - e^{-x}}{e^x + e^{-x}}, \tag{8.28}$$

$$\frac{d}{dx}\tanh(x) = 1 - \tanh^2(x), \tag{8.29}$$

- *Rectified linear unit* (ReLU)

$$\mathrm{ReLU}(x) = \max(0, x), \tag{8.30}$$

$$\frac{d}{dx}\mathrm{ReLU}(x) = \begin{cases} 1, & x > 0 \\ 0, & \text{otherwise}, \end{cases} \tag{8.31}$$

- *Softmax*

$$\sigma_i(\mathbf{x}) = \frac{e^{\mathbf{x}_i}}{\sum_i e^{\mathbf{x}_i}}, \tag{8.32}$$

where $\boldsymbol{\sigma}_i(.)$ and $\mathbf{x}_i$ denote the ith elements of vectors $\boldsymbol{\sigma}(.)$ and $\mathbf{x}$, respectively. Softmax is a generalization of the logistic function to multiple dimensions. It is used in the last layer of a neural network to provide a probability distribution over the predicted output classes through normalizing the output.

8.6 Backpropagation

In neural networks, *backpropagation* is the major algorithm for supervised learning. It implements the gradient descent method in the parameter space of a neural network. Regarding the fact that neural networks are general function approximators, the basic idea behind backpropagation is to efficiently compute partial derivatives of the approximating function realized by the network with respect to all adjustable parameters (weights and biases) for a given input vector [138]. Algorithm 8.4 shows the backpropagation procedure, which includes the forward propagation through a neural network to compute the loss function, and the backward computation of the gradients with respect to the parameters of each layer. These gradients are then used to update all weights and biases by any gradient-based optimization algorithm such as the stochastic gradient descent method [139].

8.7 Backpropagation Through Time

As an extension to the backpropagation algorithm, *backpropagation through time* has been proposed for training recurrent neural networks. It is derived by unfolding the temporal operation of the network into a layered feedforward network. Hence, the network architecture grows by one layer at every time instant. To use backpropagation through time in a real-time fashion, parameters must be updated while the network is running. To perform training in a computationally feasible manner, the relevant history of the input data and the network state must be taken into account only for a fixed number of time instants, called the *truncation depth*. The resulting algorithm is called the *truncated backpropagation through time* [138].

8.8 Regularization

Regularization is used to improve the *generalization* ability of the trained model in order to perform well when it encounters unseen data, which was not included in the training dataset. Regularization strategies aim at discouraging complex models and preventing overfitting. In other words, regularization is aimed at reducing the test error, even at the expense of increasing the training error. This goal can be achieved through *validation* and early stopping in the training phase. Alternatively, this issue can be addressed through modifying the objective function as [139, 147]:

$$\mathcal{L}(\boldsymbol{\theta}, \mathbf{u}, \mathbf{y}) = \mathcal{L}_e(\mathbf{y}, \hat{\mathbf{y}}) + \lambda\, \Omega(\boldsymbol{\theta}), \tag{8.33}$$

Algorithm 8.4: Backpropagation

Training data

$$(\mathbf{u}, \mathbf{y})$$

Network parameters

$$\theta = \left\{\mathbf{W}^{\ell}, \mathbf{b}^{\ell}\right\}, \quad \ell = 1, \dots, L$$

where L is the network depth.

$$\mathbf{y}^0 = \mathbf{u}$$

for $\ell = 1, \dots, L$ **do**

Compute the layer logits

$$\mathbf{a}^{\ell} = \mathbf{W}^{\ell}\mathbf{y}^{\ell-1} + \mathbf{b}^{\ell}$$

Compute the layer output

$$\mathbf{y}^{\ell} = f\left(\mathbf{a}^{\ell}\right)$$

where $f(.)$ denotes a nonlinear element-wise function.

end

Network output

$$\hat{\mathbf{y}} = \mathbf{y}^L$$

Compute the loss

$$\mathcal{L}(\theta, \mathbf{u}, \mathbf{y}) = \mathcal{L}_e(\mathbf{y}, \hat{\mathbf{y}}) + \lambda\, \Omega(\theta)$$

where $\mathcal{L}_e(\mathbf{y}, \hat{\mathbf{y}})$ is an error measure, and $\Omega(\theta)$ is a regularizer.

Compute the gradient w.r.t. the network output

$$\nabla_{\mathbf{y}^L}\mathcal{L}(\theta, \mathbf{u}, \mathbf{y}) = \nabla_{\hat{\mathbf{y}}}\mathcal{L}(\theta, \mathbf{u}, \mathbf{y}) = \nabla_{\hat{\mathbf{y}}}\mathcal{L}_e(\mathbf{y}, \hat{\mathbf{y}})$$

for $\ell = L, L-1, \dots, 1$ **do**

Compute gradients w.r.t. weights and biases

$$\nabla_{\mathbf{a}^{\ell}}\mathcal{L}(\theta, \mathbf{u}, \mathbf{y}) = \nabla_{\mathbf{y}^{\ell}}\mathcal{L}(\theta, \mathbf{u}, \mathbf{y}) \odot f'\left(\mathbf{a}^{\ell}\right)$$

$$\nabla_{\mathbf{b}^{\ell}}\mathcal{L}(\theta, \mathbf{u}, \mathbf{y}) = \nabla_{\mathbf{a}^{\ell}}\mathcal{L}(\theta, \mathbf{u}, \mathbf{y}) + \lambda\nabla_{\mathbf{b}^{\ell}}\Omega(\theta)$$

$$\nabla_{\mathbf{W}^{\ell}}\mathcal{L}(\theta, \mathbf{u}, \mathbf{y}) = \nabla_{\mathbf{a}^{\ell}}\mathcal{L}(\theta, \mathbf{u}, \mathbf{y})\left(\mathbf{y}^{\ell-1}\right)^T + \lambda\nabla_{\mathbf{W}^{\ell}}\Omega(\theta)$$

Compute gradient w.r.t. the previous layer's output

$$\nabla_{\mathbf{y}^{\ell-1}}\mathcal{L}(\theta, \mathbf{u}, \mathbf{y}) = \left(\mathbf{W}^{\ell}\right)^T \nabla_{\mathbf{a}^{\ell}}\mathcal{L}(\theta, \mathbf{u}, \mathbf{y})$$

end

where $\mathcal{L}_e(\mathbf{y}, \hat{\mathbf{y}})$ is an error measure based on the difference between the target output, $\mathbf{y}$, and the network output, $\hat{\mathbf{y}}$, for a given input, $\mathbf{u}$. The second term is used for regularization, where $\Omega(\boldsymbol{\theta})$ is the regularizer, and λ, which is a hyper-parameter, determines the regularization strength. Common regularizers are [138, 139]:

- *2-norm* or *Euclidean norm*:

$$\|\theta\|_2 = \sqrt{\sum_i \theta_i^2}. \tag{8.34}$$

  For every parameter in the network, θ_i, the term $\frac{1}{2}\lambda\theta_i^2$ is added to the objective function. This regularization, which is also known as *Tikhonov* regularization, is equivalent to *weight decay*.
- *1-norm*:

$$\|\theta\|_1 = \sum_i |\theta_i|. \tag{8.35}$$

  For every parameter in the network, θ_i, the term $\lambda|\theta_i|$ is added to the objective function. This regularization encourages *sparsity*.
- ∞*-norm*:

$$\|\theta\|_\infty = \max_i \; |\theta_i|. \tag{8.36}$$

  This regularization controls the growth of parameter values by enforcing an upper bound on the absolute value of the parameters.

Dropout, which is a regularization strategy of a different nature, is the most popular method in the context of deep learning. Dropout is implemented by allowing a neuron to be active with probability p and inactive with probability $1 - p$, where p is a hyper-parameter. A neuron is deactivated by setting its output to zero. As a result, the trained network will not be too dependent on a small combination of neurons, and will be able to provide fairly accurate predictions even in the absence of certain information. During the training phase, for each minibatch of data, dropout deactivates each neuron in the network with probability $1 - p$. This process can be viewed as approximately combining many different network architectures. If a neuron output prior to dropout is $\hat{y}$, the expected value of the output considering dropout will be:

$$p\,\hat{y} + (1 - p) \times 0 = p\,\hat{y}. \tag{8.37}$$

The outputs of neurons during the test phase are desired to be equivalent to their corresponding expected values during the training phase. This calls for scaling the neuron outputs. Since scaling of neuron outputs at test time is undesirable, they are scaled during training. The output of any neuron that has not been silenced is divided by p before propagation to the next layer. Hence, the expected value of the output will be:

$$p\,\frac{\hat{y}}{p} + (1 - p) \times 0 = \hat{y}. \tag{8.38}$$

This procedure is known as the *inverted dropout*. Mathematically speaking, dropout can be viewed as approximate Bayesian inference in *deep Gaussian processes* [148].

8.9 Initialization

To avoid the *vanishing gradient* and the *exploding gradient* problems during training, it is desired that the neural network's activations satisfy the following two conditions:

- Their means remain zero.
- Their variances remain the same across every layer.

Under these conditions, the gradient is backpropagated all the way to the input layer without vanishing or exploding. To satisfy these conditions, *Xavier initialization* recommends that for each layer ℓ of the network with n^{ℓ} neurons and hyperbolic tangent activation functions [149]:

- Biases are initialized to zeros, $\mathbf{b}^{\ell} = \mathbf{0}$.
- Weights are initialized according to the normal distribution $\mathbf{W}^{\ell}_{ij} \sim \mathcal{N}\left(0, \frac{1}{n^{\ell-1}}\right)$ or $\mathbf{W}^{\ell}_{ij} \sim \mathcal{N}\left(0, \frac{2}{n^{\ell-1}+n^{\ell}}\right)$.

For ReLU activation function, the variances of the aforementioned weight distributions are multiplied by 2 [150].

8.10 Convolutional Neural Network

A *convolutional neural network* (CNN) is a feedforward neural network with one or more layers that perform convolution instead of matrix multiplication. Different deep neural-network architectures, which are built around CNNs, deploy different combinations of convolutional, *pooling*, and fully-connected layers. CNN variants are proper tools for processing data with known grid-like topology. The idea behind using a convolutional layer is to apply multiple filters to extract different local features and generate feature maps. Each filter is represented by a kernel and a set of adjustable weights, which are spatially shared. Therefore, convolution is a patchy operation that is proper for handling correlated data with variable size, and leverages sparse interactions, parameter sharing, and equivariant representations. Convolutional layers may be followed by a pooling layer that performs *downsampling* on each feature map [139, 147].

Regarding the two-dimensional convolution, a filter kernel, which is represented by a low-dimensional matrix, is passed over the input matrix, which represents the input image for the first layer or the feature map from the previous layer for other layers. The element-wise product of the kernel with a submatrix of the input matrix that coincides with the kernel is computed, and then, sum of the elements of the resulting product matrix is obtained. This sum provides the value for the corresponding element in the output matrix of the convolution operation. Let the $n_f \times n_f$-dimensional matrix $\mathcal{K}$ represent the filter kernel and the $n_h \times n_w$-dimensional matrix $\mathcal{I}$ represent the input. Then, the elements of the feature map are calculated as:

$$(\mathcal{I} * \mathcal{K})_{mn} = \sum_{i=1}^{n_f} \sum_{j=1}^{n_f} \mathcal{K}_{ij} \mathcal{I}_{(m+i-1)(n+j-1)}, \tag{8.39}$$

where $*$ denotes the convolution operation. Assuming that there are n_c channels, where $n_c = 3$ for RGB (red, blue, and green) images, $\mathcal{K}$ and $\mathcal{I}$ will be $n_f \times n_f \times n_c$- and $n_h \times n_w \times n_c$-dimensional tensors, respectively. Then, the feature map for each channel is obtained as in (8.39).

Valid convolution refers to restricting the output to positions, where the kernel entirely lies within the image [139]. In performing convolution, the whole input matrix of a convolutional layer is covered by sliding the filter kernel on that matrix. While sliding the filter, the distance in terms of pixels between two consecutive computations of (8.39) is referred to as *stride*, where a stride of 1 leads to full convolution. The dimension of the resulting feature map at the output of a convolutional layer will be lower than the dimension its input matrix. The input image or feature map to a convolutional layer can be padded with an additional border of zeroes to guarantee that the output feature map will have the same dimension as the input matrix. This process in referred to as *zero padding* [151]. The required padding width, n_p, to maintain the dimension is obtained as:

$$n_p = \frac{n_f - 1}{2}. \tag{8.40}$$

Regarding padding, n_p, and a proper stride, n_s, dimensions of the input, n_{in}, and the output, n_{out}, matrices of a convolutional layer are related to each other as [151]:

$$n_h^{out} = \frac{n_h^{in} + 2n_p - n_f}{n_s} + 1, \tag{8.41}$$

$$n_w^{out} = \frac{n_w^{in} + 2n_p - n_f}{n_s} + 1, \tag{8.42}$$

where subscripts h and w refer to height and width, respectively.

The *max-pooling* layer, which follows a convolutional layer, reduces the dimension of the feature map and sharpens the features. Max-pooling is performed by dividing the feature map into equally sized tiles, and then generating a cell for each tile that maintains the maximum cell value in that tile. As a result, max-pooling is locally invariant. In this way, a condensed feature map is produced. *Average pooling* is another pooling method, which assigns the average of the cell values in a tile to the corresponding cell in the resulting condensed feature map. A pooling layer can be described by the spatial extent of the corresponding tile, n_e, and the stride, n_s. Regarding these two parameters, the resulting pooling may be overlapping or non-overlapping. For a proper stride, n_s, there is the following relation between dimensions of the input and the output matrices of a pooling layer [151]:

$$n_h^{out} = \frac{n_h^{in} - n_e}{n_s} + 1, \tag{8.43}$$

$$n_w^{out} = \frac{n_w^{in} - n_e}{n_s} + 1. \tag{8.44}$$

8.11 Long Short-Term Memory

Deploying feedback loops is the distinguishing feature of *recurrent neural networks* (RNNs) from feedforward neural networks [138]. Such feedback loops make RNNs proper tools for sequence modeling with applications in robot control, time-series analysis, language modeling, machine translation, grammar learning, speech recognition and synthesis, image captioning, music composition, etc. [152]. In sequence modeling, the desired design criteria can be summarized as follows [147]:

- Handling variable-length sequences
- Tracking long-term dependencies
- Maintaining information about order
- Sharing parameters across the sequence

As a class of RNNs, *long short-term memory* (LSTM) networks have been very successful in applications involving sequence modeling. The LSTM architecture provides a memory cell whose output is a filtered version of its state. Such a memory cell regulates its state over time using three nonlinear gating units [153]. In order to address the mentioned design issues in sequence modeling, LSTM controls the flow of information through the following gating mechanisms [154]:

- *Input gate* decides what new information must be incorporated in the cell state. A sigmoid layer determines the values that must be updated, $\mathbf{i}_k \in [0,1]^N$, based on the current input, $\mathbf{u}_k$, and the previous output, $\mathbf{y}_{k-1}$. A hyperbolic tangent layer forms a vector of new candidate values, $\mathbf{z}_k$, that could be incorporated in the cell state. To update the cell state, the element-wise product of these two vectors, $\mathbf{z}_k \odot \mathbf{i}_k$, will be added to the percentage of the information about the previous state that passes through the forget gate.

- *Forget gate* decides what percentage of the information included in the cell state must be maintained. The forget gate uses a layer of sigmoid functions to provide a vector whose elements are between zero and one, $\mathbf{f}_k \in [0,1]^N$, based on the current input, $\mathbf{u}_k$, and the previous output, $\mathbf{y}_{k-1}$. Then, the element-wise product of the previous state and the output of the forget gate, $\mathbf{x}_{k-1} \odot \mathbf{f}_k$, determines the percentage of the information about the previous state variables that will be kept in the current state variables. The extreme cases of $\mathbf{f}_k = 1$ and $\mathbf{f}_k = 0$ represent complete maintenance and complete ignorance, respectively.
- *Output gate* decides what information must be included in the output. A sigmoid layer determines what components of the cell state must be included in the output, $\mathbf{o}_k$, based on the current input, $\mathbf{u}_k$, and the previous output, $\mathbf{y}_{k-1}$. The cell state is passed through a hyperbolic tangent layer to obtain a filtered version of state vector through mapping the values of its elements to the interval $[-1,1]$. Then, the output is computed as $\tanh\left(\mathbf{x}_k\right) \odot \mathbf{o}_k$.

While the sigmoid or logistic function, $\sigma(x) = \frac{1}{1+e^{-x}}$, is used for gate activation, the hyperbolic tangent, $\tanh(x) = \frac{e^x - e^{-x}}{e^x + e^{-x}} = \frac{e^{2x}-1}{e^{2x}+1}$, is used for the block input and output activation. Adding *peephole connections* to the LSTM architecture allows the gate layers to take account of the cell state [155, 156].

Among different LSTM variants, the *vanilla LSTM* is the most popular one [153]. Let $\mathbf{u}_k$, $\mathbf{y}_k$, and $\mathbf{x}_k$ denote the input, the output, and the cell state vectors at time instant k, respectively. For a vanilla LSTM layer with N LSTM blocks and n_u inputs, the gating mechanisms can be mathematically expressed as follows [153]:

- *Block input*

$$\mathbf{z}_k = \tanh\left(\mathbf{W}_k^z \mathbf{u}_k + \mathbf{R}_k^z \mathbf{y}_{k-1} + \mathbf{b}_k^z\right), \tag{8.45}$$

where $\mathbf{W}_k^z \in \mathbb{R}^{N \times n_u}$, $\mathbf{R}_k^z \in \mathbb{R}^{N \times N}$, and $\mathbf{b}_k^z \in \mathbb{R}^N$ denote the input, the recurrent, and the bias weights, respectively.

- *Input gate*

$$\mathbf{i}_k = \sigma\left(\mathbf{W}_k^i \mathbf{u}_k + \mathbf{R}_k^i \mathbf{y}_{k-1} + \mathbf{p}_k^i \odot \mathbf{x}_{k-1} + \mathbf{b}_k^i\right), \tag{8.46}$$

where $\mathbf{W}_k^i \in \mathbb{R}^{N \times n_u}$, $\mathbf{R}_k^i \in \mathbb{R}^{N \times N}$, $\mathbf{p}_k^i \in \mathbb{R}^N$, and $\mathbf{b}_k^i \in \mathbb{R}^N$ denote the input, the recurrent, the peephole, and the bias weights, respectively.

- *Forget gate*

$$\mathbf{f}_k = \sigma\left(\mathbf{W}_k^f \mathbf{u}_k + \mathbf{R}_k^f \mathbf{y}_{k-1} + \mathbf{p}_k^f \odot \mathbf{x}_{k-1} + \mathbf{b}_k^f\right), \tag{8.47}$$

where $\mathbf{W}_k^f \in \mathbb{R}^{N \times n_u}$, $\mathbf{R}_k^f \in \mathbb{R}^{N \times N}$, $\mathbf{p}_k^f \in \mathbb{R}^N$, and $\mathbf{b}_k^f \in \mathbb{R}^N$ denote the input, the recurrent, the peephole, and the bias weights, respectively.

- *Cell state*

$$\mathbf{x}_k = \mathbf{z}_k \odot \mathbf{i}_k + \mathbf{x}_{k-1} \odot \mathbf{f}_k. \tag{8.48}$$

- *Output gate*

$$\mathbf{o}_k = \sigma\left(\mathbf{W}_k^o \mathbf{u}_k + \mathbf{R}_k^o \mathbf{y}_{k-1} + \mathbf{p}_k^o \odot \mathbf{x}_{k-1} + \mathbf{b}_k^o\right), \tag{8.49}$$

where $\mathbf{W}_k^i \in \mathbb{R}^{N \times n_u}$, $\mathbf{R}_k^i \in \mathbb{R}^{N \times N}$, $\mathbf{p}_k^i \in \mathbb{R}^N$, and $\mathbf{b}_k^i \in \mathbb{R}^N$ denote the input, the recurrent, the peephole, and the bias weights, respectively.

- *Block output*

$$\mathbf{y}_k = \tanh\left(\mathbf{x}_k\right) \odot \mathbf{o}_k. \tag{8.50}$$

In the aforementioned equations, both $\sigma(.)$ and $\tanh(.)$ denote element-wise functions.

The *gated recurrent unit* (GRU) has been proposed as a simplified variant of the LSTM [157]. Compared with the vanilla LSTM, the GRU architecture does not include peephole connections and output activation functions. Furthermore, the input and the forget gate are combined into a single *update gate*, and the output gate, which is called the *reset gate*, passes only the recurrent connections to the block input.

8.12 Hebbian Learning

Hebbian learning, which is the basis of many unsupervised learning algorithms, was inspired by the following postulate [158]:

Definition 8.2 ***(Hebb's postulate of learning)*** *"When an axon of cell A is near enough to excite a cell B and repeatedly or persistently takes part in firing it, some growth process or metabolic changes take place in one or both cells such that A's efficiency as one of the cells firing B is increased."*

Hebb's postulate of learning can be interpreted as follows [138]:

- If two connected neurons are simultaneously (synchronously) activated, then the connection between them will be strengthened.
- If two connected neurons are asynchronously activated, then the connection between them will be weakened.

In mathematical terms, let us consider a synaptic weight w^{ij}, which is connected to neuron i with presynaptic and postsynaptic signals x^j and y^i, respectively. The synaptic weight w^{ij} can be adjusted as:

$$w_{k+1}^{ij} = w_k^{ij} + \Delta w_k^{ij}, \tag{8.51}$$

$$\Delta w_k^{ij} = \eta y_k^i x_k^j, \tag{8.52}$$

where η denotes the *learning rate*. Equation (8.52), which is known as the *activity product rule*, reflects the correlative nature of a Hebbian synapse. Assuming that there are m synapses associated with the ith neuron (m is the number of inputs to the neuron), the output of the ith neuron at time instant k is computed as:

$$y_k^i = \sum_j^m w_k^{ij} x_k^j. \tag{8.53}$$

The learning rule in (8.52) allows for unlimited growth of the synaptic weight w^{ij}, which is unacceptable in real networks. This issue can be addressed through incorporating a form of normalization in the learning rule:

$$w_{k+1}^{ij} = \frac{w_k^{ij} + \eta y_k^i x_k^j}{\left(\sum_j^m \left(w_k^{ij} + \eta y_k^i x_k^j\right)^2\right)^{\frac{1}{2}}}. \tag{8.54}$$

Normalization provides a mechanism for stabilization by introducing competition among the synapses of a neuron over limited resources. For normalized weights, we have:

$$\sum_j^m \left(w_k^{ij}\right)^2 = \| \mathbf{w}_k^i \|^2 = 1, \quad \forall i, k. \tag{8.55}$$

Assuming that η is small, the denominator of (8.54) can be approximated as:

$$\begin{aligned}
\left(\sum_j^m \left(w_k^{ij} + \eta y_k^i x_k^j\right)^2\right)^{\frac{1}{2}} &\approx \left(\sum_j^m \left(\left(w_k^{ij}\right)^2 + 2\eta w_k^{ij} y_k^i x_k^j\right)\right)^{\frac{1}{2}} + O\left(\eta^2\right) \\
&= \left(\sum_j^m \left(w_k^{ij}\right)^2 + 2\eta y_k^i \sum_j^m w_k^{ij} x_k^j\right)^{\frac{1}{2}} + O\left(\eta^2\right) \\
&= \left(1 + 2\eta \left(y_k^i\right)^2\right)^{\frac{1}{2}} + O\left(\eta^2\right) \\
&\approx 1 + \eta \left(y_k^i\right)^2 + O\left(\eta^2\right).
\end{aligned} \tag{8.56}$$

In (8.54), replacing the denominator by its approximation of (8.56), we obtain:

$$
\begin{aligned}
w_{k+1}^{ij} &= \frac{w_k^{ij} + \eta y_k^i x_k^j}{1 + \eta \left(y_k^i\right)^2 + O\left(\eta^2\right)}. \\
&= \left(w_k^{ij} + \eta y_k^i x_k^j\right)\left(1 + \eta \left(y_k^i\right)^2 + O\left(\eta^2\right)\right)^{-1} \\
&\approx \left(w_k^{ij} + \eta y_k^i x_k^j\right)\left(1 - \eta \left(y_k^i\right)^2\right) + O\left(\eta^2\right) \\
&= w_k^{ij} + \eta y_k^i x_k^j - \eta w_k^{ij} \left(y_k^i\right)^2 + O\left(\eta^2\right) \\
&= w_k^{ij} + \eta y_k^i \left(x_k^j - w_k^{ij} y_k^i\right) + O\left(\eta^2\right).
\end{aligned} \tag{8.57}
$$

Ignoring the higher-order terms, $O\left(\eta^2\right)$, in the aforementioned equation, the weight-update rule is obtained as follows:

$$
w_{k+1}^{ij} = w_k^{ij} + \eta y_k^i \underbrace{\left(x_k^j - w_k^{ij} y_k^i\right)}_{\text{effective input}}. \tag{8.58}
$$

Compared with the Hebbian learning rule in (8.51) and (8.52), $w_{k+1}^{ij} = w_k^{ij} + \eta y_k^i x_k^j$, in the aforementioned equation, the input x_k^j has been modified as $x_k^j - w_k^{ij} y_k^i$, which can be regarded as an *effective input*. This input modification occurs through the negative feedback term $-w_k^{ij} y_k^i$, which stabilizes the synaptic weight as it evolves across time [138].

8.13 Gibbs Sampling

Regarding a set of variables represented by a vector, $\mathbf{x}$, given a joint sample of the variables, $\mathbf{x}_k$, *Gibbs sampler* generates a new sample $\mathbf{x}_{k+1}$ by sampling each variable in turn, based on the most recent values of other variables. There is no need to sample visible variables, because their values can be measured [77, 138]. Algorithm 8.5 summarizes the Gibbs sampling procedure [159]. Gibbs sampling is used in training Boltzmann machines, which will be discussed in Section 8.14.

8.14 Boltzmann Machine

Boltzmann machine is an *energy-based model* that allows for discovering features, which represent the regularities in the data [160]. Hebb's rule is used to train Boltzmann machines based on unlabeled datasets. A Boltzmann machine consists of two sets of visible and hidden stochastic binary units, which are symmetrically coupled [161]. In a Boltzmann machine, each node is connected to all other nodes.

Algorithm 8.5: Gibbs sampling

Initialization

$$\mathbf{x}_0 = \left[(\mathbf{x}_0)_1 \ (\mathbf{x}_0)_2 \ \cdots \ (\mathbf{x}_0)_{n_x}\right]^T$$

where $(\mathbf{x}_0)_i$ denotes the ith element of vector $\mathbf{x}_0$.

for $k = 0, 1, \ldots,$ **do**

Pick index

$$i \sim \mathcal{U}(1, \ldots, n_x)$$

where $\mathcal{U}$ denotes the uniform distribution.

Draw a sample

$$a \sim p\left((\mathbf{x}_k)_i \,\middle|\, (\mathbf{x}_k)_{-i}\right)$$

where $(\mathbf{x}_k)_{-i}$ denotes all elements of vector $\mathbf{x}_k$ except for the ith element.

Update

$$\mathbf{x}_{k+1} = \left[(\mathbf{x}_k)_1 \ \cdots \ (\mathbf{x}_k)_{i-1} \ a \ (\mathbf{x}_k)_{i+1} \ \cdots \ (\mathbf{x}_k)_{n_x}\right]^T$$

end

To be more precise, a Boltzmann machine can be viewed as a fully-connected network with two layers; one layer of visible nodes and one layer of hidden nodes. Every node in each layer is connected to all of the nodes in the other layer. Furthermore, this network is fully-laterally connected, and every node in each layer is connected to all other nodes in that layer too. Such a network can be viewed as an undirected graph, where each edge of the graph (connection between two nodes) represents the joint probability between the two stochastic binary variables associated with the nodes. In a *probabilistic graphical model*, dependency structure between random variables is represented by a graph. The undirected and directed edges respectively represent the joint and conditional probabilities between the random variables associated with the corresponding nodes [162].

Let us denote the visible and hidden stochastic binary units by the vectors $\mathbf{v} \in \{0,1\}^{n_v}$ and $\mathbf{h} \in \{0,1\}^{n_h}$, respectively. Then, the set of state variables is defined by the elements of $\mathbf{v}$ and $\mathbf{h}$. Ignoring the bias terms for simplicity of notation, the energy associated with a particular state $(\mathbf{v}, \mathbf{h})$ is defined as [161]:

$$\mathcal{E}_\theta(\mathbf{v}, \mathbf{h}) = -\frac{1}{2}\mathbf{v}^T\mathbf{W}^{\mathbf{vv}}\mathbf{v} - \frac{1}{2}\mathbf{h}^T\mathbf{W}^{\mathbf{hh}}\mathbf{h} - \mathbf{v}^T\mathbf{W}^{\mathbf{vh}}\mathbf{h}, \tag{8.59}$$

where the set of parameters $\boldsymbol{\theta} = \{\mathbf{W}^{\mathbf{vv}}, \mathbf{W}^{\mathbf{hh}}, \mathbf{W}^{\mathbf{vh}}\}$ includes the visible-to-visible, $\mathbf{W}^{\mathbf{vv}}$, the hidden-to-hidden, $\mathbf{W}^{\mathbf{hh}}$, and the visible-to-hidden, $\mathbf{W}^{\mathbf{vh}}$, symmetric interaction terms. The diagonal elements of $\mathbf{W}^{\mathbf{vv}}$ and $\mathbf{W}^{\mathbf{hh}}$ are zero. The assigned probability by the model to a visible vector $\mathbf{v}$ is computed as:

$$p_{\theta}(\mathbf{v}) = \frac{\sum_{\mathbf{h}} e^{-\mathcal{E}_{\theta}(\mathbf{v},\mathbf{h})}}{\sum_{\mathbf{v}}\sum_{\mathbf{h}} e^{-\mathcal{E}_{\theta}(\mathbf{v},\mathbf{h})}}. \tag{8.60}$$

The conditional distributions over hidden and visible units are computed as:

$$p_{\theta}(\mathbf{v}_i = 1|\mathbf{v}_{-i}, \mathbf{h}) = \sigma\left(\sum_{\ell=1,\neq i}^{n_v} \mathbf{W}^{\mathbf{vv}}_{i\ell}\mathbf{v}_{\ell} + \sum_{\ell=1}^{n_h} \mathbf{W}^{\mathbf{vh}}_{i\ell}\mathbf{h}_{\ell}\right), \tag{8.61}$$

$$p_{\theta}(\mathbf{h}_j = 1|\mathbf{v}, \mathbf{h}_{-j}) = \sigma\left(\sum_{\ell=1,\neq j}^{n_h} \mathbf{W}^{\mathbf{hh}}_{\ell j}\mathbf{h}_{\ell} + \sum_{\ell=1}^{n_v} \mathbf{W}^{\mathbf{vh}}_{\ell j}\mathbf{v}_{\ell}\right), \tag{8.62}$$

where σ denotes the sigmoid or logistic function, $\mathbf{v}_{-i}$ refers to all elements of vector $\mathbf{v}$ except for the ith element, and $\mathbf{h}_{-j}$ refers to all elements of vector $\mathbf{h}$ except for the jth element. Model parameters are updated as follows:

$$\mathbf{W}^{\mathbf{vv}}_{k+1} = \mathbf{W}^{\mathbf{vv}}_{k} + \eta\left(\mathbb{E}_{p_{\text{data}}}\left[\mathbf{v}_k\mathbf{v}_k^T\right] - \mathbb{E}_{p_{\text{model}}}\left[\mathbf{v}_k\mathbf{v}_k^T\right]\right), \tag{8.63}$$

$$\mathbf{W}^{\mathbf{hh}}_{k+1} = \mathbf{W}^{\mathbf{hh}}_{k} + \eta\left(\mathbb{E}_{p_{\text{data}}}\left[\mathbf{h}_k\mathbf{h}_k^T\right] - \mathbb{E}_{p_{\text{model}}}\left[\mathbf{h}_k\mathbf{h}_k^T\right]\right), \tag{8.64}$$

$$\mathbf{W}^{\mathbf{vh}}_{k+1} = \mathbf{W}^{\mathbf{vh}}_{k} + \eta\left(\mathbb{E}_{p_{\text{data}}}\left[\mathbf{v}_k\mathbf{h}_k^T\right] - \mathbb{E}_{p_{\text{model}}}\left[\mathbf{v}_k\mathbf{h}_k^T\right]\right), \tag{8.65}$$

where η denotes the learning rate, p_{model} denotes the distribution defined by the model as in (8.60), and p_{data} denotes the completed data distribution:

$$p_{\text{data}}(\mathbf{v}, \mathbf{h}) = p_{\theta}(\mathbf{h}|\mathbf{v})\, p_{\text{data}}(\mathbf{v}), \tag{8.66}$$

with $p_{\text{data}}(\mathbf{v})$ representing the empirical distribution. Algorithm 8.6 shows the Boltzmann machine learning procedure [161].

A Boltzmann machine is reduced to the *restricted Boltzmann machine* by removing the connections between the visible nodes, $\mathbf{W}^{\mathbf{vv}} = \mathbf{0}$, as well as the connections between the hidden nodes, $\mathbf{W}^{\mathbf{hh}} = \mathbf{0}$. In other words, in a restricted Boltzmann machine, each visible node is connected to all hidden nodes, and each hidden node is connected to all visible nodes, but there are no visible-to-visible and hidden-to-hidden node connections. The restricted Boltzmann machine has one layer of visible nodes and one layer of hidden nodes. This architecture can be extended by adding more layers of hidden nodes. The resulting undirected probabilistic graphical model, which is known as the *deep Boltzmann machine*, has one layer of visible nodes and multiple layers of hidden nodes. Another relevant architecture is the *deep belief network*, in which the top two layers of hidden nodes form an undirected graph, while the lower layers of hidden nodes plus the layer of visible nodes form a directed graph [161].

Algorithm 8.6: Boltzmann machine learning

Training set

$$\mathbf{v}^i,\ i = 1, \ldots, M$$

Initialization

$$\text{Parameters} : \theta_0 = \{\mathbf{W}_0^{\mathbf{vv}}, \mathbf{W}_0^{\mathbf{hh}}, \mathbf{W}_0^{\mathbf{vh}}\}$$
$$\text{Particles} : \{\widetilde{\mathbf{v}}_0^i, \widetilde{\mathbf{h}}_0^i\}, i = 1, \ldots, N$$
$$\text{Learning rate} : \eta_0$$

for $k = 0, 1, \ldots,$ **do**

foreach *training example,* $\mathbf{v}^i,\ i = 1, \ldots, M,$ **do**

Randomly initialize μ

while *not converged* **do**

Mean-field update

$$\left(\mu_k^i\right)_j = \sigma\left(\sum_{\ell=1}^{n_v} \left(\mathbf{W}_k^{\mathbf{vh}}\right)_{\ell j} \mathbf{v}_\ell^i + \sum_{\ell=1,\neq j}^{n_h} \left(\mathbf{W}_k^{\mathbf{hh}}\right)_{\ell j} \mu_\ell^i\right)$$

where σ denotes the sigmoid or logistic function, and $\left(\mathbf{W}_k^{\mathbf{vh}}\right)_{\ell j}$ denotes the element (ℓ, j) of matrix $\mathbf{W}_k^{\mathbf{vh}}$ at iteration k, and $\left(\mu_k^i\right)_j$ denotes the jth element of vector μ_k^i at iteration k.

end

end

foreach *particle,* $i = 1, \ldots, N,$ **do**

State update $\left(\widetilde{\mathbf{v}}_{k+1}^i, \widetilde{\mathbf{h}}_{k+1}^i\right)$ **by Gibbs sampling using**

$$p_{\theta_k}\left(\left(\widetilde{\mathbf{v}}_k^i\right)_m = 1 \middle| \left(\widetilde{\mathbf{v}}_k^i\right)_{-m}, \widetilde{\mathbf{h}}_k^i\right)$$
$$= \sigma\left(\sum_{\ell=1,\neq m}^{n_v} \left(\mathbf{W}_k^{\mathbf{vv}}\right)_{m\ell} \left(\widetilde{\mathbf{v}}_k^i\right)_\ell + \sum_{\ell=1}^{n_h} \left(\mathbf{W}_k^{\mathbf{vh}}\right)_{m\ell} \left(\widetilde{\mathbf{h}}_k^i\right)_\ell\right)$$
$$p_{\theta_k}\left(\left(\widetilde{\mathbf{h}}_k^i\right)_j = 1 \middle| \widetilde{\mathbf{v}}_k^i, \left(\widetilde{\mathbf{h}}_k^i\right)_{-j}\right)$$
$$= \sigma\left(\sum_{\ell=1,\neq j}^{n_h} \left(\mathbf{W}_k^{\mathbf{hh}}\right)_{\ell j} \left(\widetilde{\mathbf{h}}_k^i\right)_\ell + \sum_{\ell=1}^{n_v} \left(\mathbf{W}_k^{\mathbf{vh}}\right)_{\ell j} \left(\widetilde{\mathbf{v}}_k^i\right)_\ell\right)$$

where $\left(\widetilde{\mathbf{v}}_k^i\right)_{-m}$ denotes all elements of the ith particle, $\widetilde{\mathbf{v}}_k^i$, except for the mth element and $\left(\widetilde{\mathbf{h}}_k^i\right)_{-j}$ denotes all elements of the ith particle, $\widetilde{\mathbf{h}}_k^i$, except for the jth element at iteration k.

end

Parameter update

$$\mathbf{W}^{\mathbf{vv}}_{k+1} = \mathbf{W}^{\mathbf{vv}}_{k} + \eta_k \left(\frac{1}{M} \sum_{i=1}^{M} \mathbf{v}^i \left(\mathbf{v}^i\right)^T - \frac{1}{N} \sum_{i=1}^{N} \widetilde{\mathbf{v}}^i_{k+1} \left(\widetilde{\mathbf{v}}^i_{k+1}\right)^T \right)$$

$$\mathbf{W}^{\mathbf{hh}}_{k+1} = \mathbf{W}^{\mathbf{hh}}_{k} + \eta_k \left(\frac{1}{M} \sum_{i=1}^{M} \mu^i \left(\mu^i\right)^T - \frac{1}{N} \sum_{i=1}^{N} \widetilde{\mathbf{h}}^i_{k+1} \left(\widetilde{\mathbf{h}}^i_{k+1}\right)^T \right)$$

$$\mathbf{W}^{\mathbf{vh}}_{k+1} = \mathbf{W}^{\mathbf{vh}}_{k} + \eta_k \left(\frac{1}{M} \sum_{i=1}^{M} \mathbf{v}^i \left(\mu^i\right)^T - \frac{1}{N} \sum_{i=1}^{N} \widetilde{\mathbf{v}}^i_{k+1} \left(\widetilde{\mathbf{h}}^i_{k+1}\right)^T \right)$$

Decrease η_k

end

8.15 Autoencoder

Autoencoders allow for learning efficient representations of data in an unsupervised manner. Each autoencoder consists of two key components:

- *Encoder* learns a mapping from the observation $\mathbf{y}$ to the latent space $\mathbf{x}$.
- *Dencoder* learns an inverse mapping from the latent space $\mathbf{x}$ to a reconstructed version of the observation $\hat{\mathbf{y}}$.

Autoencoders can be implemented using different neural-network architectures. The autoencoder is trained through minimizing the following reconstruction loss function without requiring any labeled data:

$$\mathcal{L}(\mathbf{y}, \hat{\mathbf{y}}) = \| \mathbf{y} - \hat{\mathbf{y}} \|^2. \tag{8.67}$$

Using this loss function, which reflects the reconstruction quality, the latent representation will encode as much information as possible about the data [139, 147]. If dimension of the latent space is smaller than the observation space, the code layer (output of the encoder and input of the decoder) can be viewed as a bottleneck hidden layer, which is associated with *data compression*. On the other hand, in *overcomplete representation* exemplified by *sparse coding* [163], dimension of the latent space is larger than that of the observation space.

In a probabilistic framework, *variational autoendoders* (VAEs) have been proposed as directed probabilistic graphical models. Filters can be designed based on variational autoencoders. In this case, the autoencoder will have an asymmetric structure. While the encoder's input consists of observations $\mathbf{y}$, its output represents the state $\mathbf{x}$, hence, the encoder can be viewed as a *discreminative model*. Similarly, the decoder's input is the state $\mathbf{x}$ and its output is a reconstructed version of the observation $\hat{\mathbf{y}}$, hence, the decoder can be viewed as a *generative model*. In this framework, the encoder plays the role of a filter and performs state estimation through approximating the posterior. If a variational autoencoder is implemented by a neural network, the encoder and the decoder will compute the parameterized probability density functions $q_{\phi}(\mathbf{x}|\mathbf{y})$ and $p_{\theta}(\mathbf{y}|\mathbf{x})$, respectively. In a control system, these probability density functions will be conditional on the input $\mathbf{u}$ as well:

$$q_{\phi}(\mathbf{x}|\mathbf{u},\mathbf{y}), \tag{8.68}$$

$$p_{\theta}(\mathbf{y}|\mathbf{x},\mathbf{u}). \tag{8.69}$$

The corresponding loss function, $\mathcal{L}(\boldsymbol{\phi},\boldsymbol{\theta},\mathbf{y},\hat{\mathbf{y}})$, will consist of two terms, which are associated with reconstruction and regularization. Either (8.67) or a probabilistic measure such as log likelihood can be used for reconstruction. For regularization, the Kullback–Leibler divergence (KLD) between the inferred latent distribution and a fixed prior on latent distribution can be used [164]:

$$\mathcal{L}(\boldsymbol{\phi},\boldsymbol{\theta},\mathbf{y},\hat{\mathbf{y}}) = \underbrace{\mathbb{E}_{q_{\phi}}\left[\log p_{\theta}(\mathbf{y}|\mathbf{x},\mathbf{u})\right]}_{\text{reconstruction}} - \underbrace{\text{KLD}\left(q_{\phi}(\mathbf{x}|\mathbf{u},\mathbf{y})||p_{\theta}(\mathbf{x}|\mathbf{u})\right)}_{\text{regularization}}. \tag{8.70}$$

Regularization ensures that close points in the latent space will be decoded in a similar and meaningful way. For instance, a normal distribution can be used for the fixed prior, which guarantees that encodings will be evenly distributed around the center of the latent space, and prevents memorizing the data by penalizing the network if it tries to cluster points in specific regions. Different components of $\mathbf{x}$ may encode different interpretable latent features, which can be discovered by slightly perturbing one latent variable and keeping all others unchanged. Regarding the definition of state, the latent variables must be independent in order to represent state variables. This condition, which is referred to as *disentanglement*, is achieved through imposing a diagonal prior on the latent variables. In order to achieve disentanglement through using a diagonal prior, the loss function in (8.70) is modified as follows:

$$\mathcal{L}(\boldsymbol{\phi},\boldsymbol{\theta},\mathbf{y},\hat{\mathbf{y}}) = \underbrace{\mathbb{E}_{q_{\phi}}\left[\log p_{\theta}(\mathbf{y}|\mathbf{x},\mathbf{u})\right]}_{\text{reconstruction}} - \underbrace{\beta \times \text{KLD}\left(q_{\phi}(\mathbf{x}|\mathbf{u},\mathbf{y})||p_{\theta}(\mathbf{x}|\mathbf{u})\right)}_{\text{regularization}}, \tag{8.71}$$

where $\beta > 1$. This architecture is known as *β-VAE* [147].

8.16 Generative Adversarial Network

The main idea behind the *generative adversarial network* (GAN) is to generate new instances through sampling without explicitly modeling the underlying probability density function [147, 165]. GAN deploys two neural networks, which compete against each other in a zero-sum game [166]:

- *Generator* produces fake samples from noise that imitates real samples. Generator tries to trick the discriminator in a way that it cannot distinguish between the real samples and the fake ones.
- *Discriminator* tries to distinguish between the real samples and the fake ones, which are synthesized by the generator.

GAN is trained such that the generator learns to reproduce the true data distribution. During the training phase, the following optimization problem is solved [166]:

$$\min_G \max_D \mathbb{E}_{\mathbf{y} \sim p_{\text{data}}(\mathbf{y})} \left[\log D(\mathbf{y}) \right] + \mathbb{E}_{\mathbf{z} \sim p_{\mathbf{z}}(\mathbf{z})} \left[\log \left(1 - D\left(G(\mathbf{z}) \right) \right) \right], \tag{8.72}$$

where G and D denote the generator and the discriminator, respectively. $\mathbf{y}$ and $\mathbf{z}$ denote data and noise, respectively, and $p_{\mathbf{z}}(\mathbf{z})$ denotes the prior over noise variables $\mathbf{z}$ used as inputs to the generator.

8.17 Transformer

Transformer and its variants known as *X-formers* are multilayer architectures, which were originally proposed as sequence transduction models [167]. Since transformers encode relational structure as data, they can be easily applied to data with nonlinear relational structures [168]. A transformer transforms a set of n_B objects $\mathbf{u}^i \in \mathbb{R}^d$ to another set of n_B objects $\mathbf{y}^i \in \mathbb{R}^d$. Hence, a *transformer block*, which is the main building block of transformers, represents a parameterized function $\mathbf{y}^i = \mathbf{f}_\theta(\mathbf{u}^i), i = 1, \dots, n_B$. The transformer block includes the following components [167–169]:

- *Multi-head self-attention mechanism*: As an attention mechanism, self-attention relates different positions in a sequence in order to facilitate computing a representation of the sequence. In mathematical terms, the multi-head self-attention mechanism can be described as follows:

$$\mathbf{q}^{hi} = \mathbf{W}^{hq}\mathbf{u}^i, \tag{8.73}$$

$$\mathbf{k}^{hi} = \mathbf{W}^{hk}\mathbf{u}^i, \tag{8.74}$$

$$\mathbf{v}^{hi} = \mathbf{W}^{hv}\mathbf{u}^i, \tag{8.75}$$

where $\mathbf{W}^{hq}, \mathbf{W}^{hk}, \mathbf{W}^{hv} \in \mathbb{R}^{d \times \frac{d}{n_h}}$, are the weight matrices for query, key, and value projections of the input $\mathbf{u}_i$, for $h = 1, \dots, n_h$, with n_h denoting the number of heads in the multi-head self-attention mechanism. Using the softmax function $\boldsymbol{\sigma}(.)$:

$$\mathcal{A}_{ij}^{h} = \sigma_j\left(\frac{\left\langle \mathbf{q}^{hi}, \mathbf{k}^{hj} \right\rangle}{\sqrt{d}}\right), \tag{8.76}$$

where $\boldsymbol{\sigma}_j(.)$ denotes the jth element, the multi-head mechanism combines outputs of different heads as:

$$\tilde{\mathbf{z}}_i = \sum_{h=1}^{n_h} \mathbf{W}^{hc} \sum_{j=1}^{n_B} \mathcal{A}_{ij}^{h} \mathbf{v}^{hj}, \tag{8.77}$$

where $\mathbf{W}^{hc} \in \mathbb{R}^{d \times \frac{d}{n_h}}$.

- *Residual connectors*: Residual networks deploy *skip connections*, which are shortcuts that bypass a few layers. Such connections prevent the problem of vanishing gradients, and remedy the *degradation problem* (accuracy saturation), which refers to increasing the training error by adding more layers to a deep neural network [170]. In mathematical terms, the skip connection adds the input of the previous layers, $\mathbf{u}^i$, to their final output, $\tilde{\mathbf{z}}^i$. Hence, the following signal will be passed to the next layer:

$$\tilde{\mathbf{z}}^i + \mathbf{u}^i. \tag{8.78}$$

- *Layer normalization modules*: This module is mathematically described as:

$$\mathbf{z}^i = \text{LayerNorm}\left(\tilde{\mathbf{z}}^i + \mathbf{u}^i | \boldsymbol{\gamma}^a, \boldsymbol{\beta}^a\right), \tag{8.79}$$

where each element of $\mathbf{z}^i$ is computed as:

$$z_j^i = \gamma_j^a \frac{\left(\tilde{\mathbf{z}}^i + \mathbf{u}^i\right) - \mu_j^{\tilde{\mathbf{z}}+\mathbf{u}}}{\sigma_j^{\tilde{\mathbf{z}}+\mathbf{u}}} + \beta_j^a, \tag{8.80}$$

for $j = 1, \dots, d$, with $\boldsymbol{\gamma}^a, \boldsymbol{\beta}^a \in \mathbb{R}^d$, and

$$\boldsymbol{\mu}^{\tilde{\mathbf{z}}+\mathbf{u}} = \frac{1}{n_B} \sum_{i=1}^{n_B} \left(\tilde{\mathbf{z}}^i + \mathbf{u}^i\right), \tag{8.81}$$

$$\boldsymbol{\sigma}^{\tilde{\mathbf{z}}+\mathbf{u}} = \sqrt{\frac{1}{n_B} \sum_{i=1}^{n_B} \left(\left(\tilde{\mathbf{z}}^i + \mathbf{u}^i\right) - \boldsymbol{\mu}^{\tilde{\mathbf{z}}+\mathbf{u}}\right)^2}. \tag{8.82}$$

- *Position-wise feedforward network*: Position-wise network is a two-layer feedforward network with ReLU activation functions that independently operates on each position:

$$\tilde{\mathbf{y}}^i = \mathbf{W}^b \text{ ReLU}\left(\mathbf{W}^a \mathbf{z}^i\right), \tag{8.83}$$

where $\mathbf{W}^a \in \mathbb{R}^{m \times d}$ and $\mathbf{W}^b \in \mathbb{R}^{d \times m}$, with m being a hyper-parameter.

- *Residual connectors*:

$$\tilde{\mathbf{y}}^i + \mathbf{z}^i. \tag{8.84}$$

- *Layer normalization modules*:

$$\mathbf{y}^i = \text{LayerNorm}\left(\tilde{\mathbf{y}}^i + \mathbf{z}^i | \boldsymbol{\gamma}^b, \boldsymbol{\beta}^b\right), \tag{8.85}$$

  where $\boldsymbol{\gamma}^b, \boldsymbol{\beta}^b \in \mathbb{R}^d$.

The output, $\mathbf{Y} = \begin{bmatrix}\mathbf{y}^1 & \dots & \mathbf{y}^{n_B}\end{bmatrix} \in \mathbb{R}^{d \times n_B}$, has the same structure as the input, $\mathbf{U} = \begin{bmatrix}\mathbf{u}^1 & \dots & \mathbf{u}^{n_B}\end{bmatrix} \in \mathbb{R}^{d \times n_B}$. A transformer is formed by stacking L transformer blocks on top of each other, where each block is parameterized with a separate set of parameters [168]:

$$\mathbf{f}_{\theta_L} \circ \dots \circ \mathbf{f}_{\theta_1}(\mathbf{U}). \tag{8.86}$$

Transformers can be used in the following modes [169]:

- Encoder-only
- Decoder-only
- Encoder-decoder

8.18 Concluding Remarks

Combining representation learning with complex reasoning leads to major breakthroughs in artificial intelligence [171]. One way to design more sophisticated learning machines is through adding memory [172] and attention [167] to the network architecture. Following this line of thinking, the notion of the *cognitive dynamic system* [173] has been proposed based on the *Fuster's paradigm* of cognition [174]. Cognitive dynamic systems provide an abstract model for cognition based on five fundamental building blocks:

- Perception-action cycle
- Memory
- Attention
- Intelligence
- Language

In a such a system, memory consists of perceptual memory, executive memory, and working memory. Similarly, attention consists of perceptual attention and executive attention. Intelligence emerges through the interaction between perception-action cycle, memory, and attention. Language is categorized as internal language and external language. The internal language, which is used for communication between different components of a cognitive system, can be viewed as a tool of thought. The external language, which is used for communication between different cognitive systems, plays a key role in the context of a network of cognitive systems. Furthermore, fairness, accountability, transparency, and ethics (FATE) in artificial intelligence are issues of concern, which call for careful investigation.

9

Deep Learning-Based Filters

9.1 Introduction

Different classes of Bayesian filters, which were discussed in Chapters 5 and 6, rely on the availability of a fairly accurate state-space model. However, without domain knowledge, data-driven approaches must be followed to build state-space models from raw data. Learning filtering models using supervised learning, which relies on labeled datasets, may be impractical. Hence, generative models are required with the ability of learning and inferring hidden structures and latent states directly from data. This chapter explores the potential of variational inference for learning different classes of Bayesian filters. Furthermore, deep neural networks are deployed to construct state-space models from data. Such models can be tractably learned by optimizing a lower bound on the likelihood of data. Such deep learning-based filters will be able to handle severe nonlinearities with temporal and spatial dependencies. Deep learning-based filters will be very helpful for *counterfactual analysis*, which refers to inferring the probability of occurring an event given circumstances that are different from the empirically observed ones.

In what follows, after reviewing variational inference and amortized variational inference, different deep learning-based filtering algorithms are presented, which use supervised or unsupervised learning. In the formulation of filters, let us consider a nonlinear dynamic system with the latent state vector $\mathbf{x}_k \in \mathcal{X} \subset \mathbb{R}^{n_x}$, the control input vector $\mathbf{u}_k \in \mathcal{U} \subset \mathbb{R}^{n_u}$, and the measurement output vector $\mathbf{y}_k \in \mathcal{Y} \subset \mathbb{R}^{n_y}$ at time instant k. Let $\mathbf{U}_k = \{\mathbf{u}_i | i = 0, \dots, k\} \equiv \mathbf{u}_{0:k}$ and $\mathbf{Y}_k = \{\mathbf{y}_i | i = 0, \dots, k\} \equiv \mathbf{y}_{0:k}$ denote the input and the measurement sequences of length $k + 1$ corresponding to time instant k, respectively. Then, the information set at time instant k will be $I_k = \{\mathbf{U}_k, \mathbf{Y}_k\}$. Observations may be high-dimensional sensory data such as high-resolution images with complex non-Markovian

Nonlinear Filters: Theory and Applications, First Edition. Peyman Setoodeh, Saeid Habibi, and Simon Haykin.

transitions. In such cases, $\mathbf{y}_k$ will be a lower-dimensional representation of the high-dimensional observation, $\mathbf{o}_k$. A neural network-based observation encoder provides these lower-dimensional representations.

9.2 Variational Inference

A dynamic latent-variable model, which is aimed at capturing the relationship between a sequence of measurements (observations), $\mathbf{y}_{0:k+1}$, and a sequence of states (latent variables), $\mathbf{x}_{0:k+1}$, can be expressed by a parameterized joint probability distribution:

$$p_\theta(\mathbf{x}_{0:k+1}, \mathbf{y}_{0:k+1} | \mathbf{u}_{0:k+1}), \tag{9.1}$$

where θ denotes the set of parameters. It is usually assumed that at each time instant, this joint distribution can be factorized into conditional joint distributions as:

$$p_\theta(\mathbf{x}_{k+1}, \mathbf{y}_{k+1} | \mathbf{x}_{0:k}, \mathbf{u}_{0:k+1}, \mathbf{y}_{0:k}), \tag{9.2}$$

which are conditioned on preceding variables. Imposing such a structure on the model leads to the following autoregressive formulation:

$$\begin{aligned} p_\theta(\mathbf{x}_{0:k+1}, \mathbf{y}_{0:k+1} | \mathbf{u}_{0:k+1}) &\\ &= p_\theta(\mathbf{x}_0, \mathbf{y}_0 | \mathbf{u}_0) \prod_{i=0}^{k} p_\theta(\mathbf{x}_{i+1}, \mathbf{y}_{i+1} | \mathbf{x}_{0:i}, \mathbf{u}_{0:i+1}, \mathbf{y}_{0:i}) \\ &= p_\theta(\mathbf{x}_0)\, p_\theta(\mathbf{y}_0 | \mathbf{x}_0, \mathbf{u}_0) \\ &\quad \times \prod_{i=0}^{k} p_\theta(\mathbf{x}_{i+1} | \mathbf{x}_{0:i}, \mathbf{u}_{0:i}, \mathbf{y}_{0:i})\, p_\theta(\mathbf{y}_{i+1} | \mathbf{x}_{0:i+1}, \mathbf{u}_{0:i+1}, \mathbf{y}_{0:i}), \end{aligned} \tag{9.3}$$

where $p_\theta(\mathbf{x}_{k+1} | \mathbf{x}_{0:k}, \mathbf{u}_{0:k}, \mathbf{y}_{0:k})$ and $p_\theta(\mathbf{y}_k | \mathbf{x}_{0:k}, \mathbf{u}_{0:k}, \mathbf{y}_{0:k-1})$ represent the state-transition and the observation models, respectively. The dynamic latent-variable model in (9.3) has a general form, and allows both the dynamic and the observation models to be arbitrary functions of the conditioning variables. Further structure can be imposed on the model such as Markov property or linearity assumptions to achieve tractability [175].

The posterior, $p(\mathbf{x}_{0:k+1} | \mathbf{u}_{0:k+1}, \mathbf{y}_{0:k+1})$, can be inferred either online or offline by Bayesian filtering or smoothing, respectively [176], and the model parameters, θ, can be learned by maximum likelihood estimation. Since inference may be intractable, especially for general model classes, *variational inference* is deployed to estimate an approximate posterior, $q(\mathbf{x}_{0:k+1} | \mathbf{u}_{0:k+1}, \mathbf{y}_{0:k+1})$, by reformulating inference as an optimization problem [177]. The optimization problem aims at minimizing the Kullback–Leibler divergence (KLD) between the true and the approximate posteriors. To compute the KLD, let us start with the marginal likelihood:

$$
\begin{aligned}
&\log p_\theta(\mathbf{y}_{0:k+1}|\mathbf{u}_{0:k+1}) \\
&= \int q(\mathbf{x}_{0:k+1}|\mathbf{u}_{0:k+1},\mathbf{y}_{0:k+1}) \log p_\theta(\mathbf{y}_{0:k+1}|\mathbf{x}_{0:k+1},\mathbf{u}_{0:k+1}) \mathrm{d}\mathbf{x}_{0:k+1} \\
&= \int q(\mathbf{x}_{0:k+1}|\mathbf{u}_{0:k+1},\mathbf{y}_{0:k+1}) \log \frac{p_\theta(\mathbf{x}_{0:k+1},\mathbf{y}_{0:k+1}|\mathbf{u}_{0:k+1})}{p_\theta(\mathbf{x}_{0:k+1}|\mathbf{u}_{0:k+1},\mathbf{y}_{0:k+1})} \mathrm{d}\mathbf{x}_{0:k+1} \\
&= \int \Bigg(q(\mathbf{x}_{0:k+1}|\mathbf{u}_{0:k+1},\mathbf{y}_{0:k+1}) \\
&\quad \times \log \frac{p_\theta(\mathbf{x}_{0:k+1},\mathbf{y}_{0:k+1}|\mathbf{u}_{0:k+1})\; q(\mathbf{x}_{0:k+1}|\mathbf{u}_{0:k+1},\mathbf{y}_{0:k+1})}{p_\theta(\mathbf{x}_{0:k+1}|\mathbf{u}_{0:k+1},\mathbf{y}_{0:k+1})\; q(\mathbf{x}_{0:k+1}|\mathbf{u}_{0:k+1},\mathbf{y}_{0:k+1})} \Bigg) \mathrm{d}\mathbf{x}_{0:k+1} \\
&= \int \Bigg(q(\mathbf{x}_{0:k+1}|\mathbf{u}_{0:k+1},\mathbf{y}_{0:k+1}) \\
&\quad \times \log \frac{p_\theta(\mathbf{x}_{0:k+1},\mathbf{y}_{0:k+1}|\mathbf{u}_{0:k+1})\; q(\mathbf{x}_{0:k+1}|\mathbf{u}_{0:k+1},\mathbf{y}_{0:k+1})}{q(\mathbf{x}_{0:k+1}|\mathbf{u}_{0:k+1},\mathbf{y}_{0:k+1})\; p_\theta(\mathbf{x}_{0:k+1}|\mathbf{u}_{0:k+1},\mathbf{y}_{0:k+1})} \Bigg) \mathrm{d}\mathbf{x}_{0:k+1} \\
&= \int q(\mathbf{x}_{0:k+1}|\mathbf{u}_{0:k+1},\mathbf{y}_{0:k+1}) \log \frac{p_\theta(\mathbf{x}_{0:k+1},\mathbf{y}_{0:k+1}|\mathbf{u}_{0:k+1})}{q(\mathbf{x}_{0:k+1}|\mathbf{u}_{0:k+1},\mathbf{y}_{0:k+1})} \mathrm{d}\mathbf{x}_{0:k+1} \\
&\quad + \int q(\mathbf{x}_{0:k+1}|\mathbf{u}_{0:k+1},\mathbf{y}_{0:k+1}) \log \frac{q(\mathbf{x}_{0:k+1}|\mathbf{u}_{0:k+1},\mathbf{y}_{0:k+1})}{p_\theta(\mathbf{x}_{0:k+1}|\mathbf{u}_{0:k+1},\mathbf{y}_{0:k+1})} \mathrm{d}\mathbf{x}_{0:k+1} \\
&= \mathbb{E}_q \left[\log \frac{p_\theta(\mathbf{x}_{0:k+1},\mathbf{y}_{0:k+1}|\mathbf{u}_{0:k+1})}{q(\mathbf{x}_{0:k+1}|\mathbf{u}_{0:k+1},\mathbf{y}_{0:k+1})} \right] \\
&\quad + \mathrm{KLD}\left(q(\mathbf{x}_{0:k+1}|\mathbf{u}_{0:k+1},\mathbf{y}_{0:k+1}) || p_\theta(\mathbf{x}_{0:k+1}|\mathbf{u}_{0:k+1},\mathbf{y}_{0:k+1}) \right). \qquad (9.4)
\end{aligned}
$$

Therefore, the KLD is obtained as:

$$
\begin{aligned}
&\mathrm{KLD}\left(q(\mathbf{x}_{0:k+1}|\mathbf{u}_{0:k+1},\mathbf{y}_{0:k+1}) || p_\theta(\mathbf{x}_{0:k+1}|\mathbf{u}_{0:k+1},\mathbf{y}_{0:k+1}) \right) \\
&\quad = \log p_\theta(\mathbf{y}_{0:k+1}|\mathbf{u}_{0:k+1}) - \mathbb{E}_q \left[\log \frac{p_\theta(\mathbf{x}_{0:k+1},\mathbf{y}_{0:k+1}|\mathbf{u}_{0:k+1})}{q(\mathbf{x}_{0:k+1}|\mathbf{u}_{0:k+1},\mathbf{y}_{0:k+1})} \right]. \qquad (9.5)
\end{aligned}
$$

The second term in the aforementioned equation is called the *variational free energy* [175]:

$$
\mathcal{F} \equiv -\mathbb{E}_q \left[\log \frac{p_\theta(\mathbf{x}_{0:k+1},\mathbf{y}_{0:k+1}|\mathbf{u}_{0:k+1})}{q(\mathbf{x}_{0:k+1}|\mathbf{u}_{0:k+1},\mathbf{y}_{0:k+1})} \right], \qquad (9.6)
$$

which is also known as the (negative) *evidence lower bound* (ELBO).

Since $\log p_\theta(\mathbf{y}_{0:k+1}|\mathbf{u}_{0:k+1})$ does not depend on $q(\mathbf{x}_{0:k+1}|\mathbf{u}_{0:k+1},\mathbf{y}_{0:k+1})$, minimizing the KLD can theoretically provide the true posterior. In other words, approximate inference can be performed by minimizing $\mathcal{F}$ with respect to the approximate posterior, $q(\mathbf{x}_{0:k+1}|\mathbf{u}_{0:k+1},\mathbf{y}_{0:k+1})$. Regarding the fact that KLD is non-negative, it can be concluded that free energy provides an upper bound for the negative log likelihood. Therefore, upon minimizing $\mathscr{F}$ with respect to $q(\mathbf{x}_{0:k+1}|\mathbf{u}_{0:k+1}, \mathbf{y}_{0:k+1})$, the gradient $\nabla_{\theta}\mathscr{F}$ can be used to learn the model parameters $\boldsymbol{\theta}$.

9.3 Amortized Variational Inference

Inference optimization using the *stochastic gradient descent* method may require a relatively large number of inference iterations, which makes it a computationally demanding process. *Amortizing inference* across examples can improve the efficiency by learning to map data examples to approximate posterior estimates through a separate inference model [178]. Let λ^q denote the distribution parameters of the approximate posterior, q, (mean and variance for a Gaussian distribution). Denoting the inference model by $\mathbf{f}$, and deploying a neural network with the set of adjustable parameters $\boldsymbol{\phi}$ to perform the inference, we will have [175]:

$$\lambda^q \leftarrow \mathbf{f}_{\phi}(\mathbf{u}, \mathbf{y}). \tag{9.7}$$

Although such a model is computationally efficient, it cannot take account of empirical priors that occur from one latent variable to another. This is due to the fact that this model only receives data as input. Empirical priors are formed across time steps in dynamic models and across levels in hierarchical models. To address this issue, rather than performing inference optimization by iteratively encoding approximate posterior estimates and gradients, *iterative inference models* are deployed to directly consider empirical priors [179]:

$$\lambda^q \leftarrow \mathbf{f}_{\phi}(\lambda^q, \nabla_{\lambda^q}\mathscr{F}), \tag{9.8}$$

where $\mathscr{F}$ is the variational free energy or ELBO as defined in (9.6). The gradient $\nabla_{\lambda^q}\mathscr{F}$ can be estimated using black-box models [180] and *reparameterization trick*, which refers to applying noise as the input to a learnable generative model [181–183]. Inference convergence can be improved by additionally encoding the data along with equation (9.8) as the encoding form for an iterative inference model [175]. Learning to perform inference optimization by iterative inference models is called *learning to infer* [184] in analogy with *learning to learn* or meta-learning [185–187].

9.4 Deep Kalman Filter

Inspired by variational autoencoders (VAEs), the main idea behind the *deep Kalman filter* (DKF) algorithm is to use a neural network known as the *recognition model* to approximate the posterior, where the *stochastic gradient variational*

Bayes (SGVB) method is used to jointly perform identification and inference [188]. Due to the system nonlinearity, computation of the posterior distribution $p(\mathbf{x}_{0:k+1}|\mathbf{u}_{0:k+1}, \mathbf{y}_{0:k+1})$ may be intractable. DKF addresses this issue by approximating the posterior using the recognition model.

Let the joint distribution $p(\mathbf{x}, \mathbf{y}|\mathbf{u})$ be written as:

$$p(\mathbf{x}, \mathbf{y}|\mathbf{u}) = p(\mathbf{x}|\mathbf{u})p_{\theta}(\mathbf{y}|\mathbf{x}, \mathbf{u}), \tag{9.9}$$

where $p(\mathbf{x}|\mathbf{u})$ is the prior distribution on $\mathbf{x}$, and $p_{\theta}(\mathbf{y}|\mathbf{x}, \mathbf{u})$ is the generative model parameterized by the set of parameters $\boldsymbol{\theta}$. In general, for such a model, the posterior distribution $p_{\theta}(\mathbf{x}|\mathbf{u}, \mathbf{y})$ may be intractable. Based on the variational principle, an approximate posterior distribution $q_{\phi}(\mathbf{x}|\mathbf{u}, \mathbf{y})$ is considered, which is parameterized by the set of parameters $\boldsymbol{\phi}$. A deep neural network can be used for parameterizing q_{ϕ} such that the set of parameters $\boldsymbol{\phi}$ would be the adjustable parameters of the neural network. This approximate posterior distribution is known as the recognition model. Using the Jensen's inequality [26]:

$$f(\mathbb{E}[z]) \geq \mathbb{E}[f(z)], \tag{9.10}$$

where f is a concave function such as logarithm and z is a random variable, the following lower bound can be computed for the marginal likelihood:

$$\begin{aligned}
\log p_{\theta}(\mathbf{y}|\mathbf{u}) &= \log \int p_{\theta}(\mathbf{y}|\mathbf{x}, \mathbf{u})p(\mathbf{x}|\mathbf{u})d\mathbf{x} \\
&= \log \int \frac{q_{\phi}(\mathbf{x}|\mathbf{u}, \mathbf{y})}{q_{\phi}(\mathbf{x}|\mathbf{u}, \mathbf{y})} p_{\theta}(\mathbf{y}|\mathbf{x}, \mathbf{u})p(\mathbf{x}|\mathbf{u})d\mathbf{x} \\
&\geq \int q_{\phi}(\mathbf{x}|\mathbf{u}, \mathbf{y}) \log \frac{p_{\theta}(\mathbf{y}|\mathbf{x}, \mathbf{u})p(\mathbf{x}|\mathbf{u})}{q_{\phi}(\mathbf{x}|\mathbf{u}, \mathbf{y})} d\mathbf{x} \\
&= \mathbb{E}_{q_{\phi}}\left[\log p_{\theta}(\mathbf{y}|\mathbf{x}, \mathbf{u})\right] - \mathrm{KLD}\left(q_{\phi}(\mathbf{x}|\mathbf{u}, \mathbf{y}) || p(\mathbf{x}|\mathbf{u})\right) \\
&= \mathcal{L}(\boldsymbol{\phi}, \boldsymbol{\theta}).
\end{aligned} \tag{9.11}$$

VAEs aim at maximizing the lower bound $\mathcal{L}(.)$ using the parametric model q_{ϕ} conditioned on the input $\mathbf{u}$. According to (9.11), the first term in the lower bound $\mathcal{L}(.)$ is an expectation with respect to q_{ϕ}, which in turn, depends on the network parameters $\boldsymbol{\phi}$. The corresponding optimization problem can be solved using the *stochastic backpropagation* method. In this method, Monte Carlo estimates of the gradients of both the expectation and the KLD terms are obtained with respect to the network parameters $\boldsymbol{\phi}$. Following this line of thinking, DKF aims at learning a generative model for a sequence of actions and observations (inputs and outputs) assuming that each observation vector corresponds to a latent state vector, which evolves over time generally in a nonlinear manner with known actions.

In the DKF algorithm, it is assumed that latent states $\mathbf{x}_{k+1}$ have Gaussian distributions with mean vectors and covariance matrices, which are nonlinear functions of the previous latent states $\mathbf{x}_k$ and actions $\mathbf{u}_k$:

$$\mathbf{x}_0 \sim \mathcal{N}(\boldsymbol{\mu}_0, \boldsymbol{\Sigma}_0), \tag{9.12}$$

$$\mathbf{x}_{k+1} \sim \mathcal{N}\left(\boldsymbol{\mu}_{k+1}(\mathbf{x}_k, \mathbf{u}_k), \boldsymbol{\Sigma}_{k+1}(\mathbf{x}_k, \mathbf{u}_k)\right). \tag{9.13}$$

The multivariate normal or Gaussian distribution, $\mathbf{x} \sim \mathcal{N}(\boldsymbol{\mu}, \boldsymbol{\Sigma})$, is mathematically described as:

$$p(\mathbf{x}) = \frac{1}{\sqrt{(2\pi)^{n_x}|\boldsymbol{\Sigma}|}} e^{-\frac{1}{2}(\mathbf{x}-\boldsymbol{\mu})^T\boldsymbol{\Sigma}^{-1}(\mathbf{x}-\boldsymbol{\mu})}, \tag{9.14}$$

where $|\boldsymbol{\Sigma}|$ denotes the determinant of the covariance matrix $\boldsymbol{\Sigma}$. Similarly, the corresponding parameters λ_k of the distributions of observations $\mathbf{y}_k$ are functions of both latent states $\mathbf{x}_k$ and actions $\mathbf{u}_k$:

$$\mathbf{y}_k \sim \Pi\left(\lambda_k(\mathbf{x}_k, \mathbf{u}_k)\right). \tag{9.15}$$

In equations (9.13) and (9.15), any differentiable parametric function can be used for $\boldsymbol{\mu}_k$, $\boldsymbol{\Sigma}_k$, and λ_k. For instance, they can be parameterized by deep neural networks. Hence, this generic model subsumes a large family of linear and nonlinear state-space models. For instance, classic Kalman filter is obtained by choosing:

$$\boldsymbol{\mu}_{k+1}(\mathbf{x}_k, \mathbf{u}_k) = \mathbf{A}_k\mathbf{x}_k + \mathbf{B}_k\mathbf{u}_k, \tag{9.16}$$

$$\boldsymbol{\Sigma}_{k+1}(\mathbf{x}_k, \mathbf{u}_k) = \mathbf{P}_{k+1}, \tag{9.17}$$

$$\lambda_k(\mathbf{x}_k, \mathbf{u}_k) = \mathbf{C}_k\mathbf{x}_k + \mathbf{D}_k\mathbf{u}_k. \tag{9.18}$$

Distributions involved in the formulation of $\mathcal{L}(.)$ can be factorized due to the Markov property of the system model. In this way, a structure can be imposed on the learned model. Structured factorization of the corresponding distributions will be discussed in Section 9.8. In (9.11), if KLD $\left(q_{\boldsymbol{\phi}}(\mathbf{x}|\mathbf{u},\mathbf{y})||p(\mathbf{x}|\mathbf{u})\right) = 0$, then the lower bound will reduce to $\mathcal{L}(\boldsymbol{\phi}, \boldsymbol{\theta}) = \log p_{\boldsymbol{\theta}}(\mathbf{y}|\mathbf{u})$, and optimizing the objective function (9.11) will be a maximum likelihood problem. Algorithm 9.1 depicts different stages for learning the DKF [188].

9.5 Backpropagation Kalman Filter

The *backpropagation Kalman filter* (Backprop KF) aims at training discriminative state estimators by directly optimizing parameters of the latent state distribution as a deterministic computation graph [189]. In other words, rather than performing inference based on a probabilistic latent-variable model, Backprop KF constructs

---

Algorithm 9.1: Deep Kalman filter

State-space model:

$$\mathbf{x}_0 \sim \mathcal{N}(\boldsymbol{\mu}_0, \boldsymbol{\Sigma}_0)$$

$$\mathbf{x}_{k+1} \sim \mathcal{N}\left(\boldsymbol{\mu}_{k+1}(\mathbf{x}_k, \mathbf{u}_k), \boldsymbol{\Sigma}_{k+1}(\mathbf{x}_k, \mathbf{u}_k)\right)$$

$$\mathbf{y}_k \sim \Pi\left(\lambda_k(\mathbf{x}_k, \mathbf{u}_k)\right)$$

while *not Converged* **do**

Sample minibatch

$$\{\mathbf{u}_{0:k+1}, \mathbf{y}_{0:k+1}\}$$

Perform inference and estimate likelihood

$$\mathbf{x} \sim q_{\boldsymbol{\phi}}(\mathbf{x}_{0:k+1} | \mathbf{u}_{0:k+1}, \mathbf{y}_{0:k+1})$$

$$\mathbf{y} \sim p_{\boldsymbol{\theta}}(\mathbf{y}_{0:k+1} | \mathbf{x}_{0:k+1}, \mathbf{u}_{0:k+1})$$

$$\mathcal{L}(\boldsymbol{\phi}, \boldsymbol{\theta}) = \mathbb{E}_{q_{\boldsymbol{\phi}}}\left[\log p_{\boldsymbol{\theta}}(\mathbf{y}|\mathbf{x}, \mathbf{u})\right] - \text{KLD}\left(q_{\boldsymbol{\phi}}(\mathbf{x}|\mathbf{u}, \mathbf{y}) || \, p(\mathbf{x}|\mathbf{u})\right)$$

Compute $\nabla_{\boldsymbol{\theta}}\mathcal{L}$ and $\nabla_{\boldsymbol{\phi}}\mathcal{L}$

Update $\boldsymbol{\theta}$ and $\boldsymbol{\phi}$ using the Adam optimizer

end

---

a deterministic computation graph, which is trained in an end-to-end manner using backpropagation and gradient descent methods. Hence, Backprop KF can be viewed as a recurrent neural network with a correspondence between architecture of the network and structure of the state estimator. Backprop KF allows for embedding nonlinear components into the state-transition and observation models. Moreover, it provides a connection between discriminative probabilistic state estimators and recurrent computation graphs.

In systems with high-dimensional observations, $\mathbf{o}_k$, such as high-resolution images, it is preferred to find a low-dimensional *pseudo-measurement* vector $\mathbf{y}_k$, which captures dependency of the observation on the system state. Let $\mathbf{x}_k$ denote the system state and $\mathbf{x}_k^*$ denote the state label (target) that the designed estimator should be able to infer from $\mathbf{o}_k$. In this case, first, a feedforward model $\mathbf{y}_k = \mathbf{r}_{\boldsymbol{\theta}}(\mathbf{o}_k)$, which is parameterized by $\boldsymbol{\theta}$, is trained to predict $\mathbf{y}_k$ from $\mathbf{o}_k$. For instance, different variations of convolutional neural networks can be used to build such a feedforward model. Then, a filter is trained, which uses the predicted $\mathbf{y}_k$ as a measurement to estimate $\mathbf{x}_k$. In this framework, the filter can be viewed as a computation graph, which is unrolled through time. This computation graph is augmented with an output function $\lambda_k(\mathbf{x}_k)$ that provides parameters of the corresponding distribution over labels $\mathbf{x}_k^*$. Assuming that latent states $\mathbf{x}_k$ have Gaussian distributions, similar to the DKF algorithm, relations (9.12)–(9.15) will be valid, and classic Kalman filter can be obtained as in (9.16)–(9.18).

Let $\mathbf{x}_{k+1} = \boldsymbol{\kappa}(\mathbf{x}_k, \mathbf{u}_{k:k+1}, \mathbf{y}_{k+1})$ represent the operation performed by the filter to estimate the state, and $\mathcal{L}\left(\lambda_k(\mathbf{x}_k)\right)$ be a loss function on the output distribution of the computation graph. Then, the total loss for a sequence will be:

$$\mathcal{L}(\theta) = \sum_{i=0}^{k} \mathcal{L}\left(\lambda_i(\mathbf{x}_i)\right), \tag{9.19}$$

which can be expressed as a function of the set of parameters $\boldsymbol{\theta}$. Using the chain rule, gradient of the total loss with respect to parameters can be computed as:

$$\nabla_\theta \mathcal{L}(\theta) = \sum_{i=0}^{k} \frac{\partial \mathcal{L}}{\partial \mathbf{x}_i} \frac{\partial \mathbf{x}_i}{\partial \mathbf{y}_i} \frac{\partial \mathbf{y}_i}{\partial \theta}, \tag{9.20}$$

where

$$\frac{\partial \mathcal{L}}{\partial \mathbf{x}_i} = \frac{\partial \mathcal{L}}{\partial \lambda_i} \frac{\partial \lambda_i}{\partial \mathbf{x}_i} + \frac{\partial \mathcal{L}}{\partial \mathbf{x}_{i+1}} \frac{\partial \mathbf{x}_{i+1}}{\partial \mathbf{x}_i}, \tag{9.21}$$

$$\frac{\partial \mathbf{x}_i}{\partial \mathbf{y}_i} = \nabla_{\mathbf{y}_i} \boldsymbol{\kappa}(\mathbf{x}_{i-1}, \mathbf{u}_{i-1:i}, \mathbf{y}_i), \tag{9.22}$$

$$\frac{\partial \mathbf{y}_i}{\partial \theta} = \nabla_\theta \mathbf{r}_\theta(\mathbf{o}_k). \tag{9.23}$$

The required partial derivatives in (9.21) to calculate $\frac{\partial \mathcal{L}}{\partial \mathbf{x}_i}$ are obtained as follows:

$$\frac{\partial \mathcal{L}}{\partial \lambda_i} = \nabla_{\lambda_i} \mathcal{L}\left(\lambda_i(\mathbf{x}_i)\right), \tag{9.24}$$

$$\frac{\partial \lambda_i}{\partial \mathbf{x}_i} = \nabla_{\mathbf{x}_i} \lambda_i(\mathbf{x}_i), \tag{9.25}$$

$$\frac{\partial \mathbf{x}_{i+1}}{\partial \mathbf{x}_i} = \nabla_{\mathbf{x}_i} \boldsymbol{\kappa}(\mathbf{x}_i, \mathbf{u}_{i:i+1}, \mathbf{y}_{i+1}). \tag{9.26}$$

The set of parameters $\boldsymbol{\theta}$ can be optimized by backpropagation through time using the aforementioned gradients. For implementing the Backprop KF algorithm, any variation of the Kalman filter can be used as $\mathbf{x}_{k+1} = \boldsymbol{\kappa}(\mathbf{x}_k, \mathbf{u}_{k:k+1}, \mathbf{y}_{k+1})$ [189].

9.6 Differentiable Particle Filter

Differentiable particle filter (DPF) has been proposed as an end-to-end differentiable implementation of the particle filter [190]. Due to end-to-end differentiability, models can be learned for both the state dynamics and the measurement process (state and measurement equations of the state-space model) using variations of the gradient descent algorithm. Regarding the following probabilistic state-space model:

$$p(\mathbf{x}_{k+1}|\mathbf{x}_k, \mathbf{u}_k), \tag{9.27}$$

$$p(\mathbf{y}_k|\mathbf{x}_k, \mathbf{u}_k), \tag{9.28}$$

both the state-transition and the measurement models are trained by optimizing the state estimator's overall performance in an end-to-end manner, instead of focusing on model accuracy as an intermediate objective. Similar to the Backprop KF, DPF can be viewed as a recurrent neural network, which encodes the algorithmic prior from particle filters in the network architecture. DPF relies on both a learnable state-space model and backpropagation of the gradient of the loss function through the particle filter algorithm to optimize the model.

Uncertainty can be taken into account by estimating a probability distribution over the current state $\mathbf{x}_k$ conditioned on the history of inputs (actions), $\mathbf{u}_{0:k}$, and measurements (observations), $\mathbf{y}_{0:k}$:

$$\mathbf{b}(\mathbf{x}_k) = p(\mathbf{x}_k|\mathbf{u}_{0:k}, \mathbf{y}_{0:k}), \tag{9.29}$$

which is called *belief*. As a special class of Bayesian filters, particle filters use a set of particles (samples), $\mathbf{x}_k^i$, $i = 1, \dots, n_p$, to represent such a probability distribution function. As mentioned previously, recursive state estimation is performed in two steps of state prediction and update. According to Algorithm 6.1, the state-prediction step is performed by sampling from the importance posterior density, q, based on the state-space model in (9.27) and (9.28) as:

$$\mathbf{x}_{k+1}^i \sim q(\mathbf{x}_{k+1}^i|\mathbf{x}_k^i, \mathbf{u}_{k:k+1}, \mathbf{y}_{k+1}). \tag{9.30}$$

The corresponding normalized weights, w_{k+1}^i, are computed as follows:

$$\tilde{w}_{k+1}^i = \frac{p(\mathbf{y}_{k+1}|\mathbf{x}_{k+1}^i, \mathbf{u}_{k+1})p(\mathbf{x}_{k+1}^i|\mathbf{x}_k^i, \mathbf{u}_k)}{q(\mathbf{x}_{k+1}^i|\mathbf{x}_k^i, \mathbf{u}_{k:k+1}, \mathbf{y}_{k+1})} w_k^i, \tag{9.31}$$

$$w_{k+1}^i = \frac{\tilde{w}_{k+1}^i}{\sum_{i=1}^{n_p} \tilde{w}_{k+1}^i}. \tag{9.32}$$

Resampling is performed by randomly drawing particles $\mathbf{x}_k^i$ proportional to their corresponding weights w_k^i, if

$$N_{eff} = \frac{1}{\sum_{i=1}^{n_p} (w_{k+1}^i)^2} \ll n_p, \tag{9.33}$$

where N_{eff} is the effective sample size.

DPF uses such a set of weighted particles to represent the belief over states as:

$$\mathbf{b}(\mathbf{x}_k) = (\mathbf{S}_k, \mathbf{w}_k), \tag{9.34}$$

where $\mathbf{S}_k \in \mathbb{R}^{n_x \times n_p}$ describes n_p particles in the n_x-dimensional state space with the corresponding weights $\mathbf{w}_k \in \mathbb{R}^{n_p}$ at time instant k. At each time step, belief, $\mathbf{b}(\mathbf{x}_k)$, is updated based on the input and the measurement to obtain $\mathbf{b}(\mathbf{x}_{k+1})$.

To implement the state-prediction step in the DPF algorithm, the state-transition model is built on two components: an *action sampler* and a particle-transition model. The action sampler, $\mathbf{f}^{\mathbf{u}}_{\theta_a}$, which is a neural network with the set of parameters θ_a, generates noisy versions of the input, $\tilde{\mathbf{u}}^i_k$, in order to apply a different version of the input to each particle. Hence, state dynamics will be governed by:

$$\tilde{\mathbf{u}}^i_k = \mathbf{u}_k + \mathbf{f}^{\mathbf{u}}_{\theta_a}(\mathbf{u}_k, \epsilon^i \sim \mathcal{N}), \tag{9.35}$$

$$\mathbf{x}^i_{k+1} = \mathbf{x}^i_k + \mathbf{f}^{\mathbf{x}}_{\theta_s}(\mathbf{x}^i_k, \tilde{\mathbf{u}}^i_k), \tag{9.36}$$

where $\epsilon^i \in \mathbb{R}^{n_x}$ is a noise vector sampled from a normal distribution. Applying the noise vector ϵ^i as the input to the learnable generative model, $\mathbf{f}^{\mathbf{u}}_{\theta_a}$, is known as the *reparameterization trick* [181]. Another neural network with the set of parameters θ_s, which represents the nonlinear function $\mathbf{f}^{\mathbf{x}}_{\theta_s}$, can be used to learn the underlying dynamics from data.

To implement the state-update step in the DPF algorithm, measurements are used to compute particle weights as in (9.31) and (9.32). The DPF measurement model includes three components: an *observation encoder* $\mathbf{r}_{\theta_m}$, which encodes an observation $\mathbf{o}_k$ into a lower-dimensional pseudo-measurement vector $\mathbf{y}_k$, a *particle proposer* $\mathbf{s}_{\theta_p}$, which generates new particles, and an *observation likelihood estimator* l_{θ_w}, which assigns a weight to each particle based on the observation:

$$\mathbf{y}_k = \mathbf{r}_{\theta_m}(\mathbf{o}_k), \tag{9.37}$$

$$\mathbf{x}^i_k = \mathbf{s}_{\theta_p}(\mathbf{u}_{k-1:k}, \mathbf{y}_k, \delta^i \sim \mathcal{B}), \tag{9.38}$$

$$\mathbf{w}^i_k = l_{\theta_w}(\mathbf{x}^i_k, \mathbf{u}_k, \mathbf{y}_k), \tag{9.39}$$

where $\mathbf{r}_{\theta_m}$, $\mathbf{s}_{\theta_p}$, and l_{θ_w} are represented by neural networks with the sets of parameters θ_m, θ_p, and θ_w, respectively. Input δ^i denotes a dropout vector sampled from a *Bernoulli distribution*, which is used as a source of randomness for sampling different particles from the same encoded $\mathbf{y}_k$.

DPF is initialized by proposing particles from measurement, instead of uniformly sampling from the state space, which may produce few initial particles near the true state. During filtering, a gradual move occurs from hypotheses represented by the proposal distribution toward tracking the true state due to resampling. A hyperparameter $\gamma < 1$ is introduced to adjust the ratio of the proposed to the resampled particles as γ^{k-1}. Sequences of labeled data including input (action) $\mathbf{u}_{0:k}$, output (observation) $\mathbf{y}_{0:k}$, and true state (label) $\mathbf{x}^*_{0:k}$ are used to train the DPF by maximizing the belief at the true state. In order to use a reasonable metric in the training phase, different dimensions of the state space are scaled by the average step size, $\mathbb{E}\left[|\mathbf{x}^*_k - \mathbf{x}^*_{k-1}|\right]$. To estimate $\mathbf{b}(\mathbf{x}^*_k)$ from a set of particles, each particle can be viewed as a Gaussian in a mixture model with the corresponding weights $\mathbf{w}_k$. In the DPF algorithm, learning can be performed in an end-to-end manner. Alternatively, state-transition and measurement models can be learned individually:

- *End-to-end learning*: DPF is applied on overlapping subsequences and belief is maximized at all true states along the sequence as:

$$\theta^* = \arg\min_{\theta} \left(-\log \mathbb{E}\left[\mathbf{b}_\theta(\mathbf{x}^*_k)\right]\right), \tag{9.40}$$

  where $\theta = \{\theta_s, \theta_a, \theta_m, \theta_p, \theta_w\}$ is the set of all DPF parameters.
- *Individual learning of the state-transition model*: Given $\mathbf{x}^*_{k-1}$ and $\mathbf{u}_{k-1}$, particles, $\mathbf{x}^i_k$, are obtained using the state-transition model in (9.35) and (9.36). Then, parameters of the model, $\theta = \{\theta_s, \theta_a\}$, are optimized as:

$$\theta^* = \arg\min_{\theta} \left(-\log p_\theta\left(\mathbf{x}^*_k|\mathbf{x}^*_{k-1}, \mathbf{u}_{k-1}\right)\right). \tag{9.41}$$

- *Individual learning of the measurement model*: Given $\mathbf{o}_k$, particles, $\mathbf{x}^i_k$, are obtained using (9.37) and (9.38). Then, the particle proposer, $\mathbf{s}_{\theta_p}$, is trained and θ^*_p is obtained by maximizing the corresponding Gaussian mixture at $\mathbf{x}^*_k$. Both the observation encoder, $\mathbf{r}_{\theta_m}$, and the observation likelihood estimator, l_{θ_w}, are trained by simultaneously maximizing the likelihood of observations at their corresponding states and minimizing their likelihood at other states as:

$$\begin{aligned}\theta^* = \arg\min_{\theta} \Big(&-\log \mathbb{E}_k\left[\boldsymbol{l}_{\theta_w}\left(\mathbf{r}_{\theta_m}(\mathbf{o}_k), \mathbf{x}^i_k, \mathbf{u}_k\right)\right] \\ &-\log\left(1 - \mathbb{E}_{k_1,k_2}\left[\boldsymbol{l}_{\theta_w}\left(\mathbf{r}_{\theta_m}(\mathbf{o}_{k_1}), \mathbf{x}^i_{k_2}, \mathbf{u}_{k_2}\right)\right]\right)\Big),\end{aligned} \tag{9.42}$$

  where $\theta = \{\theta_m, \theta_w\}$ and $k_1 \neq k_2$.

The resampling process, which is used in particle filters, is not differentiable. Therefore, gradient computation cannot be continued after one iteration of the filtering loop; hence, the computed gradient ignores the effect of the previous state prediction and update steps on the current belief. This reduces the objective to predicting the Markov state at each time step, which can be fulfilled by supervised learning. Deploying supervised learning in the DPF algorithm facilitates prediction by taking advantage of the Markov property of the state space. However, supervised learning relies on the availability of the true state, $\mathbf{x}^*_k$, as the label, which is a challenge on its own. In real-world systems, it may be impossible to access the true state, and simulation would be the only valid choice for collecting labeled datasets for supervised learning. Therefore, filtering algorithms that deploy unsupervised learning would be more appealing, which will be covered in Section 9.8. The fact that the measurement process may not be Markovian calls for using backpropagation through multiple time steps in unsupervised learning. To address this issue, two approaches have been proposed for achieving differentiable resampling [190]:

- *Partial resampling*: Out of n_p particles, only $m_p < n_p$ particles are sampled at each time step, and the remaining $n_p - m_p$ particles are kept from the previous time step. In this way, gradient can flow backwards through the $n_p - m_p$ particles from the previous step.
- *Proxy gradients*: A proxy gradient is defined for the weight of a resampled particle, which is tied to the particle it was sampled from. In this way, gradient can flow through the connection between a particle and the particle it was sampled from.

A deep learning-based implementation of the *Rao-Blackwellized particle filter* (RBPF) is presented in Section 9.7.

9.7 Deep Rao–Blackwellized Particle Filter

A stochastic linear time-invariant system with Gaussian noise (linear Gaussian system) is described by the following state-space model based on the first-order Markov assumption:

$$\mathbf{x}_{k+1} = \mathbf{A}\mathbf{x}_k + \mathbf{B}\mathbf{u}_k + \mathbf{v}_k, \tag{9.43}$$

$$\mathbf{y}_k = \mathbf{C}\mathbf{x}_k + \mathbf{D}\mathbf{u}_k + \mathbf{w}_k, \tag{9.44}$$

$$\mathbf{v}_k \sim \mathcal{N}(\mathbf{0}, \mathbf{Q}_k), \tag{9.45}$$

$$\mathbf{w}_k \sim \mathcal{N}(\mathbf{0}, \mathbf{R}_k), \tag{9.46}$$

where $\mathbf{v}_k$ and $\mathbf{w}_k$ denote the process noise and the measurement noise, respectively. Inference and likelihood computation would be tractable for such a system. Alternatively, equations (9.43) and (9.44) can be represented as:

$$p(\mathbf{x}_{k+1}|\mathbf{x}_k, \mathbf{u}_k), \tag{9.47}$$

$$p(\mathbf{y}_k|\mathbf{x}_k, \mathbf{u}_k). \tag{9.48}$$

Using Kalman filter, inference and likelihood computation are performed by alternating between a state-prediction step:

$$p(\mathbf{x}_{k+1}|\mathbf{u}_{0:k}, \mathbf{y}_{0:k}) = \int p(\mathbf{x}_{k+1}|\mathbf{x}_k, \mathbf{u}_k)p(\mathbf{x}_k|\mathbf{u}_{0:k}, \mathbf{y}_{0:k})d\mathbf{x}_k, \tag{9.49}$$

and a state-update step based on the current measurement:

$$p(\mathbf{x}_{k+1}|\mathbf{u}_{0:k+1}, \mathbf{y}_{0:k+1}) \propto p(\mathbf{x}_{k+1}|\mathbf{u}_{0:k}, \mathbf{y}_{0:k})\, p(\mathbf{y}_{k+1}|\mathbf{x}_{k+1}, \mathbf{u}_{k+1}). \tag{9.50}$$

For nonlinear systems, the posterior distribution, $p(\mathbf{x}_k|\mathbf{u}_{0:k}, \mathbf{y}_{0:k})$, may be intractable, and therefore, must be approximated. Particle filter as a sequential Monte Carlo algorithm approximates the posterior distribution by a set of particles, $\mathbf{x}_k^i, i = 1, \dots, n_p$, with the corresponding weights w_k^i at each time step k. Particle filter deploys a combination of importance sampling and resampling. The filter distribution is approximated as:

$$p(\mathbf{x}_k|\mathbf{u}_{0:k}, \mathbf{y}_{0:k}) \approx \sum_{i=1}^{n_p} w_k^i \delta(\mathbf{x}_k - \mathbf{x}_k^i), \tag{9.51}$$

where $\delta(\mathbf{x}_k^i)$ denotes the Dirac delta function. The normalized weights w_k^i are computed using (9.31) and (9.32).

Switching linear Gaussian systems form a class of nonlinear state-space models in which model parameters vary with an additional latent variable $\mathbf{s}_k$ known as the switch variable. The switch variable allows the system to switch between different regimes. Such systems are also known as conditional or mixture conditional linear Gaussian systems, which are described by the following state-space model [191]:

$$\mathbf{x}_{k+1} = \mathbf{A}(\mathbf{s}_k)\mathbf{x}_k + \mathbf{B}(\mathbf{s}_k)\mathbf{u}_k + \mathbf{v}_k(\mathbf{s}_k), \tag{9.52}$$

$$\mathbf{y}_k = \mathbf{C}(\mathbf{s}_k)\mathbf{x}_k + \mathbf{D}(\mathbf{s}_k)\mathbf{u}_k + \mathbf{w}_k(\mathbf{s}_k), \tag{9.53}$$

$$\mathbf{v}_k(\mathbf{s}_k) \sim \mathcal{N}\left(\mathbf{0}, \mathbf{Q}(\mathbf{s}_k)\right), \tag{9.54}$$

$$\mathbf{w}_k(\mathbf{s}_k) \sim \mathcal{N}\left(\mathbf{0}, \mathbf{R}(\mathbf{s}_k)\right). \tag{9.55}$$

The switch variable $\mathbf{s}_k$ can be a categorical variable for indexing different linear state-space models. The probabilistic model is factored as:

$$\begin{aligned} p(\mathbf{x}_{0:k+1}, \mathbf{s}_{0:k+1}, \mathbf{y}_{0:k+1}|\mathbf{u}_{0:k+1}) &= p(\mathbf{x}_0)\, p(\mathbf{s}_0)\, p(\mathbf{y}_0|\mathbf{x}_0, \mathbf{s}_0, \mathbf{u}_0) \\ &\times \prod_{i=0}^{k} p(\mathbf{x}_{i+1}|\mathbf{x}_i, \mathbf{s}_i, \mathbf{u}_i)\, p(\mathbf{s}_{i+1}|\mathbf{s}_i, \mathbf{u}_i)\, p(\mathbf{y}_{i+1}|\mathbf{x}_{i+1}, \mathbf{s}_{i+1}, \mathbf{u}_{i+1}). \end{aligned} \tag{9.56}$$

Although the complete model is nonlinear due to switching, given the switch variable, the model will be conditionally linear for the corresponding sample trajectory. This paves the way for efficient approximate inference and likelihood estimation. The RBPF takes advantage of the conditional linearity to approximate the required expectations in filtering and likelihood computation [192]. The posterior distribution can be factorized as:

$$p(\mathbf{x}_{0:k+1}, \mathbf{s}_{0:k+1} | \mathbf{u}_{0:k+1}, \mathbf{y}_{0:k+1}) = p(\mathbf{x}_{0:k+1} | \mathbf{s}_{0:k+1}, \mathbf{u}_{0:k+1}, \mathbf{y}_{0:k+1}) \times p(\mathbf{s}_{0:k+1} | \mathbf{u}_{0:k+1}, \mathbf{y}_{0:k+1}). \tag{9.57}$$

Kalman filter can be used to compute $p(\mathbf{x}_{0:k+1} | \mathbf{s}_{0:k+1}, \mathbf{u}_{0:k+1}, \mathbf{y}_{0:k+1})$, where only the most recent state is required, $p(\mathbf{x}_{k+1} | \mathbf{s}_{0:k+1}, \mathbf{u}_{0:k+1}, \mathbf{y}_{0:k+1})$. Using Bayes' rule, $p(\mathbf{s}_{0:k+1} | \mathbf{u}_{0:k+1}, \mathbf{y}_{0:k+1})$ can be recursively computed as:

$$p(\mathbf{s}_{0:k+1} | \mathbf{u}_{0:k+1}, \mathbf{y}_{0:k+1}) \propto p(\mathbf{y}_{k+1} | \mathbf{s}_{0:k+1}, \mathbf{u}_{0:k+1}, \mathbf{y}_{0:k}) \times p(\mathbf{s}_{k+1} | \mathbf{s}_k, \mathbf{u}_k)\, p(\mathbf{s}_{0:k} | \mathbf{u}_{0:k}, \mathbf{y}_{0:k}), \tag{9.58}$$

where the predictive distribution $p(\mathbf{y}_{k+1} | \mathbf{s}_{0:k+1}, \mathbf{u}_{0:k+1}, \mathbf{y}_{0:k})$ is calculated as:

$$p(\mathbf{y}_{k+1} | \mathbf{s}_{0:k+1}, \mathbf{u}_{0:k+1}, \mathbf{y}_{0:k}) = \int p(\mathbf{y}_{k+1} | \mathbf{x}_{k+1}, \mathbf{s}_{k+1}, \mathbf{u}_{k+1})\, p(\mathbf{x}_{k+1} | \mathbf{s}_{0:k}, \mathbf{u}_{0:k}, \mathbf{y}_{0:k}) d\mathbf{x}_{k+1}. \tag{9.59}$$

Since all of the constituting terms of $p(\mathbf{s}_{0:k+1} | \mathbf{u}_{0:k+1}, \mathbf{y}_{0:k+1})$ can be computed, this distribution can be approximated by a set of particles using the sequential Monte Carlo method as in (9.51). The importance weights are updated as [191]:

$$\tilde{w}^i_{k+1} = \frac{p(\mathbf{y}_{k+1} | \mathbf{s}^i_{0:k+1}, \mathbf{u}_{0:k+1}, \mathbf{y}_{0:k})\, p(\mathbf{s}^i_{k+1} | \mathbf{s}^i_k, \mathbf{u}_k)}{q(\mathbf{s}^i_{k+1} | \mathbf{s}^i_{0:k}, \mathbf{u}_{0:k})} w^i_k, \tag{9.60}$$

$$w^i_{k+1} = \frac{\tilde{w}^i_{k+1}}{\sum_{i=1}^{n_p} \tilde{w}^i_{k+1}}. \tag{9.61}$$

The joint filter distribution for state and switch variables is approximated by a mixture of Gaussians [191]:

$$p(\mathbf{x}_k, \mathbf{s}_k | \mathbf{u}_{0:k}, \mathbf{y}_{0:k}) \approx \sum_{i=1}^{n_p} w^i_k \delta(\mathbf{s}_k - \mathbf{s}^i_k) \mathcal{N}\left(\mathbf{x}_k | \boldsymbol{\mu}_k(\mathbf{s}^i_{0:k}), \boldsymbol{\Sigma}_k(\mathbf{s}^i_{0:k})\right), \tag{9.62}$$

where mean $\boldsymbol{\mu}_k(\mathbf{s}^i_{0:k})$ and covariance $\boldsymbol{\Sigma}_k(\mathbf{s}^i_{0:k})$ of the filtered state $\mathbf{x}_k$ depend on the whole switch-variable trajectory $\mathbf{s}^i_{0:k}$. However, from an algorithmic perspective, in (9.60), only the current value of the switch variable, $\mathbf{s}^i_k$, is needed to update the importance weights.

Although a trade-off can be achieved between model complexity and efficient inference through using the switching linear Gaussian system, this model has two major limitations:

- Switching transitions are assumed to be independent of the states.
- Conditionally linear measurement models may not be suitable for multivariate observations with complex nonlinear dependencies such as video streams.

Independence of switching variables, $\mathbf{s}_{0:k}$, form the states, $\mathbf{x}_{0:k}$, leads to open-loop switch dynamics. However, state feedback allows for approximating a nonlinear system by a conditionally linear system via linearization at the current state, similar to the extended Kalman filter algorithm. To overcome the mentioned limitations of the switching linear Gaussian systems, it has been proposed to use the *auxiliary variable* recurrent switching linear Gaussian system as well as the proposal distribution, which is structurally similar to the minimum-variance proposal distribution [191]. Using Gaussian switch variables with a learnable Gaussian prior $p(\mathbf{s}_0)$ and conditionally linear state-to-switch transformations, we will have:

$$\mathbf{s}^i_{k+1} = \mathbf{F}(\mathbf{s}^i_k)\mathbf{x}^i_k + \mathbf{f}(\mathbf{s}^i_k) + \epsilon_k(\mathbf{s}^i_k), \tag{9.63}$$

$$\epsilon_k(\mathbf{s}^i_k) \sim \mathcal{N}\left(\mathbf{0}, \mathbf{S}(\mathbf{s}^i_k)\right), \tag{9.64}$$

where $\mathbf{F}(\mathbf{s}^i_k)$ is the transition matrix, $\mathbf{f}(\mathbf{s}^i_k)$ is a nonlinear function that can be represented by a neural network, and $\mathbf{S}(\mathbf{s}^i_k)$ is the covariance matrix predicted based on the switch variable at the previous time step. The states can be marginalized as:

$$\begin{aligned} p\left(\mathbf{s}_{k+1}|\mathbf{s}^i_{0:k}\right) = \mathcal{N}\big(\mathbf{s}_{k+1} \,|\, &\mathbf{F}(\mathbf{s}^i_k)\boldsymbol{\mu}_k(\mathbf{s}^i_{0:k}) + \mathbf{f}(\mathbf{s}^i_k), \\ &\mathbf{F}(\mathbf{s}^i_k)\boldsymbol{\Sigma}_k(\mathbf{s}^i_{0:k})\mathbf{F}^T(\mathbf{s}^i_k) + \mathbf{S}(\mathbf{s}^i_k)\big). \end{aligned} \tag{9.65}$$

Given Gaussians $\mathbf{s}^i_{k+1}$ and $\mathbf{s}^i_k$, neural networks can be used to predict parameters of the switching linear Gaussian system, $\{\mathbf{A}, \mathbf{B}, \mathbf{C}, \mathbf{D}, \mathbf{Q}, \mathbf{R}, \mathbf{F}, \mathbf{S}\}$, in two ways [191]:

- As direct outputs of a neural network.
- As a weighted average of a set of base matrices with weights being predicted by a neural network.

Raw data obtained from observations such as high-resolution images or point clouds would be high-dimensional and possibly non-Gaussian with complex dependencies on state variables. To model such data, instead of directly using them as measurements, $\mathbf{y}_k$, in (9.44) or (9.53), $\mathbf{y}_k$ is considered as a low-dimensional representation of the raw-data observation, $\mathbf{o}_k$. In this way, as a set of Gaussian latent variables, $\mathbf{y}_k$ decouples the observations, $\mathbf{o}_k$, from the switching linear Gaussian system through the following conditional-independence relation [191]:

$$p(\mathbf{x}_{0:k}|\mathbf{s}_{0:k}, \mathbf{u}_{0:k}, \mathbf{y}_{0:k}, \mathbf{o}_{0:k}) = p(\mathbf{x}_{0:k}|\mathbf{s}_{0:k}, \mathbf{u}_{0:k}, \mathbf{y}_{0:k}). \tag{9.66}$$

In this formulation, $\mathbf{y}_k$ plays the role of an *auxiliary variable*, and its samples can be viewed as pseudo-measurements (pseudo-observations) in the switching linear

Gaussian system. The corresponding joint distribution is computed as:

$$p(\mathbf{x}_{0:k+1}, \mathbf{s}_{0:k+1}, \mathbf{y}_{0:k+1}, \mathbf{o}_{0:k+1}|\mathbf{u}_{0:k+1}) = p(\mathbf{o}_0|\mathbf{y}_0)p(\mathbf{y}_0|\mathbf{x}_0, \mathbf{s}_0, \mathbf{u}_0)p(\mathbf{x}_0)p(\mathbf{s}_0)$$
$$\times \prod_{i=0}^{k} p(\mathbf{o}_{i+1}|\mathbf{y}_{i+1})p(\mathbf{y}_{i+1}|\mathbf{x}_{i+1}, \mathbf{s}_{i+1}, \mathbf{u}_{i+1})p(\mathbf{x}_{i+1}|\mathbf{x}_i, \mathbf{s}_i, \mathbf{u}_i)p(\mathbf{s}_{i+1}|\mathbf{x}_i, \mathbf{s}_i, \mathbf{u}_i). \tag{9.67}$$

Parameters of the conditional distribution $p(\mathbf{o}_k|\mathbf{y}_k)$ are learned by a neural network. This measurement model, which is based on introducing the auxiliary variables, is similar to the additional latent variable in the *Kalman variational autoencoder* (KVAE) that will be discussed later.

The RBPF can be extended to infer all the latent variables, $\{\mathbf{x}, \mathbf{s}, \mathbf{y}\}$. In the *deep Rao–Blackwellized particle filter* (DRBPF) algorithm, neural networks can be trained to learn model parameters including the base matrices of the recurrent switching linear Gaussian system as well as the parameters of the state prior, the auxiliary variable encoder, the switch prior, and the switch-transition distribution. Similar to (9.57), the posterior can be factored as [191]:

$$p(\mathbf{x}_{0:k+1}, \mathbf{s}_{0:k+1}, \mathbf{y}_{0:k+1}|\mathbf{u}_{0:k+1}, \mathbf{o}_{0:k+1})$$
$$= p(\mathbf{x}_{0:k+1}|\mathbf{s}_{0:k+1}, \mathbf{u}_{0:k+1}, \mathbf{y}_{0:k+1})\; p(\mathbf{s}_{0:k+1}, \mathbf{y}_{0:k+1}|\mathbf{u}_{0:k+1}, \mathbf{o}_{0:k+1}). \tag{9.68}$$

While the first term in (9.68) can be computed in a straightforward manner, the second term is approximated using the sequential Monte Carlo method. The importance-weights are updated as:

$$\tilde{w}^i_{k+1} = \frac{p(\mathbf{o}_{k+1}|\mathbf{y}^i_{k+1})p(\mathbf{y}^i_{k+1}|\mathbf{s}^i_{0:k+1}, \mathbf{u}_{0:k+1}, \mathbf{y}^i_{0:k})p(\mathbf{s}^i_{k+1}|\mathbf{s}^i_{0:k}, \mathbf{u}_{0:k}, \mathbf{y}^i_{0:k})}{q(\mathbf{s}^i_{k+1}, \mathbf{y}^i_{k+1}|\mathbf{s}^i_{0:k}, \mathbf{u}_{0:k+1}, \mathbf{y}^i_{0:k})} w^i_k, \tag{9.69}$$

$$w^i_{k+1} = \frac{\tilde{w}^i_{k+1}}{\sum_{i=1}^{n_p} \tilde{w}^i_{k+1}}, \tag{9.70}$$

where the numerator of the right-hand side of equation (9.69) is equivalent to the conditional distribution $p(\mathbf{s}_{k+1}, \mathbf{y}_{k+1}, \mathbf{o}_{k+1}|\mathbf{s}_{0:k}, \mathbf{u}_{0:k+1}, \mathbf{y}_{0:k}, \mathbf{o}_{0:k})$. The last term in the numerator, $p(\mathbf{s}^i_{k+1}|\mathbf{s}^i_{0:k}, \mathbf{u}_{0:k}, \mathbf{y}^i_{0:k})$, shows that according to the state-to-switch dynamics, the switch transition is non-Markovian. This is due to marginalizing the filtered state:

$$p(\mathbf{s}_{k+1}|\mathbf{s}_{0:k}, \mathbf{u}_{0:k}, \mathbf{y}_{0:k}) = \int p(\mathbf{s}_{k+1}|\mathbf{s}_k, \mathbf{x}_k)\; p(\mathbf{x}_k|\mathbf{s}_{0:k}, \mathbf{u}_{0:k}, \mathbf{y}_{0:k})d\mathbf{x}_k, \tag{9.71}$$

which depends on the history of switch variables and pseudo-observations similar to equation (9.65). The second term in the numerator of the right-hand side of equation (9.69), $p(\mathbf{y}^i_{k+1}|\mathbf{s}^i_{0:k+1}, \mathbf{u}_{0:k+1}, \mathbf{y}^i_{0:k})$, which is the conditional distribution of the auxiliary variable, is computed as:

$$p(\mathbf{y}_{k+1}|\mathbf{s}^i_{0:k+1}, \mathbf{u}_{0:k+1}, \mathbf{y}^i_{0:k}) = \int p(\mathbf{y}_{k+1}|\mathbf{x}_{k+1}, \mathbf{s}^i_{k+1}, \mathbf{u}_{k+1})p(\mathbf{x}_{k+1}|\mathbf{s}^i_{0:k}, \mathbf{u}_{0:k}, \mathbf{y}^i_{0:k})\mathrm{d}\mathbf{x}_{k+1}, \tag{9.72}$$

where

$$p(\mathbf{x}_{k+1}|\mathbf{s}^i_{0:k}, \mathbf{u}_{0:k}, \mathbf{y}^i_{0:k}) = \int p(\mathbf{x}_{k+1}|\mathbf{x}_k, \mathbf{s}^i_k, \mathbf{u}_k)p(\mathbf{x}_k|\mathbf{s}^i_{0:k}, \mathbf{u}_{0:k}, \mathbf{y}^i_{0:k})\mathrm{d}\mathbf{x}_k. \tag{9.73}$$

Maximum likelihood estimation and stochastic gradient descent are used to learn the model parameters. In the sequential Monte Carlo method, an unbiased estimation of the marginal likelihood can be computed based on the unnormalized importance weights as [193]:

$$\hat{p}_\theta(\mathbf{y}_{0:k}) = \prod_{i=0}^{k}\sum_{j=1}^{n_p}\tilde{w}^j_i(\theta). \tag{9.74}$$

According to the Jensen's inequality, we have:

$$\mathbb{E}\left[\log \hat{p}_\theta(\mathbf{y}_{0:k})\right] \leq \log p_\theta(\mathbf{y}_{0:k}), \tag{9.75}$$

which provides a tractable lower bound for the log marginal likelihood as:

$$\mathcal{L}(\mathbf{y}_{0:k}, \theta) = \sum_{i=0}^{k} \log \hat{p}_\theta(\mathbf{y}_i|\mathbf{y}_{0:i-1}) \tag{9.76}$$

$$= \sum_{i=0}^{k} \log\left(\sum_{j=1}^{n_p}\tilde{w}^j_i(\theta)\right)$$

$$\leq \log p_\theta(\mathbf{y}_{0:k}). \tag{9.77}$$

This lower bound can be optimized using the stochastic gradient descent method, where the importance weights are updated according to (9.69).

The proposal distribution plays a key role in achieving reliable estimates with low variance. The minimum-variance proposal distribution is proportional to the conditional distribution $p(\mathbf{s}_{k+1}, \mathbf{y}_{k+1}, \mathbf{o}_{k+1}|\mathbf{s}_{0:k}, \mathbf{u}_{0:k+1}, \mathbf{y}_{0:k})$, which is the numerator of the right-hand side of equation (9.69). Choosing a proposal distribution with a similar structure, we obtain [191]:

$$q(\mathbf{s}_{k+1}, \mathbf{y}_{k+1}|\mathbf{s}_{0:k}, \mathbf{u}_{0:k+1}, \mathbf{y}_{0:k}, \mathbf{o}_{k+1}) \propto q(\mathbf{s}_{k+1}, \mathbf{y}_{k+1}|\mathbf{o}_{k+1})q(\mathbf{s}_{k+1}, \mathbf{y}_{k+1}|\mathbf{s}_{0:k}, \mathbf{u}_{0:k+1}, \mathbf{y}_{0:k})$$

$$= q(\mathbf{s}_{k+1}, \mathbf{y}_{k+1} | \mathbf{o}_{k+1}) q(\mathbf{s}_{k+1} | \mathbf{s}_{0:k}, \mathbf{u}_{0:k}, \mathbf{y}_{0:k}) q(\mathbf{y}_{k+1} | \mathbf{s}_{0:k+1}, \mathbf{u}_{0:k+1}, \mathbf{y}_{0:k}), \tag{9.78}$$

where in the last line, among the three probability distributions, the first one represents a Gaussian approximation of the likelihood, which can be predicted by a neural network-based encoder, the second one represents the transition probability of the switch variables, and the last one represents predictive distribution of the auxiliary variable in the generative model. Combination of these Gaussians as a *product of experts* provides a Gaussian proposal distribution. While the switch-transition probability is available from the previous time step k, the predictive distribution for the auxiliary variable can be computed after sampling $\mathbf{s}_{k+1}$. The encoder is structured in a way that dependencies have the same direction as the generative model shown in (9.78). Hence, the designed encoder will have a structure similar to the encoder in a *ladder variational autoencoder* [194]. Combination of such an encoder and the forward model allows for factorizing the proposal distribution as:

$$q(\mathbf{s}_{k+1}, \mathbf{y}_{k+1} | \mathbf{s}_{0:k}, \mathbf{u}_{0:k+1}, \mathbf{y}_{0:k}, \mathbf{o}_{k+1})$$
$$= q(\mathbf{s}_{k+1} | \mathbf{s}_{0:k}, \mathbf{u}_{0:k}, \mathbf{y}_{0:k}, \mathbf{o}_{k+1}) q(\mathbf{y}_{k+1} | \mathbf{s}_{0:k+1}, \mathbf{u}_{0:k+1}, \mathbf{y}_{0:k}, \mathbf{o}_{k+1}). \tag{9.79}$$

Reusing the predicted variables, samples are drawn in the same direction as the generative process. In case that the Gaussian encoder distribution is proportional to the likelihood, the proposal distribution will be optimal. Resampling is performed based on the effective sample-size criterion [191].

9.8 Deep Variational Bayes Filter

Deep variational Bayes filters (DVBF) were proposed for unsupervised learning of latent state-space models [2]. Let us assume a generative model with an underlying latent dynamic system, which has the following probabilistic models for state transition and measurement:

$$p(\mathbf{x}_{0:k+1} | \mathbf{x}_{0:k}, \mathbf{u}_{0:k}), \tag{9.80}$$

$$p(\mathbf{y}_{0:k} | \mathbf{x}_{0:k}, \mathbf{u}_{0:k}). \tag{9.81}$$

Both of these models must be learned by performing latent system identification. An end-to-end external description of the system from its input to its output is obtained as follows:

$$p(\mathbf{y}_{0:k+1} | \mathbf{u}_{0:k+1}) = \int p(\mathbf{y}_{0:k+1} | \mathbf{x}_{0:k+1}, \mathbf{u}_{0:k+1}) p(\mathbf{x}_{0:k+1} | \mathbf{x}_{0:k}, \mathbf{u}_{0:k}) d\mathbf{x}_{0:k}. \tag{9.82}$$

In order to perform prediction, filtering, or smoothing, the posterior distribution $p(\mathbf{x}_{0:k+1}|\mathbf{u}_{0:k+1}, \mathbf{y}_{0:k+1})$ must be estimated by efficient inference methods.

Generally, the state-transition model may be time-varying. To take account of this issue, a regularizing prior distribution can be imposed on a set of transition parameters $\xi_{0:k}$ such that:

$$p(\mathbf{y}_{0:k+1}|\mathbf{u}_{0:k+1}) = \int\int p(\mathbf{y}_{0:k+1}|\mathbf{x}_{0:k+1}, \mathbf{u}_{0:k+1})p(\mathbf{x}_{0:k+1}|\mathbf{x}_{0:k}, \mathbf{u}_{0:k}, \xi_{0:k})p(\xi_{0:k})\mathrm{d}\xi_{0:k}\,\mathrm{d}\mathbf{x}_{0:k}. \tag{9.83}$$

The corresponding state-space model is then represented by:

$$p(\mathbf{x}_{0:k+1}|\mathbf{x}_{0:k}, \mathbf{u}_{0:k}, \xi_{0:k}) = p(\mathbf{x}_0)\prod_{i=0}^{k} p(\mathbf{x}_{i+1}|\mathbf{x}_i, \mathbf{u}_i, \xi_i), \tag{9.84}$$

$$p(\mathbf{y}_{0:k}|\mathbf{x}_{0:k}, \mathbf{u}_{0:k}) = \prod_{i=0}^{k} p(\mathbf{y}_i|\mathbf{x}_i, \mathbf{u}_i). \tag{9.85}$$

As a special case, let us consider the following linear time-varying system with Gaussian process noise $\mathbf{v}_k$ and Gaussian measurement noise $\mathbf{w}_k$:

$$\mathbf{x}_{k+1} = \mathbf{A}_k\mathbf{x}_k + \mathbf{B}_k\mathbf{u}_k + \mathbf{v}_k, \tag{9.86}$$

$$\mathbf{y}_k = \mathbf{C}_k\mathbf{x}_k + \mathbf{D}_k\mathbf{u}_k + \mathbf{w}_k, \tag{9.87}$$

where $\mathbf{v}_k \sim \mathcal{N}(\mathbf{0}, \mathbf{Q}_k)$ and $\mathbf{w}_k \sim \mathcal{N}(\mathbf{0}, \mathbf{R}_k)$. In this setting, transition parameters are defined as $\xi_k = (\mathbf{v}_k, \nu_k)$, where ν_k denotes the subset of the transition parameters that are related to the $(\mathbf{A}_k, \mathbf{B}_k)$ pair. For linear systems with Gaussian noise, the Kalman filter provides the optimal solution for the inference problem. Although extensions of the Kalman filter such as extended Kalman filter (EKF), unscented Kalman filter (UKF), and cubature Kalman filter (CKF) have been successfully applied to different nonlinear systems, their domain of applicability is rather limited due to the restrictive assumptions on the system model, which may be violated in practice. Moreover, performance of such algorithms heavily relies on the accuracy of the corresponding state-space model. Therefore, statistical characteristics of the process noise and the measurement noise as well as the model parameters such as $(\mathbf{A}_k, \mathbf{B}_k, \mathbf{C}_k, \mathbf{D}_k)$ must be well-known. To address this modeling issue, it has been proposed to learn the system dynamics using neural network-based implementations of the *expectation maximization* (EM) algorithm [195]. However, the domain of applicability of the EM algorithm is restricted to problems with tractable true posterior distributions. Hence, in [2], it has been suggested to use machine learning for jointly performing identification and inference.

If in a deterministic autoencoder, the code layer is replaced with stochastic units, $\mathbf{x}$, then the resulting VAE will be capable of learning complex marginal distributions on $\mathbf{y}$ from simpler distributions in an unsupervised manner [181]:

$$p(\mathbf{y}|\mathbf{u}) = \int p(\mathbf{x}, \mathbf{y}|\mathbf{u})d\mathbf{x} = \int p(\mathbf{y}|\mathbf{x}, \mathbf{u})p(\mathbf{x}|\mathbf{u})d\mathbf{x}. \tag{9.88}$$

A neural network with adjustable parameters $\boldsymbol{\theta}$ can be used to parameterize the conditional probability $p(\mathbf{y}|\mathbf{x}, \mathbf{u})$ as $p_{\boldsymbol{\theta}}(\mathbf{y}|\mathbf{x}, \mathbf{u})$. Then, the SGVB method can be used to jointly perform identification and inference. The system model is trained by maximizing the following objective function [2]:

$$\mathcal{L}(\boldsymbol{\phi}, \boldsymbol{\theta}) = \mathbb{E}_{q_{\boldsymbol{\phi}}}\left[\log p_{\boldsymbol{\theta}}(\mathbf{y}|\mathbf{x}, \mathbf{u})\right] - \text{KLD}\left(q_{\boldsymbol{\phi}}(\mathbf{x}|\mathbf{u}, \mathbf{y})||p(\mathbf{x}|\mathbf{u})\right), \tag{9.89}$$

where $q_{\boldsymbol{\phi}}(\mathbf{x}|\mathbf{u}, \mathbf{y})$ is the *recognition model* or approximate posterior, which is parameterized using a neural network with adjustable parameters $\boldsymbol{\phi}$. This objective function is a lower bound for the marginal data log likelihood:

$$\mathcal{L}(\boldsymbol{\phi}, \boldsymbol{\theta}) \leq \log p(\mathbf{y}|\mathbf{u}). \tag{9.90}$$

Maximizing the objective function $\mathcal{L}(\boldsymbol{\phi}, \boldsymbol{\theta})$ is equivalent to minimizing the KLD between the approximate posterior $q_{\boldsymbol{\phi}}(\mathbf{x}|\mathbf{u}, \mathbf{y})$ and the true posterior $p(\mathbf{x}|\mathbf{u}, \mathbf{y})$, which may be intractable.

An efficient filtering algorithm must take account of two key problems: reconstruction and prediction. Learning a state-space model calls for inferring a proper latent space that satisfies the corresponding model assumptions. The idea behind DVBF is to satisfy the state-space model assumptions while preserving full information in the latent states through forcing the latent space to fit the corresponding state transitions. Focusing on transitions allows DVBF to avoid getting stuck in local optima with only good reconstruction properties. Such local optima are usually learned by stationary autoencoders, which aim at extracting as much information as possible from a single measurement without considering the entire sequence.

To learn the transitions, they must be the driving force for shaping the latent space. Hence, instead of adjusting the transition to the recognition model's latent space, gradient paths are established through transitions over time. In the DVBF algorithm, the state transition

$$\mathbf{x}_{k+1} = \mathbf{f}(\mathbf{x}_k, \mathbf{u}_k, \boldsymbol{\xi}_k) \tag{9.91}$$

is made differentiable with respect to both the last sate, $\mathbf{x}_k$, and the corresponding parameters, $\boldsymbol{\xi}_k$. Then, the recognition model, $q_{\boldsymbol{\phi}}(\mathbf{x}_{0:k+1}|\mathbf{u}_{0:k+1}, \mathbf{y}_{0:k+1})$, is used to infer the transition parameters, $\boldsymbol{\xi}_k$, instead of the latent state, $\mathbf{x}_k$. In other words, the recognition model is prevented from directly drawing samples of the latent state $\mathbf{x}_k$. According to (9.91), the gradient $\frac{\partial \mathbf{x}_{k+1}}{\partial \mathbf{x}_k}$ is well-defined, and the gradient information can be backpropagated through the transition. Regarding the stochastic nature of the set of parameters, ξ_k, transitions will be stochastic, and errors in reconstruction of $\mathbf{y}_k$ from $\mathbf{x}_k$ and $\mathbf{u}_k$ are backpropagated through time as well.

As a corrective offset term, the process noise $\mathbf{v}_k$ in (9.86) highlights the role of the recognition model as a filter. The set of stochastic parameters, ξ_k, which includes the process noise, $\mathbf{v}_k$, allows for regularizing the transition with meaningful priors. In addition to preventing overfitting, this kind of regularization allows for enforcing meaningful manifolds in the latent space via *transition priors*. To be more precise, in case that the model tends to ignore the transitions over time, it will face large penalties from these priors. In (9.86), the process noise vector $\mathbf{v}_k$ is viewed as a set of sample-specific parameters, while ν_k is treated as a set of universal sample-independent parameters, which are inferred from data during the training phase. To form proper transition priors, the stochastic parameters can be considered as $\xi_k = (\mathbf{v}_k, \nu_k)$, which is inspired by the idea of weight uncertainty in the sense described in [196]. Then, the recognition model can be factorized as:

$$q_\phi(\xi_{0:k}|\mathbf{y}_{0:k}) = q_\phi(\mathbf{v}_{0:k}|\mathbf{y}_{0:k})q_\phi(\nu_{0:k}). \tag{9.92}$$

In the absence of input, while $\mathbf{v}_{0:k}$ is drawn from the prior, the universal transition parameters ν_k are drawn from $q_\phi(\nu_{0:k})$, using the trained model for generative sampling.

Similar to (9.90), a lower bound can be derived for the marginal data likelihood $p(\mathbf{y}_{0:k}|\mathbf{u}_{0:k})$. Substituting (9.84) and (9.85) in (9.83), we obtain:

$$\begin{aligned} &p(\mathbf{y}_{0:k+1}|\mathbf{u}_{0:k+1}) \\ &= \int\int p(\xi_{0:k})\prod_{i=-1}^{k} p_\theta(\mathbf{y}_{i+1}|\mathbf{x}_{i+1}, \mathbf{u}_{i+1})\prod_{i=0}^{k} p(\mathbf{x}_{i+1}|\mathbf{x}_i, \mathbf{u}_i, \xi_i)\mathrm{d}\xi_{0:k}\, \mathrm{d}\mathbf{x}_{0:k}, \end{aligned} \tag{9.93}$$

which is simplified as:

$$\begin{aligned} p(\mathbf{y}_{0:k+1}|\mathbf{u}_{0:k+1}) &= \int p(\xi_{0:k})\prod_{i=-1}^{k} p_\theta(\mathbf{y}_{i+1}|\mathbf{x}_{i+1}, \mathbf{u}_{i+1})|_{\mathbf{x}_{i+1}=\mathbf{f}(\mathbf{x}_i,\mathbf{u}_i,\xi_i)}\mathrm{d}\xi_{0:k} \\ &= \int p(\xi_{0:k})p_\theta(\mathbf{y}_{0:k+1}|\mathbf{x}_{0:k+1}, \mathbf{u}_{0:k+1})\mathrm{d}\xi_{0:k}, \end{aligned} \tag{9.94}$$

where $p_\theta(\mathbf{y}_{0:k+1}|\mathbf{x}_{0:k+1}, \mathbf{u}_{0:k+1})$ depends on both $\mathbf{u}_{0:k+1}$ and $\xi_{0:k}$. A lower bound on the data likelihood (9.94) is calculated as follows:

$$\begin{aligned} &\log p(\mathbf{y}_{0:k+1}|\mathbf{u}_{0:k+1}) \\ &= \log \int p(\xi_{0:k})p_\theta(\mathbf{y}_{0:k+1}|\mathbf{x}_{0:k+1}, \mathbf{u}_{0:k+1})\mathrm{d}\xi_{0:k} \end{aligned}$$

$$
\begin{aligned}
&= \log \int p(\xi_{0:k}) p_\theta(\mathbf{y}_{0:k+1}|\mathbf{x}_{0:k+1}, \mathbf{u}_{0:k+1}) \frac{q_\phi(\xi_{0:k}|\mathbf{y}_{0:k+1}, \mathbf{u}_{0:k+1})}{q_\phi(\xi_{0:k}|\mathbf{y}_{0:k+1}, \mathbf{u}_{0:k+1})} d\xi_{0:k} \\
&\geq \int q_\phi(\xi_{0:k}|\mathbf{y}_{0:k+1}, \mathbf{u}_{0:k+1}) \log \left(\frac{p(\xi_{0:k}) p_\theta(\mathbf{y}_{0:k+1}|\mathbf{x}_{0:k+1}, \mathbf{u}_{0:k+1})}{q_\phi(\xi_{0:k}|\mathbf{y}_{0:k+1}, \mathbf{u}_{0:k+1})} \right) d\xi_{0:k} \\
&= \mathbb{E}_{q_\phi} \left[\log p_\theta(\mathbf{y}_{0:k+1}|\mathbf{x}_{0:k+1}, \mathbf{u}_{0:k+1}) \right] \\
&\quad - \mathbb{E}_{q_\phi} \left[\log q_\phi(\xi_{0:k}|\mathbf{y}_{0:k+1}, \mathbf{u}_{0:k+1}) - \log p(\xi_{0:k}) \right] \\
&= \mathbb{E}_{q_\phi} \left[\log p_\theta(\mathbf{y}_{0:k+1}|\mathbf{x}_{0:k+1}, \mathbf{u}_{0:k+1}) \right] \\
&\quad - \mathrm{KLD} \left(q_\phi(\xi_{0:k}|\mathbf{y}_{0:k+1}, \mathbf{u}_{0:k+1}) || p(\xi_{0:k}) \right) \\
&= \mathcal{L}(\boldsymbol{\phi}, \boldsymbol{\theta}).
\end{aligned} \tag{9.95}
$$

This lower bound is used as the objective function. In practice, the overall performance can be improved by using an annealed version of

$$
\mathbb{E}_{q_\phi} \left[\log p_\theta(\mathbf{y}_{0:k+1}|\mathbf{x}_{0:k+1}, \mathbf{u}_{0:k+1}) - \log q_\phi(\xi_{0:k}|\mathbf{y}_{0:k+1}, \mathbf{u}_{0:k+1}) + \log p(\xi_{0:k}) \right]
$$

in (9.95), which smoothens the error landscape [2, 138]. Such an annealed version of the objective function is computed as:

$$
\begin{aligned}
&\mathbb{E}_{q_\phi} \left[c_i \log p_\theta(\mathbf{y}_{0:k+1}|\mathbf{x}_{0:k+1}, \mathbf{u}_{0:k+1}) - \log q_\phi(\xi_{0:k}|\mathbf{y}_{0:k+1}, \mathbf{u}_{0:k+1}) \right. \\
&\qquad \left. + c_i \log p(\mathbf{v}_{0:k}) + \log p(\boldsymbol{\nu}_{0:k}) \right],
\end{aligned} \tag{9.96}
$$

where $c_i = \min(1, 0.01 + i/T_A)$ has the role of an inverse temperature, which increases with the number of gradient updates, i, until it reaches the value 1 after T_A annealing iterations. The transition prior $p(\boldsymbol{\nu}_{0:k})$ can be estimated during optimization using the empirical Bayes approach [2]. Algorithm 9.2 presents the general computational architecture of the DVBF, where the stochastic transition parameters ξ_k are inferred using the recognition model. Then, the latent state is updated based on the sampled ξ_k. Finally, the updated latent state $\mathbf{x}_{k+1}$ is used to predict the output $\mathbf{y}_{k+1}$.

For locally linear state transitions such as

$$
\mathbf{x}_{k+1} = \mathbf{A}_k \mathbf{x}_k + \mathbf{B}_k \mathbf{u}_k + \mathbf{G}_k \mathbf{v}_k, \tag{9.97}
$$

the process noise $\mathbf{v}_k$ is a stochastic sample from the recognition model, and the triplet of the time-varying matrices $(\mathbf{A}_k, \mathbf{B}_k, \mathbf{G}_k)$ is a stochastic function of $\mathbf{x}_k$ and $\mathbf{u}_k$. The subset of the transition parameters related to this triplet,

$$
\boldsymbol{\nu}_k = \left\{ \left(\mathbf{A}_k^i, \mathbf{B}_k^i, \mathbf{G}_k^i \right) | i = 1, \dots, N \right\}, \tag{9.98}
$$

is drawn from $q_\phi(\boldsymbol{\nu}_k)$. Using the Bayesian paradigm, each one of these N triplets of matrices $\left(\mathbf{A}_k^i, \mathbf{B}_k^i, \mathbf{G}_k^i\right)$ can be learned as point estimates. Then, the matrices

---

Algorithm 9.2: Deep variational Bayes filter

State-space model:

$$p(\mathbf{x}_{k+1}|\mathbf{x}_k, \mathbf{u}_k, \xi_k)$$
$$p(\mathbf{y}_k|\mathbf{x}_k, \mathbf{u}_k)$$

Initialization:

$$\mathbf{x}_0$$
$$q_\phi(\mathbf{v}_0|\cdot)$$
$$q_\phi(\nu_0)$$

for $k = 0, 1, \ldots,$ **do**

Transition parameters inference

$$\xi_k \sim q_\phi(\xi_k) = q_\phi(\mathbf{v}_k|\cdot)q_\phi(\nu_k)$$

State estimate update

$$\mathbf{x}_{k+1} = \mathbf{f}(\mathbf{x}_k, \mathbf{u}_k, \xi_k)$$

Output prediction

$$p_\theta(\mathbf{y}_{k+1}|\mathbf{x}_{k+1}, \mathbf{u}_{k+1})$$

end

$(\mathbf{A}_k, \mathbf{B}_k, \mathbf{G}_k)$ are calculated as the linear combinations of the corresponding point estimates:

$$\mathbf{A}_k = \sum_{i=1}^{N} \alpha_k^i \mathbf{A}_k^i, \tag{9.99}$$

$$\mathbf{B}_k = \sum_{i=1}^{N} \alpha_k^i \mathbf{B}_k^i, \tag{9.100}$$

$$\mathbf{G}_k = \sum_{i=1}^{N} \alpha_k^i \mathbf{G}_k^i. \tag{9.101}$$

The weight vector $\boldsymbol{\alpha}_k \in \mathbb{R}^N$, which is shared between the three matrices, can be obtained from a neural network with parameters $\boldsymbol{\psi}$:

$$\boldsymbol{\alpha}_k = f_\psi(\mathbf{x}_k, \mathbf{u}_k). \tag{9.102}$$

As part of the trainable set of model parameters, $\boldsymbol{\psi}$ can be viewed as a subset of the generative parameters $\boldsymbol{\theta}$. Algorithm 9.3 presents the DVBF with the state estimate update procedure, which uses a time-varying locally linear transition model in the latent state space.

Algorithm 9.3: Deep variational Bayes filter with locally linear transition in the latent state space

---

State-space model:

$$p(\mathbf{x}_{k+1}|\mathbf{x}_k, \mathbf{u}_k, \xi_k)$$
$$p(\mathbf{y}_k|\mathbf{x}_k, \mathbf{u}_k)$$

Initialization:

$$\mathbf{x}_0$$
$$q_\phi(\mathbf{v}_0|.)$$
$$q_\phi(\nu_0)$$

for $k = 0, 1, \ldots,$ **do**

Transition parameters inference

$$\xi_k \sim q_\phi(\xi_k) = q_\phi(\mathbf{v}_k|.)q_\phi(\nu_k)$$

State estimate update

$$\alpha_k = f_\psi(\mathbf{x}_k, \mathbf{u}_k)$$
$$\left(\mathbf{A}_k, \mathbf{B}_k, \mathbf{G}_k\right) = \sum_{i=1}^{N} \alpha_k^i \left(\mathbf{A}_k^i, \mathbf{B}_k^i, \mathbf{G}_k^i\right)$$
$$\mathbf{x}_{k+1} = \mathbf{A}_k\mathbf{x}_k + \mathbf{B}_k\mathbf{u}_k + \mathbf{G}_k\mathbf{v}_k$$

Output prediction

$$p_\theta(\mathbf{y}_{k+1}|\mathbf{x}_{k+1})$$

end

The DVBF is capable of learning a state-space model in an unsupervised manner based on sequential observations from the environment. Although DVBF performs quite well on prediction tasks, its performance degrades for high-dimensional observations due to the relative difference between the dimensions of the observation space n_y and the latent state space n_x. To improve the scalability of the algorithm, this relative difference must be taken into account [197]. Observations $\mathbf{y}_k$ that include high-resolution images will be elements of a high-dimensional output space. Regarding a Markovian framework, such images cannot capture the exact system state at each time instant.

The observation distribution is governed by the following probabilistic measurement equation:

$$p_{\theta}(\mathbf{y}_{0:k+1}|\mathbf{u}_{0:k+1}) = \int p_{\theta_0}(\mathbf{x}_0, \mathbf{y}_0)\prod_{i=0}^{k} p_{\theta_m}(\mathbf{y}_{k+1}|\mathbf{x}_{k+1}, \mathbf{u}_{k+1})p_{\theta_s}(\mathbf{x}_{k+1}|\mathbf{x}_k, \mathbf{u}_k)d\mathbf{x}_{0:k}, \tag{9.103}$$

where $\boldsymbol{\theta} = \{\boldsymbol{\theta}_0, \boldsymbol{\theta}_m, \boldsymbol{\theta}_s\}$ represents the set of parameters of the initial, the measurement, and the state-transition distributions, respectively. The filtering algorithm has to perform two tasks [197]:

(i) Learning the state-space model, which includes both the state-transition model $p(\mathbf{x}_{k+1}|\mathbf{x}_k, \mathbf{u}_k)$ and the measurement model $p(\mathbf{y}_k|\mathbf{x}_k, \mathbf{u}_k)$.
(ii) Estimating the current system state given the sequence of control inputs and measurements $p(\mathbf{x}_k|\mathbf{u}_{0:k}, \mathbf{y}_{0:k})$.

The difference between dimensions n_x and n_y would be relatively large, when observations include high-resolution images. For effective compression, the observation space $\mathcal{Y} \in \mathbb{R}^{n_y}$ is projected to a lower-dimensional manifold $\mathcal{X} \in \mathbb{R}^{n_x}$. A probabilistic approximation of this projection is learned by maximizing the *ELBO* on the data log likelihood, $\mathcal{L}(\cdot)$. The learned projection must provide both the posterior distribution $p(\mathbf{x}|\mathbf{u}, \mathbf{y})$ and the generative distribution $p(\mathbf{y}|\mathbf{x}, \mathbf{u})$ for the data. The ELBO is calculated as:

$$\begin{aligned}
&\mathcal{L}(\boldsymbol{\phi}, \boldsymbol{\theta}) \\
&= \int \Bigg[q_{\boldsymbol{\phi}}(\mathbf{x}_{0:k+1}|\mathbf{u}_{0:k+1}, \mathbf{y}_{0:k+1}) \\
&\quad \log \frac{p_{\boldsymbol{\theta}}(\mathbf{y}_{0:k+1}|\mathbf{x}_{0:k+1}, \mathbf{u}_{0:k+1})p_{\boldsymbol{\theta}}(\mathbf{x}_{0:k+1}|\mathbf{x}_{0:k}, \mathbf{u}_{0:k})}{q_{\boldsymbol{\phi}}(\mathbf{x}_{0:k+1}|\mathbf{u}_{0:k+1}, \mathbf{y}_{0:k+1})}\Bigg] d\mathbf{x}_{0:k} \\
&= \underbrace{\mathbb{E}_{q_{\boldsymbol{\phi}}}\left[\log p_{\boldsymbol{\theta}}(\mathbf{y}_{0:k+1}|\mathbf{x}_{0:k+1}, \mathbf{u}_{0:k+1})\right]}_{\text{reconstruction error}} \\
&\quad - \mathrm{KLD}\left(\underbrace{q_{\boldsymbol{\phi}}(\mathbf{x}_{0:k+1}|\mathbf{u}_{0:k+1}, \mathbf{y}_{0:k+1})}_{\text{recognition model}}||\underbrace{p_{\boldsymbol{\theta}}(\mathbf{x}_{0:k+1}|\mathbf{x}_{0:k}, \mathbf{u}_{0:k})}_{\text{includes the prior}}\right) \\
&\leq \log p(\mathbf{y}_{0:k+1}|\mathbf{u}_{0:k+1}).
\end{aligned} \tag{9.104}$$

As mentioned previously, $\boldsymbol{\theta} = \{\boldsymbol{\theta}_0, \boldsymbol{\theta}_m, \boldsymbol{\theta}_s\}$ denotes the set of parameters of the initial, the measurement, and the state-transition distributions, respectively, and $\boldsymbol{\phi}$ is the parameters of the approximate posterior distribution $q_{\boldsymbol{\phi}}$. Maximizing the

ELBO with respect to $\boldsymbol{\theta}$ and $\boldsymbol{\phi}$ allows for learning a proper state-space model and estimating the latent state variables from the data sequence.

The objective function $\mathcal{L}(\cdot)$ in (9.104) is focused on the reconstruction error and the prior distribution. This lower bound can be reformulated in a way to encode the specifications of the underlying state-space model by focusing on the state-transition and the measurement models in addition to the recognition model. Following this line of thinking, the following reformulation of the ELBO guarantees that at each time instant k, the latent state represents the complete system information:

$$\begin{aligned}
&\mathcal{L}(\boldsymbol{\phi}, \boldsymbol{\theta}) \\
&= \mathbb{E}_{q_{\boldsymbol{\phi}}}\left[\sum_{i=-1}^{k} \underbrace{\log p_{\boldsymbol{\theta}_m}(\mathbf{y}_{i+1}|\mathbf{x}_{i+1}, \mathbf{u}_{i+1})}_{\text{measurement model}}\right] \\
&\quad - \text{KLD}\left(\underbrace{q_{\boldsymbol{\phi}}(\mathbf{x}_{0:k+1}|\mathbf{u}_{0:k+1}, \mathbf{y}_{0:k+1})}_{\text{recognition model}} || p_{\boldsymbol{\theta}_0}(\mathbf{x}_0) \prod_{i=0}^{k} \underbrace{p_{\boldsymbol{\theta}_s}(\mathbf{x}_{i+1}|\mathbf{x}_i, \mathbf{u}_i)}_{\text{state-transition model}}\right). \quad (9.105)
\end{aligned}$$

Additional structure can be considered for the underlying state-space model to include prior knowledge and reduce the number of parameters. Such additional structures include deploying locally linear dynamics [2], switching linear dynamics [198], and conditional Gaussian distributions. Similar to the state-transition and the measurement models, the recognition model can be decomposed as:

$$q_{\boldsymbol{\phi}}(\mathbf{x}_{0:k+1}|\mathbf{u}_{0:k+1}, \mathbf{y}_{0:k+1}) = q_{\boldsymbol{\phi}_0}(\mathbf{x}_0|\mathbf{y}_0) \prod_{i=0}^{k} q_{\boldsymbol{\phi}_f}(\mathbf{x}_{i+1}|\mathbf{x}_i, \mathbf{u}_{i:i+1}, \mathbf{y}_{i+1}), \quad (9.106)$$

where $\boldsymbol{\phi} = \{\boldsymbol{\phi}_0, \boldsymbol{\phi}_f\}$ represents the set of parameters of the initial and the filtering distributions, respectively. For online inference, $q_{\boldsymbol{\phi}_f}(\mathbf{x}_{k+1}|\mathbf{x}_k, \mathbf{u}_{k:k+1}, \mathbf{y}_{k+1})$ represents an approximation of the filtering distribution $p(\mathbf{x}_{k+1}|\mathbf{u}_{0:k+1}, \mathbf{y}_{0:k+1})$. In case that future observations are available, $q_{\boldsymbol{\phi}_f}(\mathbf{x}_{k+1}|\mathbf{x}_k, \mathbf{u}_{k:T}, \mathbf{y}_{k+1:T})$ can be used as an approximation of the smoothing distribution $p(\mathbf{x}_k|\mathbf{u}_{0:T}, \mathbf{y}_{0:T})$, where $T > k$ is the control horizon.

The lower bound $\mathcal{L}(\cdot)$ in (9.104) includes the sum of two terms: the reconstruction error for the observed data and the KLD between the prior and the approximate posterior distributions over the latent space. When the observed data includes high-resolution images, the former is calculated over the higher-dimensional observation space, while the latter is calculated over the lower-dimensional latent state space. Therefore, the two terms may have significantly different contributions to the objective function regarding their magnitudes. To address this issue, the KLD term in the ELBO is scaled by a hyper-parameter $\beta \approx \frac{n_y}{n_x} \geq 1$ as:

$$\begin{aligned}\mathcal{L}(\phi,\theta) &\\ \geq \mathbb{E}_{q_\phi} &\underbrace{\left[\log p_\theta(\mathbf{y}_{0:k+1}|\mathbf{x}_{0:k+1},\mathbf{u}_{0:k+1})\right]}_{\text{reconstruction error}} \\ &- \beta \times \mathrm{KLD}\left(\underbrace{q_\phi(\mathbf{x}_{0:k+1}|\mathbf{u}_{0:k+1},\mathbf{y}_{0:k+1})}_{\text{recognition model}}||\underbrace{p_\theta(\mathbf{x}_{0:k+1}|\mathbf{x}_{0:k},\mathbf{u}_{0:k})}_{\text{includes the prior}}\right).\end{aligned} \tag{9.107}$$

This modified version of the DVBF algorithm is called β-DVBF. To ensure a smooth training process in the β-DVBF algorithm, an appropriate annealing scheme is deployed, which multiplies the KLD term by a scaling factor that linearly increases with the training time from 0 to the chosen value for β [197].

9.9 Kalman Variational Autoencoder

KVAE provides a framework for unsupervised learning of sequential data. For instance, KVAE allows for temporal reasoning regarding a dynamically changing scene in a video signal. Reasoning is performed based on the latent space that describes nonlinear dynamics of an object in an environment rather than the pixel space that constitutes the video frames. This algorithm disentangles two latent representations, which are related to the object representation provided by a recognition model and the state variables that describe the object dynamics. This framework provides the ability of imagining the world evolution as well as imputing the missing data [199].

A linear Gaussian state-space model, which takes account of temporal correlations via a first-order Markov process on latent states, is described as follows:

$$p_{\xi_k}(\mathbf{x}_{k+1}|\mathbf{x}_k,\mathbf{u}_k) = \mathcal{N}(\mathbf{x}_k|\mathbf{A}_k\mathbf{x}_k + \mathbf{B}_k\mathbf{u}_k, \mathbf{Q}_k), \tag{9.108}$$

$$p_{\xi_k}(\mathbf{y}_k|\mathbf{x}_k,\mathbf{u}_k) = \mathcal{N}(\mathbf{y}_k|\mathbf{C}_k\mathbf{x}_k + \mathbf{D}_k\mathbf{u}_k, \mathbf{R}_k), \tag{9.109}$$

$$x_0 \sim \mathcal{N}(\mathbf{x}_0|\mathbf{0},\mathbf{\Sigma}_0), \tag{9.110}$$

where $\xi_k = (\mathbf{A}_k, \mathbf{B}_k, \mathbf{C}_k, \mathbf{D}_k)$ represents the state-space model parameters at time instant k. The joint probability distribution of this system is given by:

$$p_{\xi}(\mathbf{x}, \mathbf{y}|\mathbf{u}) = p_{\xi}(\mathbf{y}|\mathbf{x}, \mathbf{u})p_{\xi}(\mathbf{x}|\mathbf{u})$$
$$= p(\mathbf{x}_0)\prod_{i=0}^{k} p_{\xi_k}(\mathbf{x}_{k+1}|\mathbf{x}_k, \mathbf{u}_k)\prod_{i=-1}^{k} p_{\xi_{k+1}}(\mathbf{y}_{k+1}|\mathbf{x}_{k+1}, \mathbf{u}_{k+1}). \tag{9.111}$$

For linear Gaussian state-space model, the filtered posterior, $p(\mathbf{x}_{k+1}|\mathbf{u}_{0:k+1}, \mathbf{y}_{0:k+1})$, and the smoothed posterior, $p(\mathbf{x}_k|\mathbf{u}, \mathbf{y})$, can be computed by the classic Kalman filter and smoother algorithms, respectively. Moreover, missing data can be properly handled in this framework.

A VAE [181, 200] represents a generative model as:

$$p_{\theta_k}(\mathbf{y}_k|\mathbf{x}_k, \mathbf{u}_k)p(\mathbf{x}_k), \tag{9.112}$$

which can be implemented using deep neural networks. For data, $(\mathbf{u}_k, \mathbf{y}_k)$, VAE introduces a latent encoding $\mathbf{x}_k$. Given the likelihood, $p_{\theta_k}(\mathbf{y}_k|\mathbf{x}_k, \mathbf{u}_k)$, and the prior, $p(\mathbf{x}_k)$, the posterior, $p_{\theta_k}(\mathbf{x}_k|\mathbf{u}_k, \mathbf{y}_k)$, represents a stochastic map from the input (action) and output (observation) spaces to a manifold in the latent state space. Due to intractability of the posterior computation, VAE approximates the posterior with the variational distribution, $q_{\boldsymbol{\phi}}(\mathbf{x}_k|\mathbf{u}_k, \mathbf{y}_k)$, which is parameterized by $\boldsymbol{\phi}$. The approximate posterior, $q_{\boldsymbol{\phi}}(\mathbf{x}_k|\mathbf{u}_k, \mathbf{y}_k)$, is known as the recognition model or the *inference network*.

The required information for describing both the movement and the interaction between objects in a video signal usually lies on a manifold with a smaller dimension than the number of pixels in each frame. The KVAE algorithm takes advantage of this property and disentangles recognition and spatial representations by mapping the sensory inputs (observations), $\mathbf{o}_k$, to pseudo-observations, $\mathbf{y}_k$, using an observation encoder. The pseudo-observations, $\mathbf{y}_k$, lie on a manifold with a lower dimension than the observation space. In fact, $\mathbf{y}_k$ acts as a latent representation of the observation $\mathbf{o}_k$, which encodes an object's position and other visual properties of interest. In the state-space model, $\mathbf{y}_k$ is treated as the measurement vector. The corresponding generative model for a sequence of observations can be factorized as:

$$p_{\theta}(\mathbf{o}|\mathbf{y}) = \prod_{i=0}^{k} p_{\theta}(\mathbf{o}_i|\mathbf{y}_i), \tag{9.113}$$

where $p_{\theta}(\mathbf{o}_k|\mathbf{y}_k)$ can be represented by a deep neural network, which is parameterized by $\boldsymbol{\theta}$. For instance, depending on the type of the observation data, $\mathbf{o}_k$, this neural network may emit a factorized Gaussian distribution. The joint distribution for the KVAE is factorized as:

$$p(\mathbf{x}, \mathbf{y}, \mathbf{o}|\mathbf{u}) = p_{\theta}(\mathbf{o}|\mathbf{y})p_{\xi}(\mathbf{y}|\mathbf{x}, \mathbf{u})p_{\xi}(\mathbf{x}|\mathbf{u}), \tag{9.114}$$

and the data likelihood is computed as:

$$p_\xi(\mathbf{y}|\mathbf{u}) = \int p_\xi(\mathbf{y}|\mathbf{x},\mathbf{u})p_\xi(\mathbf{x}|\mathbf{u})d\mathbf{x}. \tag{9.115}$$

The latent encoding, $\mathbf{y}$, of observations, $\mathbf{o}$, allows for performing temporal reasoning and long-term prediction in the latent space, $\mathbf{x}$, without relying on an autoregressive model, which is directly dependent on the high-dimensional observations [199].

From a set of example sequences, model parameters, θ and ξ, are learned by maximizing the sum of their respective log likelihoods:

$$\mathcal{L}(\theta,\xi) = \sum_{i=0}^{k} \log p_{\theta,\xi}(\mathbf{o}_i|\mathbf{u}_i). \tag{9.116}$$

To facilitate learning and inference, the posterior, $p(\mathbf{x},\mathbf{y}|\mathbf{u},\mathbf{o})$, is approximated by the variational distribution, $q(\mathbf{x},\mathbf{y}|\mathbf{u},\mathbf{o})$, then the ELBO is computed as:

$$\begin{aligned}\log p(\mathbf{o}|\mathbf{u}) &= \log \int p(\mathbf{x},\mathbf{y},\mathbf{o}|\mathbf{u})d\mathbf{x}\, d\mathbf{y} \\ &\geq \mathbb{E}_q\left[\log \frac{p_\theta(\mathbf{o}|\mathbf{y})p_\xi(\mathbf{y}|\mathbf{x},\mathbf{u})p_\xi(\mathbf{x}|\mathbf{u})}{q_\phi(\mathbf{x},\mathbf{y}|\mathbf{u},\mathbf{o})}\right] \\ &= -\mathcal{F}(\theta,\xi,\phi).\end{aligned} \tag{9.117}$$

Instead of maximizing the sum of likelihoods, the sum of $-\mathcal{F}$'s is maximized. In order to have a tight bound, the variational distribution, q, must be specified in a way to be as close as possible to the true posterior distribution, p. However, q depends on parameters ϕ, but p depends on parameters θ and ξ. To address this issue, q is structured as follows to include the conditional posterior $p_\xi(\mathbf{x}|\mathbf{u},\mathbf{y})$, which can be computed by a Kalman smoother:

$$\begin{aligned}q(\mathbf{x},\mathbf{y}|\mathbf{u},\mathbf{o}) &= q_\phi(\mathbf{y}|\mathbf{o})p_\xi(\mathbf{x}|\mathbf{u},\mathbf{y}) \\ &= \prod_{i=0}^{k} q_\phi(\mathbf{y}_i|\mathbf{o}_i)p_\xi(\mathbf{x}|\mathbf{u},\mathbf{y}).\end{aligned} \tag{9.118}$$

A deep neural network can be used to represent $q_\phi(\mathbf{y}_k|\mathbf{o}_k)$. In the context of linear Gaussian state-space models, such a neural network maps $\mathbf{o}_k$ to the parameters of a Gaussian distribution (mean and covariance). Substituting (9.118) in (9.117), we obtain:

$$-\mathcal{F}(\theta,\xi,\phi) = \mathbb{E}_{q_\phi}\left[\log \frac{p_\theta(\mathbf{o}|\mathbf{y})}{q_\phi(\mathbf{y}|\mathbf{o})} + \mathbb{E}_{p_\xi}\left[\log \frac{p_\xi(\mathbf{y}|\mathbf{x},\mathbf{u})p_\xi(\mathbf{x}|\mathbf{u})}{p_\xi(\mathbf{x}|\mathbf{u},\mathbf{y})}\right]\right]. \tag{9.119}$$

Using Monte Carlo integration with samples $\{\tilde{\mathbf{x}}^i,\tilde{\mathbf{y}}^i\}_{i=1}^N$ drawn from q, the lower bound in (9.119) can be estimated as:

$$\begin{aligned}-\hat{\mathcal{F}}(\theta,\xi,\phi) = \frac{1}{N}\sum_{i=1}^{N} \Big(&\log p_\theta(\mathbf{o}|\tilde{\mathbf{y}}^i) + \log p_\xi(\tilde{\mathbf{x}}^i,\tilde{\mathbf{y}}^i|\mathbf{u}) \\ &- \log q_\phi(\tilde{\mathbf{y}}^i|\mathbf{o}) - \log p_\xi(\tilde{\mathbf{x}}^i|\mathbf{u},\tilde{\mathbf{y}}^i)\Big).\end{aligned} \tag{9.120}$$

Note that in (9.120),

$$\frac{p_\xi(\tilde{\mathbf{x}}^i, \tilde{\mathbf{y}}^i|\mathbf{u})}{p_\xi(\tilde{\mathbf{x}}^i|\mathbf{u}, \tilde{\mathbf{y}}^i)} = p_\xi(\tilde{\mathbf{y}}^i|\mathbf{u}). \tag{9.121}$$

However, expressing the lower bound (9.120) as a function of $\tilde{\mathbf{x}}^i$ allows for computing the stochastic gradient with respect to ξ. Drawing a sample $\tilde{\mathbf{y}} \sim q_\phi(\mathbf{y}|\mathbf{o})$ provides a measurement for the corresponding linear Gaussian state-space model. The posterior $p_\xi(\mathbf{x}|\tilde{\mathbf{y}}, \mathbf{u})$ can be tractably computed by a Kalman smoother to obtain $\tilde{\mathbf{x}} \sim p_\xi(\mathbf{x}|\tilde{\mathbf{y}}, \mathbf{u})$. Learning is performed by jointly updating parameters $\boldsymbol{\theta}$, ξ, and $\boldsymbol{\phi}$ via maximizing the ELBO by the stochastic gradient ascent [199].

Although in the linear Gaussian state-space model, the state transition model is linear, nonlinearity of the dynamics can be captured by nonlinear evolution of the time-varying model parameters, ξ_k. Model parameters, ξ_k, determine how states, $\mathbf{x}_k$, change over time. In a general setting, model parameters, ξ_k, must be learnable from pseudo-observations, $\mathbf{y}_{0:k}$, which reflect the history of the system. Hence, the joint probability distribution in (9.111) can be viewed as:

$$\begin{aligned} &p_\xi(\mathbf{x}, \mathbf{y}|\mathbf{u}) \\ &= p(\mathbf{x}_0) \prod_{i=0}^{k} p_{\xi_k(\mathbf{y}_{0:k-1})}(\mathbf{x}_{k+1}|\mathbf{x}_k, \mathbf{u}_k) \prod_{i=-1}^{k} p_{\xi_{k+1}(\mathbf{y}_{0:k})}(\mathbf{y}_{k+1}|\mathbf{x}_{k+1}, \mathbf{u}_{k+1}). \end{aligned} \tag{9.122}$$

N different state-space models, $(\mathbf{A}^i, \mathbf{B}^i, \mathbf{C}^i, \mathbf{D}^i)$, $i = 1, \dots, N$ are learned, which are then interpolated at each time step based on the information provided by the VAE encodings. A recurrent neural network that deploys long short-term memory (LSTM) cells can be used to capture the dependency of ξ_k on $\mathbf{y}_{0:k-1}$. The output of this network, which is known as the *parameter network*, is a set of normalized weights $\boldsymbol{\alpha}_k(\mathbf{y}_{0:k-1})$ for interpolating between the N different operating modes [199]:

$$\mathbf{A}_k = \sum_{i=1}^{N} \alpha_k^i(\mathbf{y}_{0:k-1})\mathbf{A}^i, \tag{9.123}$$

$$\mathbf{B}_k = \sum_{i=1}^{N} \alpha_k^i(\mathbf{y}_{0:k-1})\mathbf{B}^i, \tag{9.124}$$

$$\mathbf{C}_k = \sum_{i=1}^{N} \alpha_k^i(\mathbf{y}_{0:k-1})\mathbf{C}^i, \tag{9.125}$$

$$\mathbf{D}_k = \sum_{i=1}^{N} \alpha_k^i(\mathbf{y}_{0:k-1})\mathbf{D}^i, \tag{9.126}$$

where

$$\sum_{i=1}^{N} \alpha_k^i(\mathbf{y}_{0:k-1}) = 1. \tag{9.127}$$

The parameter network uses a softmax layer to compute these weights. The weighted sums in (9.123)–(9.126) can be interpreted as a soft mixture of N different linear Gaussian state-space models, where the time-invariant matrices of these N models are combined using the time-varying weights α_k^i to provide a time-varying linear Gaussian state-space model. Each one of the N sets of matrices $(\mathbf{A}^i, \mathbf{B}^i, \mathbf{C}^i, \mathbf{D}^i)$ represents a different dynamic system, which will have a dominant role in constructing $(\mathbf{A}_k, \mathbf{B}_k, \mathbf{C}_k, \mathbf{D}_k)$, when the corresponding α_k^i has a relatively high value. The parameter network resembles the locally-linear transitions proposed in [2, 201], which were used in the DVBF algorithm described in Section 9.8. After determining the state-space model parameters $(\mathbf{A}_k, \mathbf{B}_k, \mathbf{C}_k, \mathbf{D}_k)$ from (9.123)–(9.126), the Kalman smoothing algorithm can be used to find the conditional posterior over $\mathbf{x}$, which is needed for computing the gradient of the ELBO.

The KVAE algorithm can take advantage of Kalman smoothing based on the linear Gaussian state-space model to impute missing data using the information from both past and future. To take account of the missing data, the observation sequence $\mathbf{o}$ can be viewed as the union of two disjoint subsets, the observed and unobserved data, which are denoted by $\mathbf{o}_{\mathcal{O}}$ and $\mathbf{o}_{\overline{\mathcal{O}}}$, respectively; $\mathbf{o} = \{\mathbf{o}_{\mathcal{O}}, \mathbf{o}_{\overline{\mathcal{O}}}\}$. Here, the indices $\mathcal{O}$ and $\overline{\mathcal{O}}$ refer to the complementary sets of the observed and unobserved data, respectively. The observer encoder will not provide any pseudo-observation for the set of missing data, $\mathbf{o}_{\overline{\mathcal{O}}}$. Hence, the measurement set of the state-space model can be divided into two disjoint subsets as $\mathbf{y} = \{\mathbf{y}_{\mathcal{O}}, \mathbf{y}_{\overline{\mathcal{O}}}\}$, where $\mathbf{y}_{\overline{\mathcal{O}}}$ corresponds to the missing data. Missing data can be imputed by sampling from the joint conditional density $p(\mathbf{x}, \mathbf{y}_{\mathcal{O}}, \mathbf{y}_{\overline{\mathcal{O}}}|\mathbf{u}, \mathbf{o}_{\mathcal{O}})$, and then, generating $\mathbf{o}_{\overline{\mathcal{O}}}$ from $\mathbf{y}_{\overline{\mathcal{O}}}$. This distribution can be factorized as:

$$p(\mathbf{x}, \mathbf{y}_{\mathcal{O}}, \mathbf{y}_{\overline{\mathcal{O}}}|\mathbf{u}, \mathbf{o}_{\mathcal{O}}) = p_{\xi}(\mathbf{y}_{\overline{\mathcal{O}}}|\mathbf{x}, \mathbf{u}) p_{\xi}(\mathbf{x}|\mathbf{u}, \mathbf{y}_{\mathcal{O}}) p(\mathbf{y}_{\mathcal{O}}|\mathbf{o}_{\mathcal{O}}), \tag{9.128}$$

where sampling starts from $\mathbf{o}_{\mathcal{O}}$. Samples from $p(\mathbf{y}_{\mathcal{O}}|\mathbf{o}_{\mathcal{O}})$ can be approximated by samples from the variational distribution $q_{\phi}(\mathbf{y}_{\mathcal{O}}|\mathbf{o}_{\mathcal{O}})$. Given model parameters ξ and samples of $\mathbf{y}_{\mathcal{O}}$, an extension of the Kalman smoothing algorithm to sequences with missing data can be used to compute $p_{\xi}(\mathbf{x}|\mathbf{u}, \mathbf{y}_{\mathcal{O}})$ to provide samples of $\mathbf{x}$. Then, knowing ξ and samples of $\mathbf{x}$, samples of $\mathbf{y}_{\overline{\mathcal{O}}}$ can be drawn from $p_{\xi}(\mathbf{y}_{\overline{\mathcal{O}}}|\mathbf{x}, \mathbf{u})$.

At each time step, ξ_k depends on all of the previous encoded observations, $\mathbf{y}_{0:k-1}$, including $\mathbf{y}_{\overline{\mathcal{O}}}$, which must be estimated before ξ_k can be computed. Therefore, ξ_k is recursively estimated. Let us assume that $\mathbf{o}_{0:k-1}$ is known, but $\mathbf{o}_k$ is unknown. Using the VAE, $\mathbf{y}_{0:k-1}$ is sampled from $q_{\phi}(\mathbf{y}_{\mathcal{O}}|\mathbf{o}_{\mathcal{O}})$. Then, these samples are used to compute $\xi_{0:k}$. To compute ξ_{k+1}, the measurement $\mathbf{y}_k$ is needed, which is unknown due to missing of $\mathbf{o}_k$. Hence, instead of $\mathbf{y}_k$, its estimate, $\hat{\mathbf{y}}_k$, is used to compute ξ_{k+1}. Such an estimate can be obtained in two steps. The filtered posterior distribution, $p_\xi(\mathbf{x}_{k-1}|\mathbf{u}_{0:k-1}, \mathbf{y}_{0:k-1})$ can be computed since it depends on $\xi_{0:k-1}$. Then, given the information about input and output up to the time instant $k-1$, sample of the predicted state, $\hat{\mathbf{x}}_{k|k-1}$, at time instant k can be drawn as:

$$\begin{aligned} \hat{\mathbf{x}}_{k|k-1} &\sim p_\xi(\mathbf{x}_k|\mathbf{u}_{0:k-1}, \mathbf{y}_{0:k-1}) \\ &= \int p_{\xi_k}(\mathbf{x}_k|\mathbf{x}_{k-1}, \mathbf{u}_{k-1}) p_\xi(\mathbf{x}_{k-1}|\mathbf{u}_{0:k-1}, \mathbf{y}_{0:k-1}) d\mathbf{x}_{k-1}. \end{aligned} \tag{9.129}$$

Given $\hat{\mathbf{x}}_{k|k-1}$, sample of the predicted measurement $\hat{\mathbf{y}}_k$ at time instant k can be drawn as:

$$\begin{aligned} \hat{\mathbf{y}}_k &\sim p_\xi(\mathbf{y}_k|\mathbf{u}_{0:k}, \mathbf{y}_{0:k-1}) \\ &= \int p_{\xi_k}(\mathbf{y}_k|\mathbf{x}_k, \mathbf{u}_k) p_\xi(\mathbf{x}_k|\mathbf{u}_{0:k-1}, \mathbf{y}_{0:k-1}) d\mathbf{x}_k. \end{aligned} \tag{9.130}$$

Now, parameters of the linear Gaussian state-space model at time $k+1$ can be estimated as $\xi_{k+1}(\mathbf{y}_{0:k-1}, \hat{\mathbf{y}}_k)$. In the next step, if $\mathbf{o}_{k+1}$ is missing, the same procedure will be repeated, otherwise the observation encoder will provide $\mathbf{y}_{k+1}$. After the forward pass through the sequence, ξ is estimated and the filtered posterior is computed. Then, the smoothed posterior can be computed by the Kalman smoother's backward pass. Since the smoothed posterior relies on the estimate of ξ, which is computed during the forward pass, it is not exact. However, it enhances data imputation due to using information of the whole sequence [199].

9.10 Deep Variational Information Bottleneck

The *information bottleneck* (IB) method aims at obtaining an optimal representation in terms of a trade-off between complexity of the representation and its predictive power. In other words, the IB method aims at approximating the minimal sufficient statistics. Regarding two random variables, $\mathbf{X}$ and $\mathbf{Y}$, the IB technique tries to extract an optimally compressed representation of $\mathbf{X}$, denoted by $\hat{\mathbf{X}}$, which is relevant for predicting $\mathbf{Y}$. The compression level can be quantified by the mutual information between $\mathbf{X}$ and $\hat{\mathbf{X}}$, denoted by $I(\mathbf{X};\hat{\mathbf{X}})$. Similarly, the relevant information can be quantified by the mutual information between $\hat{\mathbf{X}}$ and $\mathbf{Y}$, denoted by $I(\hat{\mathbf{X}};\mathbf{Y})$. It is worth noting that mutual information is a symmetric measure. While $I(\mathbf{X};\hat{\mathbf{X}})$ must be minimized, $I(\hat{\mathbf{X}};\mathbf{Y})$ should be maximized. Therefore, these two objectives can be simultaneously satisfied through maximizing the following objective function [202]:

$$\mathcal{L}_{IB} = I(\hat{\mathbf{X}};\mathbf{Y}) - \beta I(\mathbf{X};\hat{\mathbf{X}}), \tag{9.131}$$

where β is the Lagrange multiplier. A trade-off can be achieved between the compression level and the preserved relevant information through adjusting the parameter $\beta > 0$. This trade-off between representation accuracy and coding cost is important, especially for real-time information processing. In this regard, for a dynamic system, the IB method allows for selectively extracting information from past that is predictive for future [202]:

$$\mathcal{L}_{IB} = I(\text{internal representation; future}) - \beta I(\text{past; internal representation}). \tag{9.132}$$

Regarding a dynamic system with input vector $\mathbf{u} \in \mathbb{R}^{n_u}$ and output vector $\mathbf{y} \in \mathbb{R}^{n_y}$, the following loss function may be considered for the past-future IB [202]:

$$\mathcal{L}_{IB} = I(\hat{\mathbf{Y}}_{future}; \mathbf{Y}_{future}) - \beta I(\mathbf{U}_{past}; \hat{\mathbf{Y}}_{future}), \tag{9.133}$$

where $\mathbf{U}$ and $\mathbf{Y}$ denote the input and the output random variables, respectively, and $\hat{\mathbf{Y}}$ denotes the predicted output random variable. In a state-space model with state vector $\mathbf{x} \in \mathbb{R}^{n_x}$, output is predicted, $\hat{\mathbf{y}}$, based on the predicted state, $\hat{\mathbf{x}}$, and the input, $\mathbf{u}$. Therefore, the objective function (9.133) can be rewritten as:

$$\mathcal{L}_{IB} = I(\mathbf{X}; \mathbf{Y}) - \beta I(\mathbf{U}; \mathbf{X}), \tag{9.134}$$

where $\mathbf{X}$ denotes the state random variable. Therefore, regarding $\mathbf{X}$ as the bottleneck variable, for any specific value of β, the IB method allows to infer an optimal state-space realization of the system under study.

A variational approximation to the IB can be obtained based on a parameterized model such as a deep neural network along with the reparameterization trick, which refers to applying noise as the input to the learnable model [203]. In such a deep neural network, the internal representation, $\mathbf{X}$, can be viewed as a stochastic encoding of the input, $\mathbf{U}$, which is provided by a parameterized encoder, $p_\theta(\mathbf{X}|\mathbf{U})$. It is desired to learn an encoding, which is optimally informative about the output, $\mathbf{Y}$. The mutual information between $\mathbf{X}$ and $\mathbf{Y}$ provides the measure of informativeness that must be maximized:

$$I_\theta(\mathbf{X}; \mathbf{Y}) = \int\int p_\theta(\mathbf{x}, \mathbf{y}) \log \frac{p_\theta(\mathbf{x}, \mathbf{y})}{p_\theta(\mathbf{x})\, p(\mathbf{y})} d\mathbf{x}\, d\mathbf{y}. \tag{9.135}$$

The maximally informative representation is obtained subject to a constraint on its complexity. Hence, to train the deep neural network, a parameterized version of the objective function in (9.134) must be optimized:

$$\mathcal{L}_{IB} = I_\theta(\mathbf{X}; \mathbf{Y}) - \beta I_\theta(\mathbf{U}; \mathbf{X}). \tag{9.136}$$

Generally speaking, computation of the mutual information is challenging except for two special cases [203, 204]:

- Random variables $\mathbf{X}$, $\mathbf{U}$, and $\mathbf{Y}$ are discrete, hence, discrete data can be clustered.
- Random variables $\mathbf{X}$, $\mathbf{U}$, and $\mathbf{Y}$ are all jointly Gaussian, which leads to linear optimal encoding and decoding maps.

In order to obtain a computationally efficient implementation of the IB method, which avoids such restrictions, variational inference can be used to compute a lower bound for the objective function in (9.136). Then, deep neural networks can be used to parameterize the corresponding distributions. Such networks can handle high-dimensional continuous data with arbitrary distributions. An unbiased estimate of the gradient can be computed using the reparameterization trick along with the Monte Carlo sampling. Hence, the objective function can be optimized by the stochastic gradient descent method. This *deep variational information bottleneck* (DVIB) method is robust against overfitting and adversarial inputs, because the representation $\mathbf{X}$ ignores details of the input $\mathbf{U}$ [203].

The joint distribution $p(\mathbf{X}, \mathbf{U}, \mathbf{Y})$ can be factorized as:

$$\begin{aligned} p(\mathbf{X}, \mathbf{U}, \mathbf{Y}) &= p(\mathbf{X}|\mathbf{U}, \mathbf{Y})\, p(\mathbf{Y}|\mathbf{U})\, p(\mathbf{U}) \\ &= p(\mathbf{X}|\mathbf{U})\, p(\mathbf{Y}|\mathbf{U})\, p(\mathbf{U}). \end{aligned} \tag{9.137}$$

Assuming $p(\mathbf{X}|\mathbf{U}, \mathbf{Y}) = p(\mathbf{X}|\mathbf{U})$ in the last line means that the representation $\mathbf{X}$ does not directly depend on the labels $\mathbf{Y}$, which paves the way for using unlabeled datasets and deploying unsupervised learning. The mutual information in (9.135) can be rewritten as:

$$I_\theta(\mathbf{X}; \mathbf{Y}) = \int\int p_\theta(\mathbf{x}, \mathbf{y}) \log \frac{p_\theta(\mathbf{y}|\mathbf{x})}{p(\mathbf{y})} \mathrm{d}\mathbf{x}\, \mathrm{d}\mathbf{y} \tag{9.138}$$

Let $q_\theta(\mathbf{y}|\mathbf{x})$ denote a variational approximation to $p_\theta(\mathbf{y}|\mathbf{x})$. Regarding the fact that

$$\text{KLD}\left(p_\theta(\mathbf{y}|\mathbf{x}) || q_\theta(\mathbf{y}|\mathbf{x})\right) \geq 0, \tag{9.139}$$

we have:

$$\int p_\theta(\mathbf{y}|\mathbf{x}) \log \frac{p_\theta(\mathbf{y}|\mathbf{x})}{q_\theta(\mathbf{y}|\mathbf{x})} \mathrm{d}\mathbf{y} \geq 0, \tag{9.140}$$

which leads to:

$$\int p_\theta(\mathbf{y}|\mathbf{x}) \log p_\theta(\mathbf{y}|\mathbf{x}) \mathrm{d}\mathbf{y} \geq \int p_\theta(\mathbf{y}|\mathbf{x}) \log q_\theta(\mathbf{y}|\mathbf{x}) \mathrm{d}\mathbf{y}. \tag{9.141}$$

Therefore, for the mutual information in (9.138), we can write:

$$\begin{aligned} I_\theta(\mathbf{X}; \mathbf{Y}) &\geq \int\int p_\theta(\mathbf{x}, \mathbf{y}) \log \frac{q_\theta(\mathbf{y}|\mathbf{x})}{p(\mathbf{y})} \mathrm{d}\mathbf{x}\, \mathrm{d}\mathbf{y} \\ &= \int\int p_\theta(\mathbf{x}, \mathbf{y}) \log q_\theta(\mathbf{y}|\mathbf{x}) \mathrm{d}\mathbf{x}\, \mathrm{d}\mathbf{y} - \int p(\mathbf{y}) \log p(\mathbf{y}) \mathrm{d}\mathbf{y} \\ &= \int\int p_\theta(\mathbf{x}, \mathbf{y}) \log q_\theta(\mathbf{y}|\mathbf{x}) \mathrm{d}\mathbf{x}\, \mathrm{d}\mathbf{y} + H(\mathbf{Y}). \end{aligned} \tag{9.142}$$

Since the entropy of the labels $H(\mathbf{Y})$ is independent of the parameters, for the optimization procedure, it can be ignored. Hence, focusing on the first term in the right hand side of the inequality (9.142) and using (9.137), we obtain:

$$I_\theta(\mathbf{X};\mathbf{Y}) \geq \int\int\int p(\mathbf{y}|\mathbf{u})\ p_\theta(\mathbf{x}|\mathbf{u})\ p(\mathbf{u}) \log q_\theta(\mathbf{y}|\mathbf{x}) \mathrm{d}\mathbf{x}\ \mathrm{d}\mathbf{u}\ \mathrm{d}\mathbf{y}. \tag{9.143}$$

Similar to (9.138), $I_\theta(\mathbf{U};\mathbf{X})$ can be written as:

$$\begin{aligned} I_\theta(\mathbf{U};\mathbf{X}) &= \int\int p_\theta(\mathbf{x},\mathbf{u}) \log \frac{p_\theta(\mathbf{x}|\mathbf{u})}{p_\theta(\mathbf{x})} \mathrm{d}\mathbf{x}\ \mathrm{d}\mathbf{u} \\ &= \int\int p_\theta(\mathbf{x},\mathbf{u}) \log p_\theta(\mathbf{x}|\mathbf{u}) \mathrm{d}\mathbf{x}\ \mathrm{d}\mathbf{u} - \int p_\theta(\mathbf{x}) \log p_\theta(\mathbf{x}) \mathrm{d}\mathbf{x} \\ &= \int\int p_\theta(\mathbf{x},\mathbf{u}) \log p_\theta(\mathbf{x}|\mathbf{u}) \mathrm{d}\mathbf{x}\ \mathrm{d}\mathbf{u} + H_\theta(\mathbf{X}). \end{aligned} \tag{9.144}$$

Let $q_\theta(\mathbf{x})$ denote a variational approximation to $p_\theta(\mathbf{x})$. Then,

$$\mathrm{KLD}\left(p_\theta(\mathbf{x})||q_\theta(\mathbf{x})\right) \geq 0, \tag{9.145}$$

leads to:

$$\int p_\theta(\mathbf{x}) \log p_\theta(\mathbf{x}) \mathrm{d}\mathbf{x} \geq \int p_\theta(\mathbf{x}) \log q_\theta(\mathbf{x}) \mathrm{d}\mathbf{x}. \tag{9.146}$$

Hence, for the mutual information in (9.144), we can write:

$$\begin{aligned} I_\theta(\mathbf{U};\mathbf{X}) &\leq \int\int p_\theta(\mathbf{x},\mathbf{u}) \log \frac{p_\theta(\mathbf{x}|\mathbf{u})}{q_\theta(\mathbf{x})} \mathrm{d}\mathbf{x}\ \mathrm{d}\mathbf{u} \\ &= \int\int p_\theta(\mathbf{x}|\mathbf{u})\ p(\mathbf{u}) \log \frac{p_\theta(\mathbf{x}|\mathbf{u})}{q_\theta(\mathbf{x})} \mathrm{d}\mathbf{x}\ \mathrm{d}\mathbf{u}. \end{aligned} \tag{9.147}$$

Regarding the two bounds in (9.143) and (9.147), we can write:

$$\begin{aligned} & I_\theta(\mathbf{X};\mathbf{Y}) - \beta I_\theta(\mathbf{U};\mathbf{X}) \\ & \geq \int\int\int p(\mathbf{y}|\mathbf{u})\ p_\theta(\mathbf{x}|\mathbf{u})\ p(\mathbf{u}) \log q_\theta(\mathbf{y}|\mathbf{x}) \mathrm{d}\mathbf{x}\ \mathrm{d}\mathbf{u}\ \mathrm{d}\mathbf{y} \\ & \quad - \beta \int\int p_\theta(\mathbf{x}|\mathbf{u})\ p(\mathbf{u}) \log \frac{p_\theta(\mathbf{x}|\mathbf{u})}{q_\theta(\mathbf{x})} \mathrm{d}\mathbf{x}\ \mathrm{d}\mathbf{u}. \end{aligned} \tag{9.148}$$

In order to compute this lower bound, $p(\mathbf{u},\mathbf{y}) = p(\mathbf{y}|\mathbf{u})\ p(\mathbf{u})$ is approximated by the empirical data distribution:

$$p(\mathbf{u},\mathbf{y}) = p(\mathbf{y}|\mathbf{u})\ p(\mathbf{u}), \tag{9.149}$$

$$\approx \frac{1}{n_p} \sum_{i=1}^{n_p} \delta(\mathbf{u}-\mathbf{u}^i)\ \delta(\mathbf{y}-\mathbf{y}^i). \tag{9.150}$$

Then, the lower bound in (9.148) can be computed as:

$$\frac{1}{n_p} \sum_{i=1}^{n_p} \left(\int \left(p_\theta(\mathbf{x}|\mathbf{u}^i) \log q_\theta(\mathbf{y}^i|\mathbf{x}) - \beta\ p_\theta(\mathbf{x}|\mathbf{u}^i) \log \frac{p_\theta(\mathbf{x}|\mathbf{u}^i)}{q_\theta(\mathbf{x})} \right) \mathrm{d}\mathbf{x} \right). \tag{9.151}$$

The *nonlinear information bottleneck* (NIB) has been proposed as an alternative to the DVIB [204]. In the NIB method, a variational lower bound similar to (9.143) is computed for $I_\theta(\mathbf{X};\mathbf{Y})$. However, instead of the variational upper bound in (9.147), a nonparametric upper bound is computed for $I_\theta(\mathbf{U};\mathbf{X})$.

9.11 Wasserstein Distributionally Robust Kalman Filter

Filtering aims at estimating the hidden state vector $\mathbf{x} \in \mathbb{R}^{n_x}$ from noisy observations $\mathbf{y} \in \mathbb{R}^{n_y}$ given inputs $\mathbf{u} \in \mathbb{R}^{n_u}$. Hence, any estimation method must take account of the inherent uncertainty in the joint distribution of $\mathbf{x}$ and $\mathbf{y}$ given $\mathbf{u}$, $p(\mathbf{x},\mathbf{y}|\mathbf{u})$. Let us define $\mathbf{z}$ as:

$$\mathbf{z} = \begin{bmatrix}\mathbf{x}\\ \mathbf{y}\end{bmatrix} \in \mathbb{R}^{n_x+n_y}. \tag{9.152}$$

Now, the distributional uncertainty can be modeled by an *ambiguity set*, $\mathcal{P}$, which is defined by a family of distributions on $\mathbb{R}^{n_x+n_y}$ that are sufficiently close to the governing distribution of $\mathbf{x}$ and $\mathbf{y}$ in light of the collected data. A robust filter can be designed by minimizing the worst-case mean-square error with respect to the distributions in the ambiguity set. The ambiguity set can be constructed using the *Wasserstein distance* [205].

Definition 9.1 ***(Wasserstein distance)*** *Type-2 Wasserstein distance between two distributions q and p on $\mathbb{R}^{n_x+n_y}$ is defined as:*

$$W_2(q,p) = \inf_{\pi\in\Pi(q,\,p)} \left\{ \left(\int_{\mathbb{R}^{(n_x+n_y)\times(n_x+n_y)}} \| \mathbf{z}^q - \mathbf{z}^p \|^2 \, \pi(d\mathbf{z}^q, d\mathbf{z}^p) \right)^{\frac{1}{2}} \right\}, \tag{9.153}$$

where $\Pi(q,p)$ denotes the set of probability distributions on $\mathbb{R}^{(n_x+n_y)\times(n_x+n_y)}$ with marginals q and p.

For Gaussian distributions, $q = \mathcal{N}\left(\boldsymbol{\mu}^q, \boldsymbol{\Sigma}^q\right)$ and $p = \mathcal{N}\left(\boldsymbol{\mu}^p, \boldsymbol{\Sigma}^p\right)$, the Wasserstein distance is computed as [206]:

$$W_2(q,p) = \sqrt{\| \boldsymbol{\mu}^q - \boldsymbol{\mu}^p \|^2 + \operatorname{trace}\left(\boldsymbol{\Sigma}^q + \boldsymbol{\Sigma}^p - 2\left(\left(\boldsymbol{\Sigma}^p\right)^{\frac{1}{2}} \boldsymbol{\Sigma}^q \left(\boldsymbol{\Sigma}^p\right)^{\frac{1}{2}} \right)^{\frac{1}{2}} \right)}. \tag{9.154}$$

At time instant k, the Wasserstein ambiguity set is defined as:

$$\mathcal{P} = \left\{ q \mid W_2(q,p) \le \rho_k \right\}, \tag{9.155}$$

which can be viewed as a ball in the space of Gaussian distributions that is centered at $p = \mathcal{N}\left(\hat{\mathbf{z}}_{k|k}, \mathbf{P}^{\mathbf{z}}_{k|k}\right)$ with radius $\rho_k \geq 0$. The robust filter is designed to have the best performance under the worst possible distribution $q \in \mathcal{P}$. The distributionally robust minimum mean-square error (MMSE) estimator is obtained by solving the following optimization problem [205]:

$$\begin{aligned}
&\sup \ \text{trace}\left(\mathbf{S}^{\mathbf{xx}}_{k+1} - \mathbf{S}^{\mathbf{xy}}_{k+1}\left(\mathbf{S}^{\mathbf{yy}}_{k+1}\right)^{-1}\mathbf{S}^{\mathbf{yx}}_{k+1}\right)\\
&\text{subject to } \mathbf{S}_{k+1} = \begin{bmatrix} \mathbf{S}^{\mathbf{xx}}_{k+1} & \mathbf{S}^{\mathbf{yx}}_{k+1} \\ \mathbf{S}^{\mathbf{xy}}_{k+1} & \mathbf{S}^{\mathbf{yy}}_{k+1} \end{bmatrix} \in \mathbb{S}^{n_x+n_y}_{+},\\
&\mathbf{S}^{\mathbf{xx}}_{k+1} \in \mathbb{S}^{n_x}_{+}, \quad \mathbf{S}^{\mathbf{yy}}_{k+1} \in \mathbb{S}^{n_y}_{+}, \quad \mathbf{S}^{\mathbf{xy}}_{k+1} = \left(\mathbf{S}^{\mathbf{yx}}_{k+1}\right)^T \in \mathbb{R}^{n_x \times n_y},\\
&\text{trace}\left(\mathbf{S}_{k+1} + \mathbf{P}^{\mathbf{z}}_{k|k} - 2\left(\left(\mathbf{P}^{\mathbf{z}}_{k|k}\right)^{\frac{1}{2}}\mathbf{S}_{k+1}\left(\mathbf{P}^{\mathbf{z}}_{k|k}\right)^{\frac{1}{2}}\right)^{\frac{1}{2}}\right) \leq \rho_k^2,\\
&\mathbf{S}_{k+1} \geq \lambda_{\min}\left(\mathbf{P}^{\mathbf{z}}_{k|k}\right)\mathbf{I}_{n_x+n_y},
\end{aligned} \tag{9.156}$$

where $\mathbb{S}_+$ denotes the cone of symmetric positive semidefinite matrices, $\mathbf{A} \geq \mathbf{B}$ means that $\mathbf{A} - \mathbf{B} \in \mathbb{S}_+$, and $\mathbf{I}$ denotes the identity matrix. Let $\mathbf{S}^{*}_{k+1}$, $\mathbf{S}^{\mathbf{xx}*}_{k+1}$, $\mathbf{S}^{\mathbf{yy}*}_{k+1}$, and $\mathbf{S}^{\mathbf{xy}*}_{k+1}$ denote the optimal values obtained by solving (9.156). Regarding the a priori output error estimate (innovation):

$$\mathbf{e}_{\mathbf{y}_{k+1|k}} = \mathbf{y}_{k+1} - \hat{\mathbf{y}}_{k+1|k}, \tag{9.157}$$

the distributionally robust MMSE estimate and the corresponding estimation error covariance matrix are computed as:

$$\hat{\mathbf{x}}_{k+1|k+1} = \hat{\mathbf{x}}_{k+1|k} + \mathbf{S}^{\mathbf{xy}*}_{k+1}\left(\mathbf{S}^{\mathbf{yy}*}_{k+1}\right)^{-1}\mathbf{e}_{\mathbf{y}_{k+1|k}}, \tag{9.158}$$

$$\mathbf{P}_{k+1|k+1} = \mathbf{S}^{\mathbf{xx}*} - \mathbf{S}^{\mathbf{xy}*}_{k+1}\left(\mathbf{S}^{\mathbf{yy}*}_{k+1}\right)^{-1}\mathbf{S}^{\mathbf{yx}*}_{k+1}. \tag{9.159}$$

Distributionally robust estimation problem can be viewed as a zero-sum game between the filter designer and a fictitious adversary (environment) that chooses q. The *Nash equilibrium* of this game provides the robust estimator as well as the *least favorable conditional distribution*:

$$q^* = \mathcal{N}\left(\begin{bmatrix} \hat{\mathbf{x}}_{k+1|k+1} \\ \mathbf{y}_{k+1} \end{bmatrix}, \mathbf{S}^{*}_{k+1}\right), \tag{9.160}$$

which will be used as the prior in the next iteration:

$$p(\mathbf{z}_{k+1}|\mathbf{u}_{0:k+1}, \mathbf{y}_{0:k+1}) = \mathcal{N}\left(\hat{\mathbf{z}}_{k+1|k+1} = \begin{bmatrix} \hat{\mathbf{x}}_{k+1|k+1} \\ \mathbf{y}_{k+1} \end{bmatrix}, \mathbf{P}^{\mathbf{z}}_{k+1|k+1} = \mathbf{S}^{*}_{k+1}\right). \tag{9.161}$$

As $\rho \to 0$, the robust update step reduces to the update step of the classic Kalman filter [205].

9.12 Hierarchical Invertible Neural Transport

Hierarchical invertible neural transport (HINT) deploys invertible neural architectures, which are built on invertible coupling blocks. Such blocks split their incoming variables into two sub-spaces. Then, one set of the variables is used to parameterize an invertible affine transformation, which is applied to the other set. Deploying recursive coupling in a hierarchical architecture, HINT allows for sampling from both a joint distribution and its corresponding posterior using a single invertible neural network [207].

Let us consider a generative model with the state vector $\mathbf{x} \in \mathbb{R}^{n_x}$ and the output vector $\mathbf{y} \in \mathbb{R}^{n_y}$. Using the *floor*, $\lfloor . \rfloor$, and the *ceiling*, $\lceil . \rceil$, functions, the state vector, $\mathbf{x}$, can be split into two halves as $\mathbf{x}_1 = \mathbf{x}_{0:\lfloor n_x/2 \rfloor}$ and $\mathbf{x}_2 = \mathbf{x}_{\lceil n_x/2 \rceil : n_x}$. Let $C(\mathbf{x}_2|\mathbf{x}_1)$ denote an invertible transform of $\mathbf{x}_2$ conditioned on $\mathbf{x}_1$. Then, a standard invertible coupling block and its inverse are respectively defined by the following functions [207]:

$$\mathbf{f}_C(\mathbf{x}) = \begin{bmatrix} \mathbf{x}_1 \\ C\left(\mathbf{x}_2|\mathbf{x}_1\right) \end{bmatrix} \tag{9.162}$$

and

$$\mathbf{f}_C^{-1}(\mathbf{x}) = \begin{bmatrix} \mathbf{x}_1 \\ C^{-1}\left(\mathbf{x}_2|\mathbf{x}_1\right) \end{bmatrix}. \tag{9.163}$$

For an affine coupling block, $C(\mathbf{x}_2|\mathbf{x}_1)$ is defined as [208]:

$$C(\mathbf{x}_2|\mathbf{x}_1) = \mathbf{x}_2 \odot e^{\mathbf{s}(\mathbf{x}_1)} + \mathbf{t}(\mathbf{x}_1), \tag{9.164}$$

where $\mathbf{s}$ and $\mathbf{t}$ represent two unconstrained feed-forward networks, and $\odot$ is the *Hadamard product* or element-wise product. For the Jacobian of such an affine coupling block, $\mathbf{J}_{\mathbf{f}_C}(\mathbf{x})$, we have:

$$\log |\mathbf{J}_{\mathbf{f}_C}(\mathbf{x})| = \log \left| \frac{\partial \mathbf{f}_C(\mathbf{x})}{\partial \mathbf{x}} \right| = \sum \mathbf{s}(\mathbf{x}_1), \tag{9.165}$$

which can be efficiently computed. In (9.165), $|\cdot|$ denotes the determinant.

A pipeline $\mathbf{y} = \mathbf{T}(\mathbf{x})$, which is constructed by a series of alternating coupling blocks and random orthogonal matrices $\mathbf{Q}$, allows for transformation of all entries of $\mathbf{x}$ and their interaction via different combinations. Regarding the fact that for orthogonal matrices, $\mathbf{Q}^{-1} = \mathbf{Q}^T$, transformations by such matrices and their inverses are respectively defined by:

$$\mathbf{f}_Q(\mathbf{x}) = \mathbf{Q}\mathbf{x} \tag{9.166}$$

and

$$\mathbf{f}_{\mathrm{Q}}^{-1}(\mathbf{x}) = \mathbf{Q}^T\mathbf{x}. \tag{9.167}$$

For the Jacobian of such transformation, $\mathbf{J}_{\mathbf{f}_{\mathrm{Q}}}(\mathbf{x})$, we have:

$$\log \left|\mathbf{J}_{\mathbf{f}_{\mathrm{Q}}}(\mathbf{x})\right| = 0. \tag{9.168}$$

The pipeline is mathematically described by:

$$\mathbf{T} = \mathbf{f}_{\mathrm{C}_1} \circ \mathbf{f}_{\mathrm{Q}_1} \circ \mathbf{f}_{\mathrm{C}_2} \circ \mathbf{f}_{\mathrm{Q}_2} \circ \dots, \tag{9.169}$$

where $\circ$ denotes the composition of functions. In $\mathbf{T}$, each term $\mathbf{f}_{\mathrm{C}_i} \circ \mathbf{f}_{\mathrm{Q}_i}$, which is a composition of a triangular map, $\mathbf{f}_{\mathrm{C}_i}$ and an orthogonal transformation, $\mathbf{f}_{\mathrm{Q}_i}$, can be viewed as a nonlinear generalization of the **QR** decomposition in linear algebra [209, 210]. While the triangular map, which is known as the *Knothe–Rosenblatt rearrangement* in the *theory of transport maps* [211, 212], allows for representing nonlinear functions, the orthogonal transformation enhances the representational power of the pipeline by ensuring that each element of the input will contribute to the output through reshuffling of the variables. This pipeline is trained by optimizing the following loss function:

$$\mathcal{L}(\mathbf{x}) = \frac{1}{2} \parallel \mathbf{T}(\mathbf{x})\parallel^2 - \log \left|\mathbf{J}_{\mathbf{T}}(\mathbf{x})\right|, \tag{9.170}$$

to create a normalizing flow by transporting $p_{\mathbf{X}}(\mathbf{x})$ to a standard normal distribution, $p_{\mathbf{Y}}(\mathbf{y}) = \mathcal{N}(\mathbf{0}, \mathbf{I})$.

According to the theory of transport maps [211], this pipeline allows for pushing a reference density toward a target density [212]. To be more precise, samples $\mathbf{x}$ can be drawn from $p_{\mathbf{X}}(\mathbf{x})$ by drawing samples $\mathbf{y}$ from $p_{\mathbf{Y}}(\mathbf{y})$, and then, passing them through the inverse model, $\mathbf{x} = \mathbf{T}^{-1}(\mathbf{y})$. In this way, for a given data point, $\mathbf{x}$, we have [207]:

$$p_{\mathbf{X}}(\mathbf{x}) = p_{\mathbf{Y}}\left(\mathbf{T}(\mathbf{x})\right) \cdot \left|\mathbf{J}_{\mathbf{T}}(\mathbf{x})\right|. \tag{9.171}$$

The objective function in (9.170) can be interpreted as the KLD between $p_{\mathbf{X}}(\mathbf{x})$ and the pull-back of $p_{\mathbf{Y}}(\mathbf{y})$ under the map $\mathbf{T}^{-1}$, which is denoted by $\mathbf{T}_{\#}^{-1}$. To be more precise, $\mathrm{KLD}(p_{\mathbf{X}} \parallel \mathbf{T}_{\#}^{-1} p_{\mathbf{Y}})$ can be written as:

$$\begin{aligned}
&\mathrm{KLD}(p_{\mathbf{X}} \parallel \mathbf{T}_{\#}^{-1} p_{\mathbf{Y}}) \\
&= \int p_{\mathbf{X}}(\mathbf{x}) \log \frac{p_{\mathbf{X}}(\mathbf{x})}{\mathbf{T}_{\#}^{-1} p_{\mathbf{Y}}(\mathbf{x})} d\mathbf{x} \\
&= -\int p_{\mathbf{X}}(\mathbf{x}) \log\left(p_{\mathbf{Y}}\left(\mathbf{T}(\mathbf{x})\right) \mathbf{J}_{\mathbf{T}}(\mathbf{x})\right) d\mathbf{x} + \int p_{\mathbf{X}}(\mathbf{x}) \log p_{\mathbf{X}}(\mathbf{x}) d\mathbf{x} \\
&= -\int p_{\mathbf{X}}(\mathbf{x}) \log\left(p_{\mathbf{Y}}\left(\mathbf{T}(\mathbf{x})\right) \mathbf{J}_{\mathbf{T}}(\mathbf{x})\right) d\mathbf{x} - H(p_{\mathbf{X}}) \\
&= \mathbb{E}_{p_{\mathbf{X}}}[\mathcal{L}(\mathbf{x})] - H(p_{\mathbf{X}}),
\end{aligned} \tag{9.172}$$

where $H(p_{\mathbf{X}})$ denotes the entropy, which is fixed but unknown.

In many problems, data consists of input and output pairs, $(\mathbf{u}, \mathbf{y})$, obtained from a stochastic forward process, $\mathbf{u} \rightarrow \mathbf{y}$. In order to perform Bayesian inference and approximate the posterior, $p(\mathbf{x}|\mathbf{u}, \mathbf{y})$, these pairs are concatenated into a single vector, $[\mathbf{u}^T, \mathbf{y}^T]^T$. Then, the transport pipeline estimates the state vector based on the joint distribution of input and output, $\mathbf{T}^{-1}\left([\mathbf{u}^T, \mathbf{y}^T]^T\right)$. Moreover, deploying conditional coupling blocks allows for conditioning the entire normalizing flow on external variables (inputs) [213, 214]. In this way, the transport $\mathbf{T}$ between $p_{\mathbf{X}}(\mathbf{x})$ and $p_{\mathbf{Y}}(\mathbf{y})$ is conditioned on the corresponding values of $\mathbf{u}$ as $\mathbf{y} = T(\mathbf{x}|\mathbf{u})$, then, the inverse $T^{-1}(\mathbf{y}|\mathbf{u})$ provides an approximation of the posterior. Now, the objective function in (9.170) can be adjusted accordingly as [207]:

$$\mathcal{L}(\mathbf{x}, \mathbf{u}) = \frac{1}{2} \parallel \mathbf{T}(\mathbf{x}|\mathbf{u})\|^2 - \log\left|\mathbf{J}_{\mathbf{T}}(\mathbf{x}|\mathbf{u})\right|, \tag{9.173}$$

where the Jacobian is taken with respect to $\mathbf{x}$:

$$\mathbf{J}_{\mathbf{T}}(\mathbf{x}|\mathbf{u}) = \nabla_{\mathbf{x}}\mathbf{T}(\mathbf{x}|\mathbf{u}). \tag{9.174}$$

This optimization problem is equivalent to minimizing the KLD between $p_{\mathbf{X}|\mathbf{U}}(\mathbf{x}|\mathbf{u})$ and the pull-back of a standard normal distribution via $\mathbf{T}^{-1}$:

$$\text{KLD}\left(p_{\mathbf{X}|\mathbf{U}}(\mathbf{x}|\mathbf{u})||\mathbf{T}_{\#}^{-1}(.|\mathbf{U})p_{\mathbf{Y}}\right) = \mathbb{E}_{p_{\mathbf{X}|\mathbf{U}}}[\mathcal{L}(\mathbf{x}, \mathbf{u})] - H(p_{\mathbf{X}|\mathbf{U}}). \tag{9.175}$$

In the following, the basic coupling block architecture is extended in two ways to build recursive coupling block and the HINT [207].

A large number of possible interactions between variables are not usually modeled. This leads to sparse Jacobian matrices, $\mathbf{J}_{\mathbf{f}_{\mathrm{C}}}$, for coupling blocks. If the Jacobian $\mathbf{J}_{\mathbf{f}_{\mathrm{C}}}$ is a lower-triangular matrix, its log determinant can be efficiently computed as shown in (9.165). According to [215], arbitrary distributions can be represented by triangular transformations. The recursive coupling scheme has been proposed to take advantage of these properties and fill the null elements below the main diagonal $\mathbf{J}_{\mathbf{f}_{\mathrm{C}}}$. This scheme allows for building more expressive architectures. In this scheme, each one of the subspaces $\mathbf{x}_1$ and $\mathbf{x}_2$ is split and transformed, where each subcoupling has its own affine coupling function C with independent parameters. This procedure can be repeated until one-dimensional subspaces are reached. The recursive coupling block and its inverse are respectively defined as [207]:

$$\mathbf{f}_{\mathrm{R}}(\mathbf{x}) = \begin{cases} \mathbf{f}_{\mathrm{C}}(\mathbf{x}), & \text{if } \mathbf{x} \in \mathbb{R}^{n_x \leq 3} \\ \begin{bmatrix} \mathbf{f}_{\mathrm{R}}(\mathbf{x}_1) \\ C\left(\mathbf{f}_{\mathrm{R}}(\mathbf{x}_2)|\mathbf{x}_1\right) \end{bmatrix}, & \text{otherwise} \end{cases} \tag{9.176}$$

and

$$\mathbf{f}_{\mathrm{R}}^{-1}(\mathbf{x}) = \begin{cases} \mathbf{f}_{\mathrm{C}}^{-1}(\mathbf{x}), & \text{if } \mathbf{x} \in \mathbb{R}^{n_x \leq 3}, \\ \begin{bmatrix} \mathbf{f}_{\mathrm{R}}^{-1}(\mathbf{x}_1) \\ \mathbf{f}_{\mathrm{R}}^{-1}\left(C^{-1}\left(\mathbf{x}_2|\mathbf{f}_{\mathrm{R}}^{-1}(\mathbf{x}_1)\right)\right) \end{bmatrix}, & \text{otherwise.} \end{cases} \tag{9.177}$$

This procedure results in a dense lower-triangular Jacobian matrix, for which the log determinant is the sum of the log determinants of all subcouplings C. If n_x is a power of two that ensures proper dimensions for all splits, then $\mathbf{f}_R$ will be a full Knothe–Rosenblatt map.

In the recursive affine coupling block, splitting continues for the inner functions $\mathbf{f}_R(\mathbf{x}_i)$ until the vectors $\mathbf{x}_i$ cannot be split any further. In this way, each block will have a triangular Jacobian. However, the final structure of the Jacobian of the transport (composition of a number of these blocks), $\mathbf{J}_\mathbf{T}$, is determined by placement of the orthogonal matrices $\mathbf{Q}$, which act as permutation matrices. The Jacobian $\mathbf{J}_\mathbf{T}$ exhibits different structures according to different ways that the recursive affine coupling block can be integrated with an orthogonal permutation matrix $\mathbf{Q}$:

(1) Without using $\mathbf{Q}$, composition of the lower-triangular maps leads to a dense lower-triangular map, which can be viewed as an autoregressive model. In other words, no permutation results in a dense lower-triangular $\mathbf{J}_\mathbf{T}$.
(2) Permutation in the lower branch leads to a recursive conditional structure in which the lower-branch variables cannot affect the upper-branch variables. In this setting, the lower lane performs a transport conditioned on intermediate representations from the upper lane. In other words, placing $\mathbf{Q}$ at the beginning of the lower branch in each recursive split leads to a block-triangular Jacobian $\mathbf{J}_\mathbf{T}$.
(3) Permutation between coupling blocks leads to a map with full Jacobian. In other words, placing $\mathbf{Q}$ at the beginning (outside) of each recursive coupling block to apply the permutation over all variables leads to a dense $\mathbf{J}_\mathbf{T}$.

Moving from a triangular to a block triangular to a dense Jacobian increases the expressive power of the map but decreases its interpretability.

HINT can be performed using hierarchical affine coupling blocks, which is inspired by the second case mentioned earlier that leads to a block-triangular Jacobian matrix. In such a hierarchical architecture, splitting the variables at different levels may have different interpretations [207]. In a system with paired data, both $\mathbf{x}$ and $\mathbf{u}$ can be treated as inputs to the generative model pipeline and the top level split is used to transform them separately at the next recursion level. At the beginning of the hierarchical affine coupling block, only one orthogonal permutation is applied to each of the $\mathbf{x}$- and $\mathbf{u}$-lanes with the corresponding permutation matrices $\mathbf{Q}_\mathbf{x}$ and $\mathbf{Q}_\mathbf{u}$. Similar to (9.170), the objective function for training the model is defined as:

$$\mathcal{L}(\mathbf{x},\mathbf{u}) = \frac{1}{2} \parallel \mathbf{T}(\mathbf{x},\mathbf{u})\|^2 - \log\left|\mathbf{J}_\mathbf{T}(\mathbf{x},\mathbf{u})\right| . \tag{9.178}$$

The joint density of $(\mathbf{x},\mathbf{u})$ is the pull-back of the output density $p_\mathbf{Y}(\mathbf{y})$ via $\mathbf{T}^{-1}$:

$$p_\mathbf{T}(\mathbf{x},\mathbf{u}) = \mathbf{T}_\#^{-1} p_\mathbf{Y}(\mathbf{y}) = \mathbf{T}_\#^{-1} \mathcal{N}(\mathbf{0}, \mathbf{I}_{|\mathbf{x}|+|\mathbf{u}|}). \tag{9.179}$$

Since in the hierarchical architecture, the **u**-lane can be evaluated independently of the **x**-lane, for a given **u**, the partial latent code associated with the **u**-lane can be fixed, while drawing samples from the **x**-part of the distribution, which yields samples from the following conditional probability density function:

$$p_{\mathbf{T}}(\mathbf{x}|\mathbf{u}) = \mathbf{T}_{\#}^{-1} \begin{bmatrix} \mathbf{T}_{\mathbf{u}}(\mathbf{u}) \\ \mathcal{N}(\mathbf{0}, \mathbf{I}_{|\mathbf{x}|}) \end{bmatrix}. \tag{9.180}$$

Therefore, HINT can provide both the joint and the conditional densities of the variables of interest.

9.13 Applications

Two case studies are reviewed in this section. The first case study is focused on healthcare and using the DKF for estimating the effect of anti-diabetic drugs. The second case study is focused on visual odometry and using the DPF for autonomous driving.

9.13.1 Prediction of Drug Effect

Application of artificial intelligence in healthcare has been facilitated through developing large-scale medical databases. *Electronic health records* include sequences of diagnoses, drug prescriptions, laboratory reports, and surgeries for a large number of patients. Learning causal generative temporal models from such datasets paves the way for answering medical questions toward finding the best course of treatment for a patient. To build such models, a representation of the patient's status must be learned, which is sensitive to the effect of treatments and evolves over time. For these models, the generalization ability is of crucial importance in order to predict the probability of an effect in circumstances, which are different from the empirically observed ones. For instance, DKF has been successfully used for predicting the effect of anti-diabetic drugs [188]. The corresponding generative model was trained on a population of 8000 diabetic and pre-diabetic patients to model the disease progression. Two variables of interest were glucose and hemoglobin A1c. The former reflects the amount of a patient's blood sugar, and the latter is an indicator of the level of diabetes. Age, gender, and ICD-9 diagnosis codes were considered in the model as well. The ICD-9 diagnosis codes are related to comorbidities of diabetes including congestive heart failure, chronic kidney disease, and obesity. Prescriptions of nine diabetic drugs including Metformin and Insulin were considered as actions, and the effect of anti-diabetic drugs on a patient's A1c and glucose levels were studied.

9.13.2 Autonomous Driving

According to the definition of SAE J3016 [216], the following levels of driving automation are considered for on-road vehicles:

- *Level 0*: No automation
- *Level 1*: Driver assistance
- *Level 2*: Partial automation
- *Level 3*: Conditional automation
- *Level 4*: High automation
- *Level 5*: Full automation

This definition, which spans from no automation to full automation, paves the way for a stepwise introduction of automated vehicles regarding standardization, testing, safety, or in-vehicle technology [217].

The KITTI Vision Benchmark Suite provides real-world computer vision benchmarks for stereo, optical flow, and visual odometry as well as three-dimensional object detection and tracking [218]. DPF was applied to the KITTI visual odometry dataset, which includes 11 trajectories of a real car driving in an urban area for 40 minutes. Images from an RGB stereo camera as well as the ground-truth position and velocity were available in the dataset. The car's position and orientation as well as its translational and angular velocity were considered as state variables. Simulations were performed by training the DPF in an end-to-end manner as well as using individual learning for the state-transition and the measurement models. The DPF was compared against the Backprop KF in terms of performance in the considered visual odometry task. The DPF algorithm was further evaluated regarding a global localization task in a maze [190].

9.14 Concluding Remarks

Combination of deep learning with Bayesian filtering has led to a wide variety of filtering algorithms, which rely on supervised or unsupervised learning for building data-driven state-space models. The majority of deep learning-based filters are trained via optimizing the ELBO. In a class of filters including the DKF algorithm, the transition occurs only in the KLD term of the objective function. Hence, the gradient from the generative model cannot be backpropagated through the transitions. To address this issue, the DVBF algorithm performs latent dynamic system identification, which allows for learning state-space models even from non-Markovian sequential data. Using the SGVB, DVBF can handle large-scale datasets and overcome intractable inference. The corresponding generative model provides stable long-term predictions beyond the sequence length used for training.

10

Expectation Maximization

10.1 Introduction

In machine learning and statistics, computing the *maximum likelihood* (ML) or *maximum a posteriori* (MAP) parameter estimate relies on the availability of complete data. However, if the model consists of latent variables or there is missing data, then ML and MAP estimation will become a challenging problem. In such cases, gradient descent methods can be used to find a local minimum of the *negative log likelihood* (NLL) [77]:

$$\mathrm{NLL}(\boldsymbol{\theta}) = -\frac{1}{N+1} \log p_{\boldsymbol{\theta}}(\mathcal{D}), \tag{10.1}$$

where $\boldsymbol{\theta}$ represents the set of model parameters and $\mathcal{D}$ denotes the set of $N+1$ observed data points for $k = 0, \dots, N$. It is often required to impose additional constraints on the model such as mixing weights must be normalized and covariance matrices must be positive definite. *Expectation maximization* (EM) algorithm paves the way for addressing this issue. As an iterative algorithm, EM enforces the required constraints and handles the missing data problem by alternating between two steps [77]:

- *E-step*: inferring the missing values given the parameters, and then,
- *M-step*: optimizing the parameters given the filled-in data.

In the following, after reviewing the EM algorithm, its machine learning-based variants will be presented.

10.2 Expectation Maximization Algorithm

Regarding a state-space model, let $\mathbf{x} \in \mathbb{R}^{n_x}$, $\mathbf{u} \in \mathbb{R}^{n_u}$, and $\mathbf{y} \in \mathbb{R}^{n_y}$ denote the latent state, the input (action), and the output (observation) vectors, respectively. The

Nonlinear Filters: Theory and Applications, First Edition. Peyman Setoodeh, Saeid Habibi, and Simon Haykin.

goal is to find the set of optimal model parameters by maximizing the log likelihood of the observed data:

$$\begin{aligned}\ell(\theta) &= \sum_{k=0}^{N} \log p_\theta(\mathbf{y}_k) \\ &= \sum_{k=0}^{N} \log\left(\sum_{\mathbf{x}_k} p_\theta(\mathbf{x}_k, \mathbf{y}_k)\right).\end{aligned} \tag{10.2}$$

Maximizing this likelihood would be difficult due to the fact that the logarithm cannot be moved inside the inner summation. To get around this problem, let us define the *complete-data log likelihood* as:

$$\ell_c(\theta) = \sum_{k=0}^{N} \log p_\theta(\mathbf{x}_k, \mathbf{y}_k), \tag{10.3}$$

which cannot be computed, since state variables are usually hidden and may not be directly measured. Now, let us define the *auxiliary function* $\mathcal{Q}(\cdot)$ as the expected complete-data log likelihood:

$$\mathcal{Q}(\theta, \theta_t) = \mathbb{E}\left[\ell_c(\theta)|\mathcal{D}, \theta_t\right], \tag{10.4}$$

where the expectation is taken with respect to the parameter values at the current iteration, θ_t, and the observed data, $\mathcal{D}$. The E-step computes the *expected sufficient statistics*, on which the ML estimation depends. The M-step optimizes $\mathcal{Q}$ with respect to θ:

$$\theta_{t+1} = \arg\max_{\theta} \ \mathcal{Q}(\theta, \theta_t). \tag{10.5}$$

MAP estimation can be performed by keeping the E-step unchanged, but modifying the M-step by adding the log prior to the aforementioned objective function:

$$\theta_{t+1} = \arg\max_{\theta} \ \mathcal{Q}(\theta, \theta_t) + \log p(\theta). \tag{10.6}$$

Regarding an arbitrary distribution $q(\mathbf{x}_k)$ over the hidden state variables, the observed data log likelihood can be written as:

$$\begin{aligned}\ell(\theta) &= \sum_{k=0}^{N} \log\left(\sum_{\mathbf{x}_k} p_\theta(\mathbf{x}_k, \mathbf{y}_k)\right) \\ &= \sum_{k=0}^{N} \log\left(\sum_{\mathbf{x}_k} q(\mathbf{x}_k)\frac{p_\theta(\mathbf{x}_k, \mathbf{y}_k)}{q(\mathbf{x}_k)}\right).\end{aligned} \tag{10.7}$$

Using *Jensen's inequality*:

$$f(\mathbb{E}[z]) \geq \mathbb{E}[f(z)], \tag{10.8}$$

where f is a concave function such as logarithm and z is a random variable, the following lower bound can be obtained:

$$
\begin{aligned}
\ell(\theta) &\geq \sum_{k=0}^{N} \sum_{\mathbf{x}_k} q(\mathbf{x}_k) \log \frac{p_\theta(\mathbf{x}_k, \mathbf{y}_k)}{q(\mathbf{x}_k)} \\
&= \sum_{k=0}^{N} \mathbb{E}_q \left[\log p_\theta(\mathbf{x}_k, \mathbf{y}_k)\right] + H(q) \\
&= \sum_{k=0}^{N} L(\theta, q) \\
&= \mathcal{Q}(\theta, q),
\end{aligned} \tag{10.9}
$$

where $H(q)$ is the entropy of q. This entropy is independent of θ. Although equation (10.9) is valid for any positive distribution q, it is desired to choose q in a way to achieve the tightest lower bound. Regarding the lower bound in (10.9), $L(\theta, q)$ can be written as:

$$
\begin{aligned}
L(\theta, q) &= \sum_{\mathbf{x}_k} q(\mathbf{x}_k) \log \frac{p_\theta(\mathbf{x}_k, \mathbf{y}_k)}{q(\mathbf{x}_k)} \\
&= \sum_{\mathbf{x}_k} q(\mathbf{x}_k) \log \frac{p_\theta(\mathbf{x}_k|\mathbf{y}_k) p_\theta(\mathbf{y}_k)}{q(\mathbf{x}_k)} \\
&= \sum_{\mathbf{x}_k} q(\mathbf{x}_k) \log \frac{p_\theta(\mathbf{x}_k|\mathbf{y}_k)}{q(\mathbf{x}_k)} + \sum_{\mathbf{x}_k} q(\mathbf{x}_k) \log p_\theta(\mathbf{y}_k) \\
&= \sum_{\mathbf{x}_k} q(\mathbf{x}_k) \log \frac{p_\theta(\mathbf{x}_k|\mathbf{y}_k)}{q(\mathbf{x}_k)} + \log p_\theta(\mathbf{y}_k) \sum_{\mathbf{x}_k} q(\mathbf{x}_i) \qquad (10.10) \\
&= -\text{KLD}\left(q(\mathbf{x}_k) || p_\theta(\mathbf{x}_k|\mathbf{y}_k)\right) + \log p_\theta(\mathbf{y}_k), \qquad (10.11)
\end{aligned}
$$

where in (10.11), KLD denotes the Kullback–Leibler divergence, and the second term was reached due to the fact that $p_\theta(\mathbf{y}_k)$ is independent of q and $\sum_{\mathbf{x}_k} q(\mathbf{x}_k) = 1$. The lower bound can be maximized by choosing $q_t(\mathbf{x}_k) = p_{\theta_t}(\mathbf{x}_k|\mathbf{y}_k)$, where θ_t represents the estimate of parameters at iteration t. Regarding (10.9), the following auxiliary function is provided by the E-step:

$$
\mathcal{Q}(\theta, q_t) = \sum_{k=0}^{N} \mathbb{E}_{q_t} \left[\log p_\theta(\mathbf{x}_k, \mathbf{y}_k)\right] + H(q_t). \tag{10.12}
$$

Since $H(q_t)$ is independent of θ, the following optimization is solved in the M-step:

$$
\begin{aligned}
\theta_{t+1} &= \arg\max_\theta \ \mathcal{Q}(\theta, \theta_t) \\
&= \arg\max_\theta \ \sum_{k=0}^{N} \mathbb{E}_{q_t} \left[\log p_\theta(\mathbf{x}_k, \mathbf{y}_k)\right].
\end{aligned} \tag{10.13}
$$

Choosing $q_t(\mathbf{x}_k) = p_{\theta_t}(\mathbf{x}_k|\mathbf{y}_k)$, the KLD in (10.11) will be zero, and we will have:

$$L(\boldsymbol{\theta}_t, q_t) = \log p_{\theta_t}(\mathbf{y}_k), \tag{10.14}$$

$$\begin{aligned}\mathcal{Q}(\boldsymbol{\theta}_t, \boldsymbol{\theta}_t) &= \sum_{k=0}^{N} \log p_{\theta_t}(\mathbf{y}_k)\\ &= \ell(\boldsymbol{\theta}_t).\end{aligned} \tag{10.15}$$

The log likelihood of the observed data for ML estimation either stays the same or monotonically increases by the EM algorithm [77]:

$$\begin{aligned}\ell(\boldsymbol{\theta}_{t+1}) &\geq \mathcal{Q}(\boldsymbol{\theta}_{t+1}, \boldsymbol{\theta}_t)\\ &= \max_{\boldsymbol{\theta}} \ \mathcal{Q}(\boldsymbol{\theta}, \boldsymbol{\theta}_t)\\ &\geq \mathcal{Q}(\boldsymbol{\theta}_t, \boldsymbol{\theta}_t) \qquad (10.16)\\ &= \ell(\boldsymbol{\theta}_t). \qquad (10.17)\end{aligned}$$

For MAP estimation, this result can be extended to the log likelihood of the observed data plus the log prior.

Section 10.3 provides a particle-based implementation of the EM algorithm.

10.3 Particle Expectation Maximization

The *particle expectation maximization* (P-EM) algorithm has been proposed for joint state and parameter estimation for a general class of stochastic nonlinear systems described by the following state-space model [219]:

$$\mathbf{x}_{k+1} = \mathbf{f}_k(\mathbf{x}_k, \mathbf{u}_k, \mathbf{v}_k, \boldsymbol{\theta}), \tag{10.18}$$

$$\mathbf{y}_k = \mathbf{g}_k(\mathbf{x}_k, \mathbf{u}_k, \mathbf{w}_k, \boldsymbol{\theta}), \tag{10.19}$$

where $\mathbf{v}_k$ denotes the process noise, $\mathbf{w}_k$ denotes the measurement noise, and $\boldsymbol{\theta}$ denotes the set of model parameters. Regarding the uncertainty due to $\mathbf{v}_k$ and $\mathbf{w}_k$, this state-space model can be expressed as the following conditional probabilities:

$$p_{\theta}(\mathbf{x}_{k+1}|\mathbf{x}_k, \mathbf{u}_k), \tag{10.20}$$

$$p_{\theta}(\mathbf{y}_k|\mathbf{x}_k, \mathbf{u}_k). \tag{10.21}$$

For ML estimation, the P-EM algorithm proceeds as follows [219]:

- The model parameters are initialized in a way to obtain a finite log likelihood, $\ell(\boldsymbol{\theta}_0)$.

- E-step performs nonlinear state estimation to obtain the following representations using particle filtering and particle smoothing, respectively:

$$p_\theta(\mathbf{x}_{k+1}|\mathbf{u}_{0:k+1},\mathbf{y}_{0:k+1}) \approx \sum_{i=1}^{n_p} w^i_{k+1}\delta(\mathbf{x}_{k+1}-\mathbf{x}^i_{k+1}), \tag{10.22}$$

$$p_\theta(\mathbf{x}_{k+1}|\mathbf{u}_{0:N},\mathbf{y}_{0:N}) \approx \sum_{i=1}^{n_p} w^i_{k+1|N}\delta(\mathbf{x}_{k+1}-\mathbf{x}^i_{k+1}), \tag{10.23}$$

where n_p is the number of particles and $k = 0, \dots, N-1$.
Particle filtering is performed using Algorithm 6.1 to obtain $\{\mathbf{x}^i_{0:N}, w^i_{0:N}\}$ for $i = 1, \dots, n_p$. The *forward-backward particle smoother* maintains the original particle locations but reweights the particles to approximate the smoothed posterior [220]. Setting $w^i_{N|N} = w^i_N$ for $i = 1, \dots, n_p$, weights $w^i_{k|N}$ are computed using the following backward recursion:

$$w^i_{k|N} = w^i_k\left(\sum_{j=1}^{n_p} w^j_{k+1|N}\frac{p_\theta(\mathbf{x}^j_{k+1}|\mathbf{x}^i_k,\mathbf{u}_k)}{\sum_{l=1}^{n_p} w^l_k\, p_\theta(\mathbf{x}^j_{k+1}|\mathbf{x}^l_k,\mathbf{u}_k)}\right). \tag{10.24}$$

Equation (10.24) has a computational complexity of $O(n_p^3)$. However, the computational complexity can be reduced to $O(n_p^2)$ by computing the denominator for each j independent of i, because the denominator does not depend on i. A more computationally efficient particle smoother has been proposed in [221] with the computational complexity of $O(n_p)$. Deploying such a particle smoother will significantly reduce the computational burden.
Defining the weights $w^{ij}_{k|N}$ as:

$$w^{ij}_{k|N} = \frac{w^i_k\, w^j_{k+1|N}\, p_{\theta_t}(\mathbf{x}^j_{k+1}|\mathbf{x}^i_k,\mathbf{u}_k)}{\sum_{l=1}^{n_p} w^l_k\, p_{\theta_t}(\mathbf{x}^j_{k+1}|\mathbf{x}^l_k,\mathbf{u}_k)}, \tag{10.25}$$

the following importance sampling approximations are computed:

$$I_1 \approx \sum_{i=1}^{n_p} w^i_{0|N}\log p_\theta(\mathbf{x}^i_0), \tag{10.26}$$

$$I_2 \approx \sum_{k=0}^{N-1}\sum_{i=1}^{n_p}\sum_{j=1}^{n_p} w^{ij}_{k|N}\log p_\theta(\mathbf{x}^j_{k+1}|\mathbf{x}^i_k,\mathbf{u}_k), \tag{10.27}$$

$$I_3 \approx \sum_{k=0}^{N}\sum_{i=1}^{n_p} w^i_{k|N}\log p_\theta(\mathbf{y}_k|\mathbf{x}^i_k,\mathbf{u}_k). \tag{10.28}$$

Now, $\mathcal{Q}(\boldsymbol{\theta},\boldsymbol{\theta}_t)$ is computed as:

$$\mathcal{Q}(\boldsymbol{\theta},\boldsymbol{\theta}_t) = I_1 + I_2 + I_3. \tag{10.29}$$

- M-step optimizes $\mathcal{Q}$ with respect to $\boldsymbol{\theta}$ in order to update the parameters:

$$\boldsymbol{\theta}_{t+1} = \arg\max_{\boldsymbol{\theta}} \mathcal{Q}(\boldsymbol{\theta}, \boldsymbol{\theta}_t). \tag{10.30}$$

- While the algorithm has not converged:

$$\mathcal{Q}(\boldsymbol{\theta}_{t+1}, \boldsymbol{\theta}_t) - \mathcal{Q}(\boldsymbol{\theta}_t, \boldsymbol{\theta}_t) > \epsilon, \tag{10.31}$$

for some $\epsilon > 0$, the E-step and the M-step will be repeated, otherwise the algorithm will terminate.

Section 10.4 discusses how to fit a mixture of Gaussians using the EM algorithm.

10.4 Expectation Maximization for Gaussian Mixture Models

From (10.3) and (10.4), the auxiliary function for M mixture components is computed as:

$$\mathcal{Q}(\boldsymbol{\theta}, \boldsymbol{\theta}_t) = \mathbb{E}\left[\sum_{k=0}^{N} \log p_{\boldsymbol{\theta}}(\mathbf{x}_k, \mathbf{y}_k)\right] \tag{10.32}$$

$$= \sum_{k=0}^{N}\sum_{j=1}^{M} p_{\boldsymbol{\theta}_t}\left(\mathbf{x}_k \sim \mathcal{N}^j | \mathbf{y}_k\right) \log\left(w^j p_{\boldsymbol{\theta}^j}(\mathbf{y}_k)\right), \tag{10.33}$$

where $r^{kj} \triangleq p_{\boldsymbol{\theta}_t}\left(\mathbf{x}_k \sim \mathcal{N}^j | \mathbf{y}_k\right)$ reflects the responsibility that cluster j with parameters $\boldsymbol{\theta}^j$ takes for data point k. The EM algorithm proceeds as follows [77]:

- E-step computes r^{kj} as:

$$r^{kj} = \frac{w^j p_{\boldsymbol{\theta}_t^j}(\mathbf{y}_k)}{\sum_{l=1}^{M} w^l p_{\boldsymbol{\theta}_t^l}(\mathbf{y}_k)}, \tag{10.34}$$

which is the same for any mixture model.

- M-step optimizes $\mathcal{Q}$ with respect to $\boldsymbol{\theta}_t$ and $\mathbf{w} = \begin{bmatrix} w^1 & \cdots & w^M \end{bmatrix}^T$. For $\mathbf{w}$, we have:

$$w^j = \frac{1}{N+1}\sum_{k=0}^{N} r^{kj}$$
$$= \frac{1}{N+1} r^j, \tag{10.35}$$

where $r^j = \sum_{k=0}^{N} r^{kj}$ is the weighted number of points assigned to cluster j. Regarding Gaussian distributions, $\mathcal{N}^j(\boldsymbol{\mu}^j, \boldsymbol{\Sigma}^j)$, the log likelihood of the observed data can be written as:

$$
\begin{aligned}
\ell(\theta^j) &= \ell(\boldsymbol{\mu}^j, \boldsymbol{\Sigma}^j) \\
&= \sum_{k=0}^{N}\sum_{j=1}^{M} r^{kj} \log p_{\theta^j}(\mathbf{y}_k) \\
&= -\frac{1}{2}\sum_{k=0}^{N} r^{kj}\left(\log|\boldsymbol{\Sigma}^j| + (\mathbf{x}_k - \boldsymbol{\mu}^j)^T(\boldsymbol{\Sigma}^j)^{-1}(\mathbf{x}_k - \boldsymbol{\mu}^j)\right),
\end{aligned} \tag{10.36}
$$

where the new parameter estimates are computed as:

$$
\boldsymbol{\mu}^j = \frac{\sum_{k=0}^{N} r^{kj}\mathbf{x}_k}{r^j}, \tag{10.37}
$$

$$
\begin{aligned}
\boldsymbol{\Sigma}^j &= \frac{\sum_{k=0}^{N} r^{kj}(\mathbf{x}_k - \boldsymbol{\mu}^j)(\mathbf{x}_k - \boldsymbol{\mu}^j)^T}{r^j} \\
&= \frac{\sum_{k=0}^{N} r^{kj}\mathbf{x}_k\mathbf{x}_k^T}{r^j} - \boldsymbol{\mu}^j(\boldsymbol{\mu}^j)^T.
\end{aligned} \tag{10.38}
$$

For each cluster, the mean is the weighted average of its assigned data points, and the covariance is proportional to the weighted empirical scatter matrix. The M-step ends by updating the parameter set, $\theta_{t+1} = \{w^j, \boldsymbol{\mu}^j, \boldsymbol{\Sigma}^j\}$ for $j = 1, \ldots, M$, after computing the new estimates. Then, the next E-step starts.

In Section 10.5, a neural network-based variant of the EM algorithm is presented that simultaneously learns how to group and represent individual entities.

10.5 Neural Expectation Maximization

Identification and manipulation of conceptual entities are part and parcel of many real-world tasks including reasoning and physical interaction. In performing such tasks, automated discovery of symbol-like representations can play an important role, which can be viewed as inference in a spatial mixture model with components that are parametrized by neural networks. Building on the EM algorithm, this framework provides a differentiable clustering technique for simultaneous learning of grouping and representing different entities in an unsupervised manner. The resulting algorithm known as *neural expectation maximization* (N-EM) can perform *perceptual grouping*, which refers to dynamic deconstruction of input into its constituent conceptual entities. Perceptual grouping allows for learning efficient representations for different clusters [222].

Training a system, which aims at providing efficient representations for the individual conceptual entities contained in the observed data, calls for deploying a notion of *entity*. In unsupervised learning, this notion relies on statistical properties of the data. To be more precise, a conceptual entity can be viewed as a common cause for multiple observations, hence, it induces a dependency structure. For instance, regarding the objects in an image, each object can be viewed as the common cause for the pixels that depict it. Then, due to the dependency between the affected pixels by an object, some of these pixels can be predicted based on the knowledge about the others. Hence, knowledge about some pixels of an object can help to partially recover the object. This is especially useful in handling sequential data such as video streams, where pixels associated with an object share a common fate (moving along a specific trajectory). However, knowledge about pixels of an object may not improve predictions about pixels of other objects.

Let us represent each object, j, by a set of parameters, $\boldsymbol{\theta}^j$. It is desirable that these parameters capture the dependency structure of the affected pixels by object j without carrying any information about the rest of the image. This property leads to modularity, and paves the way for generalization through reusing the same representation in different contexts. Modularity allows for representing novel objects as combinations of known objects. Treating each image as a composition of M objects and assigning each pixel to one of these objects, each image can be viewed as a mixture model. Therefore, the EM algorithm can be used for learning how to group different entities as well as computing the ML estimate for each entity's corresponding parameters, $\boldsymbol{\theta}^j$. For unsupervised learning of a family of distributions corresponding to an object level representation, $p_{\boldsymbol{\theta}^j}(\mathbf{y})$, this family of distributions is parameterized by a differentiable function $\mathbf{f}_{\boldsymbol{\phi}}(\boldsymbol{\theta}^j)$ such as a neural network with the set of parameters $\boldsymbol{\phi}$. This framework provides a fully differentiable version of the EM algorithm that allows for backpropagation of an outer loss to the weights of the neural network [222].

Let us represent each image as a vector of pixels, $\mathbf{y} \in \mathbb{R}^{n_y}$. Each image can be viewed as a spatial mixture of M entities, which are, respectively, parametrized by vectors $\boldsymbol{\theta}^j \in \mathbb{R}^{n_{obj}}$ for $j = 1, \dots, M$. The differentiable nonlinear function $\mathbf{f}_{\boldsymbol{\phi}}(\boldsymbol{\theta}^j)$ transforms the entity representation $\boldsymbol{\theta}^j$ into parameters $\boldsymbol{\psi}^{ij} = \mathbf{f}^i_{\boldsymbol{\phi}}(\boldsymbol{\theta}^j)$ for separate pixel-wise distributions, where i refers to pixels of the image, y_i. For Gaussian distributions, $\boldsymbol{\psi}^{ij}$ refers to mean and variance. In contrast with standard mixture models, it is assumed that for a representation, pixels are independent but not identically distributed. Let us define a matrix of binary latent variables, $\mathcal{X} \in [0,1]^{n_y \times M}$, which encodes the unknown true pixel assignments. Its elements, x_{ij}, are determined as follows:

$$x_{ij} = \mathcal{X}_{ij} = \begin{cases} 1, & \text{pixel } i \text{ is generated by entity } j \\ 0, & \text{otherwise,} \end{cases} \tag{10.39}$$

with $\sum_j x_{ij} = 1$. Given the set of parameters, $\theta = \{\theta^1, \dots, \theta^M\}$, the likelihood is computed as [222]:

$$p_\theta(\mathbf{y}) = \prod_{i=1}^{n_y} \sum_{j=1}^{M} p_{\psi^{ij}}(x_{ij}, y_i)$$
$$= \prod_{i=1}^{n_y} \sum_{j=1}^{M} \underbrace{p(x_{ij} = 1)}_{w^j} p_{\psi^{ij}}(y_i | x_{ij} = 1), \tag{10.40}$$

where the weight vector $\mathbf{w} = \begin{bmatrix} w^1 & \cdots & w^M \end{bmatrix}^T$ represents the mixing coefficients or the prior.

Optimizing $\log p_\psi(\mathbf{y})$ with respect to θ is difficult due to marginalization over $\mathcal{X}$. It would be easier to optimize $\log p_\psi(\mathcal{X}, \mathbf{y})$ using the EM algorithm, which maximizes the following lower bound:

$$\mathcal{Q}(\theta, \theta_t) = \sum_{\mathcal{X}} p_{\psi_t}(\mathcal{X} | \mathbf{y}) \log p_\psi(\mathcal{X}, \mathbf{y}). \tag{10.41}$$

This lower bound is iteratively optimized by alternating between the following two steps [222]:

- E-step: Given θ_t from the previous iteration, a new estimate of the posterior distribution over the latent variables is computed to obtain a new soft-assignment of pixels to entities (clusters):

$$\gamma^{ij} = p_{\psi_t}(x_{ij} = 1 | y_i). \tag{10.42}$$

- M-step: Using the posterior computed in the E-step, the optimal value of θ is found by solving the following optimization problem:

$$\theta_{t+1} = \arg\max_{\theta} \ \mathcal{Q}(\theta, \theta_t), \tag{10.43}$$

using the gradient ascent method as:

$$\theta_{t+1} = \theta_t + \eta \frac{\partial \mathcal{Q}}{\partial \theta}, \tag{10.44}$$

where η is the learning rate. For a fixed σ^2, assuming that:

$$p_{\psi^{ij}}(y_i | x_{ij} = 1) = \mathcal{N}(y_i | \mu = \psi_{ij}, \sigma^2), \tag{10.45}$$

we have:

$$\frac{\partial \mathcal{Q}}{\partial \theta^j} \propto \sum_{i=1}^{n_y} \gamma^{ij} (\psi^{ij} - y_i) \frac{\partial \psi^{ij}}{\partial \theta^j}. \tag{10.46}$$

Unrolling the iterations of this EM algorithm, an end-to-end differentiable clustering procedure is obtained based on the statistical model represented by $\mathbf{f}_\phi(\cdot)$.

The N-EM algorithm resembles M copies of a recurrent neural network with hidden states $\boldsymbol{\theta}^j$. Each copy j receives inputs $\gamma^{ij}(\boldsymbol{\psi}^{ij} - y_i)$ and generates outputs $\boldsymbol{\psi}^{ij}$, which are used to update the soft assignments γ^{ij} in the E-step. However, some constraints must be imposed on the architecture and parameters of the recurrent neural networks to force them to accurately mimic the M-step of the EM algorithm. In order to train each one of the M neural networks to model the structure of one object regardless of any other information in the image, the following loss function is considered [222]:

$$\mathcal{L}(\mathbf{y}) = -\sum_{i=1}^{n_y}\sum_{j=1}^{M}\left(\underbrace{\gamma^{ij}\log p_{\boldsymbol{\psi}^{ij}}(x_{ij}, y_i)}_{\text{intra-cluster loss}} - \underbrace{(1-\gamma^{ij})\mathrm{KLD}\left(p(y_i)||p_{\boldsymbol{\psi}^{ij}}(y_i|x_{ij})\right)}_{\text{inter-cluster loss}}\right). \tag{10.47}$$

Section 10.6 is focused on modeling interactions between objects by adding relational structure to the N-EM algorithm.

10.6 Relational Neural Expectation Maximization

Causal understanding would be a key component for creating artificial general intelligence. A truly intelligent agent must be able to acquire causal knowledge without access to labeled data. The *relational neural expectation maximization* (R-NEM) algorithm aims at learning to discover objects and model their physical interactions from images in an unsupervised manner. Inspired by the compositional nature of human perception, R-NEM relies on factoring interactions between object pairs. Compared to the N-EM algorithm, the goal of R-NEM is to model interactions between objects through adding relational structure without distorting the learned representations of objects [223].

The R-NEM relies on a generalized dynamic model, which includes an *interaction function*, $\boldsymbol{\Upsilon}$. As described in Section 10.5, such a dynamic model can be implemented using M copies of recurrent neural networks (RNNs). For $j = 1, \dots, M$, these RNNs compute object representations $\boldsymbol{\theta}_k^j$ at time step k, as functions of all object representations, $\boldsymbol{\theta}_{k-1} = \{\boldsymbol{\theta}_{k-1}^1, \dots, \boldsymbol{\theta}_{k-1}^M\}$, at the previous time step $k-1$, via their corresponding interaction functions $\boldsymbol{\Upsilon}^j$ [223]:

$$\boldsymbol{\theta}_k^j = \mathrm{RNN}\left(\mathbf{y}_k, \boldsymbol{\Upsilon}^j(\boldsymbol{\theta}_{k-1})\right), \tag{10.48}$$

where $\mathbf{y}_k$ denotes the observed data, which may be the output of an observation encoder. The N-EM algorithm can be viewed a special case of the R-NEM, and can be obtained from (10.48) by choosing $\boldsymbol{\Upsilon}^j(\boldsymbol{\theta}_{k-1}) = \boldsymbol{\theta}_{k-1}^j$.

Modeling assumptions regarding the interactions between objects in the environment can be incorporated in the model via the interaction function, Υ. Rewriting equation (10.48) as [223]:

$$\theta^j_k = \sigma\left(\mathbf{W}^y \mathbf{y}_k + \mathbf{W}^\Upsilon \Upsilon^j(\theta_{k-1})\right), \tag{10.49}$$

where $\mathbf{W}^y$ and $\mathbf{W}^\Upsilon$ are weight matrices and $\sigma(.)$ denotes the sigmoid activation function, reveals that Υ provides an *inductive bias* that reflects the underlying modeling assumptions. The interaction function can be parameterized to incorporate modeling assumptions and θ^j updates based on the pairwise interactions between objects. A set of neural networks are used to transform θ^j to $\hat{\theta}^j$ for $j = 1, \dots, M$. In this way, these representations will contain the information relevant to object dynamics in a more explicit way. Then, each pair $(\hat{\theta}^{j_1}, \hat{\theta}^{j_2})$ is concatenated and processed by another neural network to compute a shared embedding $\zeta^{j_1 j_2}$, which encodes the interaction between the two objects j_1 and j_2. To provide a clear separation between the *focus* object j_1 and the *context* object j_2, two other neural networks are used to compute the effect of object j_2 on object j_1, denoted by $\mathbf{e}_{j_1 j_2}$, and an attention coefficient [167], denoted by $\alpha_{j_1 j_2}$, based on the shared embedding, $\zeta^{j_1 j_2}$:

$$\hat{\theta}^j = \mathrm{NN}^{\mathrm{encode}}(\theta^j), \tag{10.50}$$

$$\zeta^{j_1 j_2} = \mathrm{NN}^{\mathrm{embed}}\left(\begin{bmatrix} \hat{\theta}^{j_1} \\ \hat{\theta}^{j_2} \end{bmatrix}\right), \tag{10.51}$$

$$\mathbf{e}_{j_1 j_2} = \mathrm{NN}^{\mathrm{effect}}(\zeta^{j_1 j_2}), \tag{10.52}$$

$$\alpha_{j_1 j_2} = \mathrm{NN}^{\mathrm{attention}}(\zeta^{j_1 j_2}). \tag{10.53}$$

The attention coefficient encodes the occurrence of interaction between two objects to select the relevant context objects. Finally, the total effect of other objects in the environment on the object, which is represented by θ^j, can be computed as a weighted sum of effects as:

$$\mathbf{E}^j = \sum_{i \neq j, i=1}^{K} \alpha_{ji}\, \mathbf{e}_{ji}, \tag{10.54}$$

where effects are multiplied by their corresponding attention coefficients. The interaction function can be computed as:

$$\Upsilon^j(\theta) = \begin{bmatrix} \hat{\theta}^j \\ \mathbf{E}^j \end{bmatrix}. \tag{10.55}$$

Section 10.7 presents the variational filtering EM algorithm.

10.7 Variational Filtering Expectation Maximization

The *variational filtering expectation* maximization (VFEM) algorithm uses a variational objective in the filtering setting. It provides a general-purpose method for performing variational inference for dynamic latent-variable models through solving an optimization problem at each time step [175]. In filtering, online approximate inference is performed to estimate the posterior at each time step based on the past and present information. Hence, the approximate posterior, $q(\mathbf{x}_{0:k+1}|\mathbf{u}_{0:k+1}, \mathbf{y}_{0:k+1})$, can be factorized as:

$$q(\mathbf{x}_{0:k+1}|\mathbf{u}_{0:k+1}, \mathbf{y}_{0:k+1}) = \prod_{i=0}^{k} q(\mathbf{x}_{i+1}|\mathbf{x}_{0:i}, \mathbf{u}_{0:i+1}, \mathbf{y}_{0:i+1}). \tag{10.56}$$

Regarding the joint distribution:

$$\begin{aligned} p_\theta(\mathbf{x}_{0:k+1}, \mathbf{y}_{0:k+1}|\mathbf{u}_{0:k+1}) \\ &= \prod_{i=0}^{k} p_\theta(\mathbf{x}_{i+1}, \mathbf{y}_{i+1}|\mathbf{x}_{0:i}, \mathbf{u}_{0:i+1}, \mathbf{y}_{0:i}) \\ &= \prod_{i=0}^{k} p_\theta(\mathbf{y}_{i+1}|\mathbf{x}_{0:i+1}, \mathbf{u}_{0:i+1}, \mathbf{y}_{0:i})\, p_\theta(\mathbf{x}_{i+1}|\mathbf{x}_{0:i}, \mathbf{u}_{0:i}, \mathbf{y}_{0:i}), \end{aligned} \tag{10.57}$$

and the approximate posterior (10.56), the *variational free energy*:

$$\mathcal{F} = -\mathbb{E}_q \left[\log \frac{p_\theta(\mathbf{x}_{0:k+1}, \mathbf{y}_{0:k+1}|\mathbf{u}_{0:k+1})}{q(\mathbf{x}_{0:k+1}|\mathbf{u}_{0:k+1}, \mathbf{y}_{0:k+1})} \right], \tag{10.58}$$

can be rewritten as:

$$\begin{aligned} \mathcal{F} &= -\mathbb{E}_{q(\mathbf{x}_{0:k+1}|\mathbf{u}_{0:k+1}, \mathbf{y}_{0:k+1})} \left[\log \prod_{i=0}^{k} \frac{p_\theta(\mathbf{x}_{i+1}, \mathbf{y}_{i+1}|\mathbf{x}_{0:i}, \mathbf{u}_{0:i+1}, \mathbf{y}_{0:i})}{q(\mathbf{x}_{i+1}|\mathbf{x}_{0:i}, \mathbf{u}_{0:i+1}, \mathbf{y}_{0:i+1})} \right] \\ &= -\mathbb{E}_{q(\mathbf{x}_{0:k+1}|\mathbf{u}_{0:k+1}, \mathbf{y}_{0:k+1})} \left[\sum_{i=0}^{k} \log \frac{p_\theta(\mathbf{x}_{i+1}, \mathbf{y}_{i+1}|\mathbf{x}_{0:i}, \mathbf{u}_{0:i+1}, \mathbf{y}_{0:i})}{q(\mathbf{x}_{i+1}|\mathbf{x}_{0:i}, \mathbf{u}_{0:i+1}, \mathbf{y}_{0:i+1})} \right] \\ &= -\sum_{i=0}^{k} \mathbb{E}_{q(\mathbf{x}_{0:i+1}|\mathbf{u}_{0:i+1}, \mathbf{y}_{0:i+1})} \left[\log \frac{p_\theta(\mathbf{x}_{i+1}, \mathbf{y}_{i+1}|\mathbf{x}_{0:i}, \mathbf{u}_{0:i+1}, \mathbf{y}_{0:i})}{q(\mathbf{x}_{i+1}|\mathbf{x}_{0:i}, \mathbf{u}_{0:i+1}, \mathbf{y}_{0:i+1})} \right] \\ &= -\sum_{i=0}^{k} \mathbb{E}_{\prod_{j=0}^{i} q(\mathbf{x}_{j+1}|\mathbf{x}_{0:j}, \mathbf{u}_{0:j+1}, \mathbf{y}_{0:j+1})} \left[\log \frac{p_\theta(\mathbf{x}_{i+1}, \mathbf{y}_{i+1}|\mathbf{x}_{0:i}, \mathbf{u}_{0:i+1}, \mathbf{y}_{0:i})}{q(\mathbf{x}_{i+1}|\mathbf{x}_{0:i}, \mathbf{u}_{0:i+1}, \mathbf{y}_{0:i+1})} \right]. \end{aligned} \tag{10.59}$$

For simplicity of notation, let us define $\mathcal{G}_{i+1}$ as:

$$\mathcal{G}_{i+1} = \log \frac{p_\theta(\mathbf{x}_{i+1}, \mathbf{y}_{i+1}|\mathbf{x}_{0:i}, \mathbf{u}_{0:i+1}, \mathbf{y}_{0:i})}{q(\mathbf{x}_{i+1}|\mathbf{x}_{0:i}, \mathbf{u}_{0:i+1}, \mathbf{y}_{0:i+1})}. \tag{10.60}$$

Now equation (10.59) is rewritten as:

$$\begin{aligned}\mathcal{F} &= -\sum_{i=0}^{k} \mathbb{E}_{\prod_{j=0}^{i} q(\mathbf{x}_{j+1}|\mathbf{x}_{0:j},\mathbf{u}_{0:j+1},\mathbf{y}_{0:j+1})} \left[\mathcal{G}_{i+1}\right] \\ &= -\sum_{i=0}^{k} \mathbb{E}_{\prod_{j=0}^{i-1} q(\mathbf{x}_{j+1}|\mathbf{x}_{0:j},\mathbf{u}_{0:j+1},\mathbf{y}_{0:j+1})} \left[\mathbb{E}_{q(\mathbf{x}_{i+1}|\mathbf{x}_{0:i},\mathbf{u}_{0:i+1},\mathbf{y}_{0:i+1})} \left[\mathcal{G}_{i+1}\right]\right].\end{aligned} \tag{10.61}$$

Let us define the *step free energy* as:

$$\begin{aligned}\tilde{\mathcal{F}}_{k+1} &= -\mathbb{E}_{q(\mathbf{x}_{k+1}|\mathbf{x}_{0:k},\mathbf{u}_{0:k+1},\mathbf{y}_{0:k+1})} \left[\mathcal{G}_{k+1}\right] \\ &= -\mathbb{E}_{q(\mathbf{x}_{k+1}|\mathbf{x}_{0:k},\mathbf{u}_{0:k+1},\mathbf{y}_{0:k+1})} \left[\log \frac{p_\theta(\mathbf{x}_{k+1}, \mathbf{y}_{k+1}|\mathbf{x}_{0:k}, \mathbf{u}_{0:k+1}, \mathbf{y}_{0:k})}{q(\mathbf{x}_{k+1}|\mathbf{x}_{0:k}, \mathbf{u}_{0:k+1}, \mathbf{y}_{0:k+1})}\right],\end{aligned} \tag{10.62}$$

which leads to:

$$\begin{aligned}\mathcal{F} &= \sum_{i=0}^{k} \mathbb{E}_{\prod_{j=0}^{i} q(\mathbf{x}_{j+1}|\mathbf{x}_{0:j},\mathbf{u}_{0:j+1},\mathbf{y}_{0:j+1})} \left[\tilde{\mathcal{F}}_{i+1}\right] \\ &= \sum_{i=0}^{k} \tilde{\mathcal{F}}_{i+1}.\end{aligned} \tag{10.63}$$

The step free energy in (10.62) can be expressed as a reconstruction term plus a KLD term:

$$\begin{aligned}\tilde{\mathcal{F}}_{k+1} =& -\mathbb{E}_{q(\mathbf{x}_{k+1}|\mathbf{x}_{0:k},\mathbf{u}_{0:k+1},\mathbf{y}_{0:k+1})} \left[\log p_\theta(\mathbf{y}_{k+1}|\mathbf{x}_{0:k+1}, \mathbf{u}_{0:k+1}, \mathbf{y}_{0:k})\right] \\ &+ \mathrm{KLD}\left(q(\mathbf{x}_{k+1}|\mathbf{x}_{0:k}, \mathbf{u}_{0:k+1}, \mathbf{y}_{0:k+1}) || p_\theta(\mathbf{x}_{k+1}|\mathbf{x}_{0:k}, \mathbf{u}_{0:k}, \mathbf{y}_{0:k})\right).\end{aligned} \tag{10.64}$$

To perform filtering variational inference, the approximate posterior, q, is found in a way that the filtering free energy summation in (10.63) is minimized. The step free energy at each time step is computed based on the expectation over the past latent sequence. At time step $k + 1$, the past expectations for $i = 0, \dots, k$ have already been evaluated, hence, minimizing $\mathcal{F} = \sum_{i=0}^{k} \tilde{\mathcal{F}}_{i+1}$ will entirely depend on $\tilde{\mathcal{F}}_{k+1}$. Therefore, the approximate posterior can be found by solving the following optimization problem:

$$q^*(\mathbf{x}_{k+1}|\mathbf{x}_{0:k}, \mathbf{u}_{0:k+1}, \mathbf{y}_{0:k+1}) = \underset{q(\mathbf{x}_{k+1}|\mathbf{x}_{0:k},\mathbf{u}_{0:k+1},\mathbf{y}_{0:k+1})}{\arg\min} \tilde{\mathcal{F}}_{k+1}. \tag{10.65}$$

The variational E-step is performed by sequentially minimizing each $\tilde{\mathcal{F}}_{k+1}$ with respect to $q(\mathbf{x}_{k+1}|\mathbf{x}_{0:k}, \mathbf{u}_{0:k+1}, \mathbf{y}_{0:k+1})$. In general, expectations are estimated via Monte Carlo samples from q, and stochastic gradient descent method is used to perform the inference optimization. Initializing $q(\mathbf{x}_{k+1}|\mathbf{x}_{0:k}, \mathbf{u}_{0:k+1}, \mathbf{y}_{0:k+1})$ at or near the prior $p_\theta(\mathbf{x}_{k+1}|\mathbf{x}_{0:k}, \mathbf{u}_{0:k}, \mathbf{y}_{0:k})$, the next observation can be predicted through the likelihood, $p_\theta(\mathbf{y}_{k+1}|\mathbf{x}_{0:k+1}, \mathbf{u}_{0:k+1}, \mathbf{y}_{0:k})$, which allows for computing $\mathcal{F}_{k+1}$. Then, the posterior estimate can be updated based on the approximate posterior gradient. This procedure resembles Bayesian filtering, where the posterior is updated from the prior prediction according to the likelihood of observations. However, prediction and update steps are repeated until convergence [175].

After inferring an optimal approximate posterior, the variational M-step is performed by minimizing the filtering free energy with respect to the model parameters, $\boldsymbol{\theta}$, based on the following gradient:

$$\nabla_\theta \mathcal{F} = \sum_{i=0}^{k} \nabla_\theta \left[\mathbb{E}_{\prod_{j=0}^{i} q(\mathbf{x}_{j+1}|\mathbf{x}_{0:j}, \mathbf{u}_{0:j+1}, \mathbf{y}_{0:j+1})} \left[\mathcal{F}_{i+1} \right] \right] \tag{10.66}$$

$$= \sum_{i=0}^{k} \nabla_\theta \tilde{\mathcal{F}}_{i+1}, \tag{10.67}$$

where such a stochastic gradient can be estimated based on a minibatch of data. Algorithm 10.1 summarizes different stages of the VFEM [175]. Section 10.8 presents a computationally efficient implementation of the algorithm known as the amortized variational filtering.

10.8 Amortized Variational Filtering Expectation Maximization

Performing approximate inference by solving the optimization problem in the E-step of Algorithm 10.1 calls for adjusting the optimizer hyper-parameters. Furthermore, it may require a large number of iterations of gradient updates. In order to improve computational efficiency of the variational filtering expectation maximization algorithm, an amortized inference model can be deployed. To be more precise, if the approximate posterior, $q(\mathbf{x}_{k+1}|\mathbf{x}_{0:k}, \mathbf{u}_{0:k+1}, \mathbf{y}_{0:k+1})$, is a parametric distribution with the set of parameters λ^q, then the inference update will be performed as:

$$\lambda^q_{k+1} \leftarrow \mathbf{f}_\phi(\lambda^q_k, \nabla_{\lambda^q_k} \tilde{\mathcal{F}}_{k+1}). \tag{10.68}$$

Using the prior to update the approximate posterior, this framework takes advantage of the model capacity for inference correction at each step. Hence, a computationally efficient implementation of the algorithm is obtained by performing inference optimization via such an iterative inference model in the filtering setting [175].

Algorithm 10.1: Variational filtering expectation maximization

State-space model:

$$p(\mathbf{x}_{k+1}|\mathbf{x}_k, \mathbf{u}_k)$$
$$p(\mathbf{y}_k|\mathbf{x}_k, \mathbf{u}_k)$$

Initialization:

$$p(\mathbf{x}_0)$$
$$\nabla_\theta \mathscr{F} = 0$$

for $k = 0, 1, \dots,$ **do**

Initialize the approximate posterior at/near the prior

$$q(\mathbf{x}_{k+1}|\mathbf{x}_{0:k}, \mathbf{u}_{0:k+1}, \mathbf{y}_{0:k+1}) \approx p_\theta(\mathbf{x}_{k+1}|\mathbf{x}_{0:k}, \mathbf{u}_{0:k}, \mathbf{y}_{0:k})$$

Compute the expectation of the step free energy

$$\widetilde{\mathscr{F}}_{k+1} = \mathbb{E}_{\prod_{i=0}^{k} q(\mathbf{x}_{i+1}|\mathbf{x}_{0:i}, \mathbf{u}_{0:i+1}, \mathbf{y}_{0:i+1})} \left[\mathscr{F}_{k+1}\right]$$

E-step

$$q(\mathbf{x}_{k+1}|\mathbf{x}_{0:k}, \mathbf{u}_{0:k+1}, \mathbf{y}_{0:k+1}) = \arg\min_q \widetilde{\mathscr{F}}_{k+1}$$

Compute the gradient

$$\nabla_\theta \mathscr{F}_{0:k+1} = \nabla_\theta \mathscr{F}_{0:k} + \nabla_\theta \widetilde{\mathscr{F}}_{k+1}$$

M-step

$$\theta_{k+1} = \theta_k - \eta \nabla_\theta \mathscr{F}_{0:k+1}$$

end

10.9 Applications

This section reviews three case studies. The particle expectation maximization algorithm has been used for joint state and parameter estimation in mathematical finance. The presented deep learning-based expectation maximization algorithms have been evaluated on sequential perceptual grounding tasks. They were able to accurately recover the constituent objects, handle occlusion, and generalize the learned knowledge to scenes with different numbers of objects.

10.9.1 Stochastic Volatility

The particle expectation maximization algorithm has been successfully applied to a *stochastic volatility* model [224]. In stochastic volatility models, variance of a random variable such as the price of a tradable financial asset, which is referred to as *security*, is treated as a random variable [225]. The P-EM algorithm was used for predicting changes in the variance (volatility) of stock prices and exchange rates. The studied stochastic volatility model is described by the following state-space model [224]:

$$x_{k+1} = a\, x_k + b\, v_k, \tag{10.69}$$

$$y_k = c\, e^{\frac{x_k}{2}}\, w_k, \tag{10.70}$$

where

$$x_0 \sim \mathcal{N}\left(0, \frac{b^2}{1-a^2}\right), \tag{10.71}$$

$$v_k \sim \mathcal{N}(0,1), \tag{10.72}$$

$$w_k \sim \mathcal{N}(0,1). \tag{10.73}$$

The P-EM algorithm jointly estimates the state and the model parameters:

$$\theta = \begin{bmatrix} a \\ b \\ c \end{bmatrix}. \tag{10.74}$$

10.9.2 Physical Reasoning

Applying the R-NEM algorithm to physical reasoning tasks for different nonlinear dynamic systems with varying degrees of visual complexity has confirmed its superiority over other unsupervised-learning methods that do not incorporate inductive bias, which reflects the real-world interactions [223]. These experiments were performed on bouncing balls with variable mass, bouncing balls with an invisible curtain, and the Arcade learning environment [226]. In the experiments, it was shown that increasing the number of components and including some additional empty groups during the training phase can improve the grouping quality. The R-NEM algorithm was able to generalize the learned knowledge about physical interactions to environments with more objects. It was demonstrated that generalization to more complex environments would heavily rely on deploying the attention mechanism [223].

10.9.3 Speech, Music, and Video Modeling

The amortized variational filtering expectation maximization algorithm has been applied to a variety of sequential data using multiple deep latent Gaussian models [175]. Experiments were performed on speech [227], music [228], and video [229] data. It was shown that the amortized variational filtering provides a general algorithm that performs well across multiple models and datasets.

10.10 Concluding Remarks

An important step towards achieving artificial general intelligence is to equip machines with common-sense physical reasoning. According to cognitive science and developmental psychology, intuitive physics and reasoning capabilities rely on the ability to perceive objects and their interactions [230]. Hence, efficient algorithms must be designed in order to discover and describe a scene in terms of objects, and incorporate inductive bias towards compositionality [231]. Such algorithms must be able to efficiently learn the nature of objects, their dynamics, and their interactions from observations. Following this line of thinking, three classes of deep learning-based algorithms were studied in this chapter:

- The neural expectation maximization algorithm emphasizes on separately representing conceptual entities contained in the data by combining neural networks and generalized EM [232] into an unsupervised-learning clustering algorithm.
- Combining object clustering with interaction functions that take account of interactions between objects allows the relational neural expectation maximization algorithm to act as an approximate simulator of the environment. The algorithm would be able to predict object movement and collision, even for occluded objects. This ability provides an important step towards unsupervised human-like learning.
- The variational filtering expectation maximization algorithm provides a general framework to design filters for dynamic systems based on deep latent-variable models. Variational filtering inference can be formulated as a sequence of optimization objectives, which are related to each other through samples across steps. Efficient implementations of the algorithm can be obtained by amortized variational filtering, which uses iterative inference models to perform inference optimization.

11

Reinforcement Learning-Based Filter

11.1 Introduction

Reinforcement learning is a branch of artificial intelligence that is focused on goal-oriented learning from interactions with an environment [233]. Reinforcement learning algorithms aim at finding effective suboptimal solutions to complex problems of sequential decision making and planning under uncertainty [234]. Optimal decision making is a core concept in many disciplines such as engineering, computer science, mathematics, economics, neuroscience, and psychology. Therefore, there is a branch of these disciplines that deals with the same problems that reinforcement learning tries to solve. In this regard, reinforcement learning can be viewed as a multi-disciplinary field with a wide range of applications.

The probabilistic models used to understand and analyze data have been enhancing in terms of complexity and scale to cope with big data, which is associated with five key attributes: volume, velocity, variety, veracity, and value (five Vs). Due to this trend, performing inference has become a more challenging task. Regarding the duality between estimation and control [235], these two fields can borrow algorithms from each other. For instance, by viewing control as an inference problem, control algorithms can be designed based on *belief propagation*. Similarly, learning the approximate posterior distribution in variational inference can be mapped to the policy optimization problem in reinforcement learning to design filters based on *reinforced variational inference* [236]. Section 11.2 provides a brief review of reinforcement learning based on references [233] and [237] to lay the groundwork for viewing variational inference as a reinforcement learning problem.

Nonlinear Filters: Theory and Applications, First Edition. Peyman Setoodeh, Saeid Habibi, and Simon Haykin.

11.2 Reinforcement Learning

In reinforcement learning, an agent learns through interaction with an environment. There is no supervisor, instead the agent receives a scalar reward signal, r_k, at time instant k from the environment, which is an indicator of how well the agent is performing at that step. Reinforcement learning is based on the *reward hypothesis* that all goals can be described as maximizing the expected cumulative reward. Therefore, the agent's goal is to maximize the cumulative reward over the *control horizon*. The agent's actions affect the subsequent data that it receives. These actions may have long-term consequences with delayed rewards, hence, the agent deals with sequential data, and there must be a trade-off between collecting immediate and long-term rewards [233].

At each time instant, k, the agent executes an action, $\mathbf{u}_k \in \mathbb{R}^{n_u}$, and receives an observation, $\mathbf{y}_k \in \mathbb{R}^{n_y}$, as well as a scalar reward, $r_k \in \mathbb{R}$. Hence, the environment receives action $\mathbf{u}_k \in \mathbb{R}^{n_u}$, and accordingly, emits observation $\mathbf{y}_{k+1} \in \mathbb{R}^{n_y}$ and scalar reward $r_{k+1} \in \mathbb{R}$. Then, k increments at environment step. The sequence of actions, observations, and rewards forms the *history*:

$$H_k = \{\mathbf{u}_0, \mathbf{y}_1, r_1, \dots, \mathbf{u}_{k-1}, \mathbf{y}_k, r_k\}, \tag{11.1}$$

which includes all observable variables up to time instant k. At each time step, the action that the agent selects as well as the observation and the reward that the environment selects depend on this history. In other words, future depends on this history. As a function of history, state, $\mathbf{x}_k = \mathbf{f}(H_k)$, refers to the information, which is used to determine what will happen next. In the context of reinforcement learning, two types of state can be considered [237]:

- The *environment state*, $\mathbf{x}_k^e$, reflects the environment's private representation or the information that the environment uses to choose the next observation and reward. In general, it is not directly accessible by the agent, moreover, it usually contains information, which is irrelevant to the task that the agent is performing.
- The *agent state*, $\mathbf{x}_k^a$, reflects the agent's internal representation or the information that the agent uses to choose the next action. Reinforcement learning algorithms are designed based on this information, which is a function of history, $\mathbf{x}_k^a = \mathbf{f}(H_k)$.

The notion of *information state* is defined as all useful information from the history, which induces the *Markov* property:

$$p(\mathbf{x}_{k+1}|\mathbf{x}_{0:k}) = p(\mathbf{x}_{k+1}|\mathbf{x}_k). \tag{11.2}$$

In other words, "*the future is independent of the past given the present*" [237]. The state can be viewed as the *sufficient statistics* of the future; hence, given the state, the history can be ignored.

Assuming that both the environment state and the history are Markovian, environments can be categorized into two groups [237]:

- *Fully observable environments*: Agent has direct access to the environment state:

$$\mathbf{y}_k = \mathbf{x}_k^e = \mathbf{x}_k^a. \tag{11.3}$$

Reinforcement learning problem is formulated as a *Markov decision process* (MDP):

$$\text{Agent state} = \text{Environment state} = \text{Information state}. \tag{11.4}$$

- *Partially observable environments*: Agent does not have direct access to the environment state and constructs its own state representation, $\mathbf{x}_k^a$:

$$\text{Agent state} \neq \text{Environment state}. \tag{11.5}$$

Reinforcement learning problem is formulated as a *partially observable Markov decision process* (POMDP). Through estimating the environment state based on the history, agent forms *beliefs* about the environment state. Recurrent neural networks can be used to model the underlying dynamics based on the history.

Reinforcement learning agents are built on the following pillars [237]:

- *Policy function*: refers to a map from state to action, which determines the agent's behavior. Policy may be deterministic, $\mathbf{u} = \boldsymbol{\pi}(\mathbf{x})$, or stochastic, $\pi(\mathbf{u}|\mathbf{x}) = p(\mathbf{u} = \mathbf{u}_k|\mathbf{x} = \mathbf{x}_k)$.
- *Value function*: evaluates the quality of each state, $V(\mathbf{x})$, or the quality of each state-action pair (selecting a specific action at a particular state), $Q(\mathbf{x}, \mathbf{u})$, in terms of the expected future rewards that can be collected from that state:

$$V(\mathbf{x}) = \mathbb{E}_\pi \left[r_{k+1} + \gamma r_{k+2} + \gamma^2 r_{k+3} + \ldots |\mathbf{x} = \mathbf{x}_k \right], \tag{11.6}$$

$$Q(\mathbf{x}, \mathbf{u}) = \mathbb{E}_\pi \left[r_{k+1} + \gamma r_{k+2} + \gamma^2 r_{k+3} + \ldots |\mathbf{x} = \mathbf{x}_k, \mathbf{u} = \mathbf{u}_k \right], \tag{11.7}$$

where $0 \leq \gamma \leq 1$ is a discount factor that reflects the uncertainty about the future, where lower and higher values of γ lead to myopic and far-sighted evaluations, respectively. The total discounted reward from time instant k to the end of the control horizon, T, which may be finite or infinite, is called *return*:

$$R_k = r_{k+1} + \gamma r_{k+2} + \gamma^2 r_{k+3} + \cdots = \sum_{i=0}^{T} \gamma^i r_{i+k+1}. \tag{11.8}$$

Actions can be selected at each state based on the value function.

- *Model*: refers to the agent's representation of the environment, which is used to predict the next state and immediate reward:

$$\mathscr{P}_{\mathbf{x}\mathbf{x}'}^a = p(\mathbf{x}_{k+1} = \mathbf{x}'|\mathbf{x} = \mathbf{x}_k, \mathbf{u} = \mathbf{u}_k), \tag{11.9}$$

$$\mathscr{R}_{\mathbf{x}}^a = \mathbb{E}\left[r_{k+1}|\mathbf{x} = \mathbf{x}_k, \mathbf{u} = \mathbf{u}_k \right]. \tag{11.10}$$

A reinforcement learning agent may include all or some of these components. Regarding policy and value functions, reinforcement learning algorithms are categorized as follows [238]:

- *Actor-only* or *policy-based*: These agents include an explicit policy function, but not a value function. These algorithms consider a parameterized family of policies, and find the optimal policy by directly applying an optimization procedure over the family. This class of algorithms, which includes *policy gradient methods*, has the following advantages and disadvantages:
  - *Advantage*: Actor-only methods are more resilient to fast-changing nonstationary environments, and the parameterized policy allows for generating a spectrum of continuous actions.
  - *Disadvantage*: Actor-only methods suffer from slow learning due to the high variance of gradient estimates.

  REINFORCE [239] is an actor-only algorithm.
- *Critic-only* or *value-based*: These agents include a value function, but not an explicit policy. These algorithms use *temporal-difference learning* (TD learning) and follow a greedy strategy to select actions with the highest expected collected rewards. This class of algorithms has the following advantages and disadvantages:
  - *Advantage*: Critic-only methods have a lower variance in gradient estimates compared to actor-only techniques. These algorithms contain extra learning rates, which provide more degrees of freedom.
  - *Disadvantage*: Critic-only methods need to use an optimization procedure at every encountered state to find the optimal action, which can be computationally demanding, especially for continuous action spaces.

  Q-learning [240, 241] and SARSA [242] are critic-only algorithms.
- *Actor-critic*: These agents include both a policy and a value function. These algorithms combine the advantages of both actor-only and critic-only methods. The parameterized actor allows for computing continuous actions without solving an optimization problem over the value function at each encountered state. The critic provides a low-variance gradient estimate, which improves the learning speed. Regarding the bias-variance trade-off, the low-variance gradient estimate is achieved at the expense of a larger bias, especially at early stages of learning, when critic estimates are less accurate [243]. Regarding convergence, actor-critic algorithms perform better than critic-only methods [244].

Both policy and value functions must be learned through interaction with the environment. In the context of deep reinforcement learning, deep neural networks are used to approximate policy function and value function, which are referred to as policy network and value network, respectively. Regarding model, reinforcement learning algorithms are categorized as [233, 237]:

- *Model-free*: Agent includes a policy function and/or a value function, but not a model.
- *Model-based*: Agent includes a policy function and/or a value function as well as a model, which can be used for planning. In planning, the agent can perform simulations using the known model of the environment to improve its policy without any external interaction with the environment [245].

Reinforcement learning can be used for both *prediction* and *control*. In prediction, agent evaluates the future based on a given policy. In control, agent optimizes the future by finding the best policy. In reinforcement learning, there must be a trade-off between *exploration* and *exploitation*. While through exploration agent learns more about the environment, through exploitation it uses the known information to maximize the expected collected reward.

11.3 Variational Inference as Reinforcement Learning

Defining the *trajectory*, τ, as the sequence of visited states and selected actions, $\tau = \{\mathbf{x}_0, \mathbf{u}_0, \dots \mathbf{x}_T, \mathbf{u}_T\}$, a distribution can be considered over the trajectories, $p_\theta(\tau)$. The goal is to maximize the expected collected reward over the control horizon [236]:

$$\begin{aligned} J(\theta) &= \mathbb{E}_{\tau \sim p_\theta}\left[\sum_{k=1}^{T} r_k\right] \\ &= \int p_\theta(\tau) R(\tau)\, \mathrm{d}\tau, \end{aligned} \tag{11.11}$$

where $R(\tau) = \sum_{k=1}^{T} r_k$ denotes the return with discount factor $\gamma = 1$. The reward at time instant k is a function of the previously visited state and the selected action at that state, $(\mathbf{x}_{k-1}, \mathbf{u}_{k-1})$. Assuming Markov property for the environment, the distribution over trajectories can be factorized as:

$$p_\theta(\tau) = p(\mathbf{x}_0)\pi_\theta(\mathbf{u}_0|\mathbf{x}_0)\prod_{k=1}^{T} p(\mathbf{x}_k|\mathbf{x}_{k-1}, \mathbf{u}_{k-1})\pi_\theta(\mathbf{u}_k|\mathbf{x}_k), \tag{11.12}$$

where π_θ denotes the parameterized policy.

The *score function* trick can be used to rewrite an integral as an expectation. Assuming that $\frac{\partial p_\theta(z)}{\partial \theta}$ exists and is continuous, according to the following equality [236]:

$$\frac{\partial \log p_\theta(z)}{\partial \theta} = \frac{\frac{\partial p_\theta(z)}{\partial \theta}}{p_\theta(z)}, \tag{11.13}$$

we have:

$$\frac{\mathrm{d}}{\mathrm{d}\theta}\int_z p_\theta(z)f(z)\mathrm{d}z = \int_z \frac{\partial p_\theta(z)}{\partial\theta}f(z)\mathrm{d}z$$
$$= \int_z p_\theta(z)\frac{\partial \log p_\theta(z)}{\partial\theta}f(z)\,\mathrm{d}z$$
$$= \mathbb{E}_{p_\theta(z)}\left[\frac{\partial \log p_\theta(z)}{\partial\theta}f(z)\right]. \tag{11.14}$$

Using the score function trick, $\nabla_\theta J(\boldsymbol{\theta})$ can be computed as:

$$\nabla_\theta J(\boldsymbol{\theta}) = \mathbb{E}_{\tau\sim p_\theta}\left[\frac{\partial}{\partial\theta}\log p_\theta(\tau)R(\tau)\right]. \tag{11.15}$$

A Monte Carlo estimate of this gradient can be obtained as:

$$\nabla_\theta J(\boldsymbol{\theta}) \approx \frac{1}{n_p}\sum_{i=1}^{n_p}\frac{\partial}{\partial\theta}\log p_\theta(\tau^i)R(\tau^i). \tag{11.16}$$

Substituting for $p_\theta(\tau^i)$ from (11.12), and considering that only policy depends on the parameters $\boldsymbol{\theta}$, the aforementioned gradient can be estimated as:

$$\nabla_\theta J(\boldsymbol{\theta}) \approx \frac{1}{n_p}\sum_{i=1}^{n_p}\sum_{k=1}^{T}\frac{\partial}{\partial\theta}\log\left(p(\mathbf{x}_k^i|\mathbf{x}_{k-1}^i,\mathbf{u}_{k-1})\pi_\theta(\mathbf{u}_k|\mathbf{x}_k^i)\right)R(\tau^i). \tag{11.17}$$

The main idea behind the REINFORCE algorithm is the aforementioned gradient estimate, which can be interpreted as follows. The agent tries different actions at visited states, and adjusts the parameters of the policy in a way to increase the chance of following high-return trajectories. The REINFORCE algorithm considers the total return for the entire trajectory rather than performing fine grained *credit assignment* to individual actions [236].

Now, let us consider variational inference for a system with the state (latent variable), input (action), and output (observable) vectors $\mathbf{x}\in\mathbb{R}^{n_x}$, $\mathbf{u}\in\mathbb{R}^{n_u}$, and $\mathbf{y}\in\mathbb{R}^{n_y}$, respectively. Given a model $p(\mathbf{x})p(\mathbf{y}|\mathbf{x},\mathbf{u})$, the goal of filtering is to find the posterior, $p(\mathbf{x}|\mathbf{u},\mathbf{y})$. Since finding the exact posterior may be computationally intractable, it is approximated by another distribution, $q(\mathbf{x}|\mathbf{u},\mathbf{y})$, that must be close to the true posterior. It is desirable that both the model p and the approximate posterior q can be decomposed into products of local conditional distributions. Markovian models have such a desired characteristic:

$$p(\mathbf{x}_{0:T},\mathbf{y}_{0:T}|\mathbf{u}_{0:T}) = p(\mathbf{x}_0)\prod_{k=1}^{T}p(\mathbf{x}_k|\mathbf{x}_{k-1},\mathbf{u}_{k-1},\mathbf{y}_{k-1})p(\mathbf{y}_k|\mathbf{x}_k,\mathbf{u}_k). \tag{11.18}$$

An approximate posterior is chosen that can be similarly factorized as:

$$q(\mathbf{x}_{0:T}|\mathbf{u}_{0:T},\mathbf{y}_{0:T}) = q(\mathbf{x}_0|\mathbf{u}_0,\mathbf{y}_0)\prod_{k=1}^{T}q(\mathbf{x}_k|\mathbf{x}_{k-1},\mathbf{u}_{k-1:k},\mathbf{y}_k) \tag{11.19}$$

Variational inference aims at maximizing the negative free energy (minimizing the free energy):

$$\mathcal{F} = -\int q(\mathbf{x}_{0:T}|\mathbf{u}_{0:T}, \mathbf{y}_{0:T}) \log \frac{p(\mathbf{x}_{0:T}, \mathbf{y}_{0:T}|\mathbf{u}_{0:T})}{q(\mathbf{x}_{0:T}|\mathbf{u}_{0:T}, \mathbf{y}_{0:T})} \, \mathrm{d}\mathbf{x}_{0:T}, \tag{11.20}$$

which can be interpreted in two ways [236]:

- Minimizing $\mathcal{F}$ is equivalent to minimizing the Kullback–Leibler divergence (KLD) between the approximate posterior, $q(\mathbf{x}|\mathbf{u}, \mathbf{y})$, and the true posterior, $p(\mathbf{x}|\mathbf{u}, \mathbf{y})$.
- Minimizing $\mathcal{F}$ is equivalent to maximizing the lower bound on the data log likelihood, $p(\mathbf{y}|\mathbf{u})$.

If the approximate posterior is chosen from a parametric family with parameters θ, then $\mathcal{F}$ can be viewed as a function of θ, therefore, it can be directly optimized with respect to θ. Learning variational approximations to posterior distributions relies on the stochastic gradient descent method.

Using the score function trick, $\nabla_\theta \mathcal{F}$ can be computed as:

$$\begin{aligned} \nabla_\theta \mathcal{F} &= -\mathbb{E}_{q_\theta}\left[\frac{\partial}{\partial \theta}\left(\log q_\theta(\mathbf{x}_{0:T}|\mathbf{u}_{0:T}, \mathbf{y}_{0:T})\right) \log \frac{p(\mathbf{x}_{0:T}, \mathbf{y}_{0:T}|\mathbf{u}_{0:T})}{q_\theta(\mathbf{x}_{0:T}|\mathbf{u}_{0:T}, \mathbf{y}_{0:T})}\right] \\ &= -\mathbb{E}_{q_\theta}\left[\frac{\partial}{\partial \theta}\left(\log q_\theta(\mathbf{x}_{0:T}|\mathbf{u}_{0:T}, \mathbf{y}_{0:T})\right)\right. \\ &\qquad \left.\log \frac{p(\mathbf{y}_{0:T}|\mathbf{x}_{0:T}, \mathbf{u}_{0:T}) p(\mathbf{x}_{0:T}|\mathbf{x}_{0:T-1}, \mathbf{u}_{0:T-1}, \mathbf{y}_{0:T-1})}{q_\theta(\mathbf{x}_{0:K}|\mathbf{u}_{0:K}, \mathbf{y}_{0:K})}\right]. \end{aligned} \tag{11.21}$$

A Monte Carlo estimate of this gradient can be obtained as:

$$\begin{aligned} \nabla_\theta \mathcal{F} \approx -\frac{1}{n_p}\sum_{i=1}^{n_p} &\left[\frac{\partial}{\partial \theta}\left(\log q_\theta(\mathbf{x}^i_{0:T}|\mathbf{u}_{0:T}, \mathbf{y}_{0:T})\right)\right. \\ &\left.\log \frac{p(\mathbf{y}_{0:T}|\mathbf{x}^i_{0:T}, \mathbf{u}_{0:T}) p(\mathbf{x}^i_{0:T}|\mathbf{x}^i_{0:T-1}, \mathbf{u}_{0:T-1}, \mathbf{y}_{0:T-1})}{q_\theta(\mathbf{x}^i_{0:T}|\mathbf{u}_{0:T}, \mathbf{y}_{0:T})}\right]. \end{aligned} \tag{11.22}$$

Comparing the objective functions (11.11) and (11.20), it can be seen that both the REINFORCE algorithm and the variational inference can be viewed as special cases of a generic expectation maximization (EM) problem that aims at maximizing an expectation in the form of $\int_z p_\theta(z) f(z) \mathrm{d}z$ with respect to the corresponding parameters, θ, of the distribution family $p_\theta(z)$. Comparison of the gradient in (11.16) or (11.17) with the gradient in (11.22) leads to the same conclusion. Table 11.1 summarizes the correspondence between different entities of these three equivalent problems [236]. Following this line of thinking, concepts and ideas from the reinforcement learning literature can be mapped to novel ideas in filtering and inference.

Table 11.1 Reinforcement learning and variational inference viewed as expectation maximization.

Generic expectation maximization	Variational inference	Reinforcement learning
θ: Optimization variable	θ: Variational parameter	θ: Policy parameter
z: Integration variable	$\mathbf{x}$: Latent state trajectory	τ: Trajectory
$p_\theta(z)$: Distribution	$q_\theta(\mathbf{x}\|\mathbf{u},\mathbf{y})$: Approximate posterior distribution	$p_\theta(\tau)$: Trajectory distribution
$f(z)$: Integrand	$\log\left(\frac{p(\mathbf{x},\mathbf{y}\|\mathbf{u})}{q_\theta(\mathbf{x}\|\mathbf{u},\mathbf{y})}\right)$: Negative free energy	$R(\tau)$: Total return

11.4 Application

Since variational inference can be formulated as reinforcement learning, many nonlinear estimation and filtering problems can be solved by reinforcement learning algorithms. In this section, battery state-of-charge (SoC) estimation is reviewed as a case study.

11.4.1 Battery State-of-Charge Estimation

Lithium-ion batteries have been extensively used as the energy source in electric vehicles due to their durability, high-energy density, safety, lack of hysteresis, and slow loss of charge in the idle mode. Accurately estimating the SoC and the state of health (SoH) of the battery is critical for battery management systems (BMSs). Different factors may have negative impact on the accuracy of the SoC estimation, including modeling imperfections, parametric uncertainties, sensor inaccuracies, and measurement noise. An equivalent circuit model is used to describe the governing dynamics of the lithium-ion cell. Then, the corresponding state-space model of the circuit is used by nonlinear filters to estimate the SoC [246]. A combination of a model-based method and a data-driven algorithm has been proposed for the SoC estimation [247]. In the proposed approach, while extended Kalman filter is used to estimate the SoC of the battery based on the equivalent circuit model, filter parameters are optimized by reinforcement learning.

11.5 Concluding Remarks

Variational inference and reinforcement learning can be viewed as instances of a generic expectation maximization problem. In this regard, inference itself can be viewed as reinforcement learning. This paves the way for developing novel filtering algorithms based on variational inference, which are inspired by the state-of-the-art (deep) reinforcement learning methods. For instance, well-established concepts in the reinforcement learning literature such as temporal difference, score function estimation, and exploration can be used in variational inference.

12

Nonparametric Bayesian Models

12.1 Introduction

In Bayesian models, any form of uncertainty is viewed as randomness, and model parameters are usually assumed to be random variables [248]. Parametric and nonparametric Bayesian models are defined on finite- and infinite-dimensional parameter spaces, respectively. While finite-dimensional probability models are described using densities, infinite-dimensional probability models are described based on concepts of measure theory. For a general type of observations known as exchangeable sequences, existence of a randomly distributed parameter is a mathematical consequence of the characteristics of the collected data rather than being a modeling assumption [249]. Nonparametric Bayesian models can be derived by starting from a parametric model and then taking the infinite limit [250]. Existence of noncomputable conditional distributions as well as limitations of using computable probability distributions for describing and computing conditional distributions would rule out the possibility of deriving generic probabilistic inference algorithms (even inefficient ones). However, the presence of additional structure such as exchangeability, which is common in Bayesian hierarchical modeling, paves the way for posterior inference [251].

12.2 Parametric vs Nonparametric Models

A parametric model uses a finite set of parameters to capture what can be known from a dataset, which is relevant for a prediction task. Mathematically speaking, given the set of parameters, θ, future predictions will be independent of the observed data, $\mathcal{D}$ [250]:

$$p(\mathbf{x}|\theta, \mathcal{D}) = p(\mathbf{x}|\theta), \tag{12.1}$$

Nonlinear Filters: Theory and Applications, First Edition. Peyman Setoodeh, Saeid Habibi, and Simon Haykin.

In this framework, a model can be viewed as an information channel from past data $\mathcal{D}$ to future predictions $\mathbf{x}$, and parameters $\boldsymbol{\theta}$ play the role of a bottleneck in this channel. Even for an unbounded amount of observed data, both complexity of a parametric model and capacity of the equivalent information channel will be bounded. Hence, the finite set of parameters in a parametric model provides limited flexibility.

On the other hand, a nonparametric model does not rely on the assumption that a finite set of parameters can determine the underlying distribution from which the observed data was sampled. This does not imply that the model does not have any parameter. On the contrary, it is built around an infinite-dimensional parameter vector $\boldsymbol{\theta}$, which is usually expressed as a function. Using the information channel analogy, there is no bottleneck in nonparametric models. To be more precise, the amount of information, which is represented by $\boldsymbol{\theta}$, increases as more data becomes available. This calls for deploying a memory to store the growing amount of information as the volume of the dataset increases. Therefore, nonparametric models provide memory-based predictions, $p(\mathbf{x}|\mathcal{D})$, and they are more flexible compared to parametric models [250]. In the following, basic concepts of measure-theoretic probability theory are briefly reviewed to lay the groundwork for construction of stochastic processes from their marginals.

12.3 Measure-Theoretic Probability

Infinite-dimensional probability models are generally described by notions of measure theoretic probability rather than densities [252]. Hence, some basic concepts of measure theory are recalled from [253] in what follows. If the outcome of an experiment cannot be certainly predicted, it is called *random*. The set of all unpredictable outcomes of such an experiment is known as the *sample space*, Ω. Subsets of the sample space corresponding to outcomes of the random experiment are called *events*. A *simple event* refers to the smallest observable outcome represented as a single-member subset of Ω, the *impossible event* refers to the empty set $\emptyset$, and the *certain event* refers to Ω itself. Possible events are associated with combinations of subsets of the sample space via set operations such as *union*, *intersection*, and *complementation*. This leads to the concept of field or algebra.

Definition 12.1 ***(Class)*** *A class, $\mathcal{F}$, is a collection of sets that is formed based on a clearly defined property, which is shared by all of its members.*

Definition 12.2 ***(σ-field or σ-algebra)*** *A* field *or* algebra *is a nonempty class $\mathcal{F}$ of subsets of a nonempty set Ω satisfying:*

(i) $\Omega \subset \mathcal{F}$
(ii) *$\mathcal{F}$ is closed under finite unions/intersections*
(iii) *$\mathcal{F}$ is closed under complementation*

$\mathcal{F}$ is a σ-field *or* σ-algebra *if instead of condition (ii), the following stronger condition holds:*

(ii') $\mathcal{F}$ is closed under countable unions/intersections

While the largest possible σ-algebra on any set Ω is the power set 2^{Ω}, which contains all subsets of Ω, the smallest σ-algebra only includes Ω and the empty set $\emptyset$.

Definition 12.3 **(Borel σ-algebra, Borel sets)** *The* Borel σ-algebra *or Borel σ-field, $\mathcal{B}(\mathbb{R})$, on the real line is the minimal σ-algebra containing the collection of open intervals and in effect therefore, all intervals. Sets in $\mathcal{B}(\mathbb{R})$ are known as Borel sets.*

The Borel σ-algebra contains all singletons and all open sets as well as all rational and irrational numbers.

Definition 12.4 **(Measurable space, measurable sets)** *The pair $(\Omega, \mathcal{F})$ is called a* measurable space, *if $\mathcal{F}$ is a σ-field or σ-algebra, then, events refer to sets $B \in \mathcal{F}$, which are known as measurable sets.*

Definition 12.5 **(Measurable map)** *Let us consider two measurable spaces $(\Omega, \mathcal{F})$ and $(\Omega_X, \mathcal{E})$, a map $X : \Omega \to \Omega_X$ is measurable if $\forall B \in \mathcal{E}, X^{-1}(B) \in \mathcal{F}$.*

Definition 12.6 **(Measure, measure space, σ-finite measure)** *For the measurable space $(\Omega, \mathcal{F})$, a map $\mu : \mathcal{F} \to [0, \infty)$ is a measure on $(\Omega, \mathcal{F})$, if μ is countably additive, and for a countable sequence of pairwise disjoint elements $B_k \in \mathcal{F}$, satisfies the σ-additivity condition:*

$$\mu\left(\cup B_k\right) = \sum \mu\left(B_k\right). \tag{12.2}$$

The triplet $(\Omega, \mathcal{F}, \mu)$ represents a measure space. *μ is a finite measure, if $\mu(\Omega) < \infty$, and it is a σ-finite measure, if there exists a partition of Ω into disjoint subsets $\{A_k\}$ such that*

(i) $\cup_k A_k = \Omega$,
(ii) $\mu(A_k) < \infty, \forall k \geq 1$.

Definition 12.7 **(Probability measure, probability space)** *For a measurable space $(\Omega, \mathcal{F})$, a probability measure $P : \mathcal{F} \to [0,1]$ is a countably additive set function satisfying:*

(i) $P(\Omega) = 1$
(ii) For a countable sequence of pairwise disjoint events $B_k \in \mathcal{F}$, the σ-additivity holds:

$$P\left(\cup B_k\right) = \sum P\left(B_k\right). \tag{12.3}$$

The triplet $(\Omega, \mathcal{F}, P)$ represents a probability space.

These two conditions imply that $P(\emptyset) = 0$. Moreover, for two events A and B, if $B \subset A$ and $P(A) = 0$, then $P(B) = 0$. Such nonempty events with zero probability are called *negligible events* or *null events*. The abstract probability space $(\Omega, \mathscr{F}, P)$ can be used to describe randomness.

Definition 12.8 ***(Random variable)*** *A random variable is a measurable real-valued function* $X : (\Omega, \mathscr{F}, P) \rightarrow (\mathbb{R}, \mathcal{B}(\mathbb{R}))$.

The probability distribution of X is the probability measure induced by X on the Borel sets $\mathcal{B}(\mathbb{R})$. For any Borel set $B \in \mathcal{B}(\mathbb{R})$, this probability distribution is defined as:

$$P_X(B) = P(X \in B). \tag{12.4}$$

The product σ-algebra of two σ-algebras $\mathscr{F}_1$ and $\mathscr{F}_2$, which is denoted by $\mathscr{F}_1 \otimes \mathscr{F}_2$, is the smallest σ-algebra that contains all rectangles $F_1 \times F_2$ with $F_1 \in \mathscr{F}_1$ and $F_2 \in \mathscr{F}_2$. The direct product of two measure spaces $(\Omega_1, \mathscr{F}_1, \mu_1)$ and $(\Omega_2, \mathscr{F}_2, \mu_2)$ defines a measure space:

$$\left(\Omega = \Omega_1 \times \Omega_2, \mathscr{F} = \mathscr{F}_1 \otimes \mathscr{F}_2, \mu = \mu_1 \times \mu_2\right), \tag{12.5}$$

where

$$\mu_1 \times \mu_2 \left(F_1 \times F_2\right) = \mu_1(F_1)\, \mu_2(F_2), \tag{12.6}$$

for $F_1 \in \mathscr{F}_1$ and $F_2 \in \mathscr{F}_2$. Such product spaces allow for formalizing multi-dimensional random variables, which can be extended to infinite-dimensional spaces.

An infinite-dimensional space, Ω_X^E, is formed by the E-fold product of Ω_X with itself, where E denotes an infinite index set. Elements of the infinite-dimensional product space Ω_X^E can be viewed as functions $f : E \rightarrow \Omega_X$. For instance, $\mathbb{R}^{\mathbb{R}}$ includes all real-valued functions on the line. A finite-dimensional subspace of Ω_X^E is obtained as Ω_X^I by choosing a finite index set I, which belongs to the set of finite subsets of E, $I \in \mathscr{F}(E)$. There would be a product Borel σ-algebra, $\mathcal{B}_X^I$, associated with each product space Ω_X^I. Then, random variables that take values on these spaces will have product structures as:

$$X^I = \otimes_{i \in I} X^i. \tag{12.7}$$

Since individual components of X^I may be dependent, the corresponding probability measure P_X^I is not necessary a product measure [252].

A *projection operator*, Π_{IJ}, can be used to restrict a vector $\mathbf{x}^J \in \Omega_X^J$ to a lower dimensional product space Ω_X^I, where $\Omega_X^I \subset \Omega_X^J$ with $I \subset J$, and the resulting image of $\mathbf{x}^J$ is denoted by $\mathbf{x}^I \in \Omega_X^I$.

Definition 12.9 ***(Cylinder set)*** *For a set $A^I \subset \Omega_X^I$, the cylinder set with base A^I is defined as the pre-image of A^I under the projection operator Π_{IJ}, which is denoted by:*

$$\Pi_{IJ}^{-1} A^I. \tag{12.8}$$

Applying a projection operator Π_{IJ} to a probability measure P_X^J is equivalent to assigning a probability to an I-dimensional set $A^I \subset \Omega_X^I$ by applying the J-dimensional measure P_X^J to the cylinder set with base A^I. The projection operator for probability measures is mathematically defined as

$$\Pi_{IJ} \circ P_X^J = P_X^J \circ \Pi_{IJ}^{-1}. \tag{12.9}$$

Hence, for an I-dimensional event $A^I \in \mathcal{B}_X^I$, we have

$$\left(\Pi_{IJ} P_X^J\right) A^I = P_X^J \left(\Pi_{IJ}^{-1} A^I\right), \tag{12.10}$$

which can be interpreted as the marginal of the measure P_X^J on the lower-dimensional subspace Ω_X^I [252].

When a family of distributions is parameterized, switching from the prior distribution to the posterior distribution would be equivalent to updating the corresponding parameters. In this regard, *conjugacy* is a property of interest for any family of parameterized distributions [44].

Definition 12.10 ***(Conjugacy)*** *A family of probability distributions, P, on the parameter space Θ is conjugate (closed under sampling) for a likelihood function $f(\mathbf{x}|\theta)$, if for every $\pi \in P$, the posterior distribution $\pi(\theta|\mathbf{x})$ belongs to P as well.*

Conjugate priors are very popular because they lead to computable posterior distributions. This topic will be further discussed in Section 12.7. The conjugacy property holds for exponential family distributions [44].

Definition 12.11 ***(Exponential family)*** *Let μ be a σ-finite measure on Ω, and Θ be the parameter space. Consider functions $C : \Omega \to \mathbb{R}_+$, $h : \Theta \to \mathbb{R}_+$, $\mathbf{R} : \Theta \to \mathbb{R}^k$, and $\mathbf{T} : \Omega \to \mathbb{R}^k$. The family of distributions with the following densities:*

$$f(\mathbf{x}|\theta) = C(\theta)h(\mathbf{x})e^{\mathbf{R}(\theta)\cdot\mathbf{T}(\mathbf{x})}, \tag{12.11}$$

is an exponential family of dimension k with respect to μ. The family is called natural, if $\Theta \subset \mathbb{R}^k$, $\Omega \subset \mathbb{R}^k$, and

$$f(\mathbf{x}|\theta) = C(\theta)h(\mathbf{x})e^{\theta\cdot\mathbf{x}}, \tag{12.12}$$

where "$\cdot$" denotes the dot product.

A generic distribution from an exponential family can be obtained by choosing $C(\theta)$ in (12.12) to be an exponential function $e^{-\psi(\theta)}$:

$$f(\mathbf{x}|\theta) = h(\mathbf{x})e^{\theta\cdot\mathbf{x}-\psi(\theta)}. \tag{12.13}$$

According to the following proposition, such a distribution allows for a conjugate family [44].

Proposition 12.1 ***(Natural conjugate distribution)*** *A conjugate family for $f(\mathbf{x}|\theta)$ is given by*

$$\pi(\theta|\boldsymbol{\mu}, \lambda) = K(\boldsymbol{\mu}, \lambda)e^{\theta\cdot\boldsymbol{\mu}-\lambda\psi(\theta)}, \tag{12.14}$$

where $K(\boldsymbol{\mu}, \lambda)$ is the normalizing constant of the density. This measure is called the natural conjugate distribution of f. The corresponding posterior distribution is $\pi(\theta|\boldsymbol{\mu}+\mathbf{x}, \lambda+1)$.

The measure defined in (12.14) is σ-finite and induces a probability distribution on Θ, if and only if $\lambda > 0$ and $\frac{1}{\lambda}\boldsymbol{\mu}$ belongs to the interior of the *natural parameter space*:

$$N = \left\{\theta \mid \int_{\mathcal{X}} e^{\theta\cdot\mathbf{x}} h(\mathbf{x})\mathrm{d}\mu(\mathbf{x}) < +\infty\right\}. \tag{12.15}$$

The family is called *regular*, if N is an open set.

Two examples of exponential family distributions are *Gaussian process* and *Dirichlet process* [44]. They are two basic building blocks for nonparametric Bayesian models, for which the conjugacy property holds [250, 252]:

- **Gaussian process**: A Gaussian distribution:

$$f(x|\mu, \sigma^2) = \frac{1}{\sqrt{2\pi\sigma}} e^{-\frac{1}{2}\left(\frac{x-\mu}{\sigma}\right)^2}, \tag{12.16}$$

is parameterized by the mean μ and the covariance σ^2 of the corresponding random variable X. Similarly, a Gaussian process is parametrized by a mean function and a covariance function, $P_X^E\left(X^E|\Theta^E\right)$, with X^E being function-valued. For instance, if $E = \mathbb{R}_+$ and $\Omega_X = \mathbb{R}$, then the product space $\Omega_X^E = \mathbb{R}^{\mathbb{R}_+}$ will include all functions $x^E : \mathbb{R}_+ \to \mathbb{R}$. While for such an infinite-dimensional space, each axis label $i \in E$ is a point on the positive real line, for a finite-dimensional space, a finite index set $I \in \mathcal{F}(E)$ determines a finite collection of points $I = \{i_1, \dots, i_m\}$. The projection of a function x^E in the infinite-dimensional space, Ω_X^E, to the finite-dimensional space, Ω_X^I, specified by index set I is the vector of function values at the points in I:

$$\mathbf{x}^I = \left[x^E(i_1), \dots, x^E(i_m)\right]^T. \tag{12.17}$$

- **Dirichlet process**: In Bayesian statistics, Dirichlet distributions are usually used as prior distributions. Conjugacy makes Dirichlet process-based inference tractable as posterior of a Dirichlet process is again a Dirichlet process. Regarding a discrete and finite space, where data points are assumed to take one of the K possible values, the Dirichlet distribution:

$$f(x|\alpha, h) = \frac{1}{Z}\prod_{k=1}^{K} x_k^{\alpha h_k - 1}, \tag{12.18}$$

is parameterized by the mean h of the corresponding random variable X and a scalar $\alpha > 0$ that controls the dispersion around the mean. In the aforementioned distribution, Z is a normalizing constant. A Dirichlet process is parametrized by a scalar concentration parameter $\alpha \in \mathbb{R}_+$ and a probability measure G_0, $P_X^E\left(X^E|\Theta^E\right)$, where the random variable X^E takes values x^E in the domain of G_0 (for instance, $\mathbb{R}$). A probability measure on $\mathbb{R}$ is a set function, whose domain is the corresponding Borel algebra $\mathcal{B}(\mathbb{R})$ and its range is [0,1]:

$$\mathcal{B}(\mathbb{R}) \rightarrow [0,1]. \tag{12.19}$$

With the choice of $E = \mathcal{B}(\mathbb{R})$ and $\Omega_X = [0,1]$, the parameter space would be $\mathbb{R}_+ \times [0,1]^{\mathcal{B}(\mathbb{R})}$.

For nonparametric models, the notion of exchangeability is a key concept, which is covered in Section 12.4.

12.4 Exchangeability

A sequence is *exchangeable*, if ordering of its elements contains no information. In such sequences, although the elements may depend on each other, they are independent of their indices. This is the case when the sequence represents a collection of objects, for which the order of collecting the objects is either unknown or irrelevant [250]. Mathematically speaking, the joint probability distribution of an exchangeable sequence is invariant under arbitrary permutations of its elements' indices. To be more precise, for the exchangeable sequence of variables $\{\mathbf{X}_0, \dots, \mathbf{X}_N\}$ over the same probability space, we have [249]:

$$P(\mathbf{X}_0 = \mathbf{x}_0, \dots, \mathbf{X}_N = \mathbf{x}_N) = P(\mathbf{X}_0 = \mathbf{x}_{\gamma(0)}, \dots, \mathbf{X}_N = \mathbf{x}_{\gamma(N)}), \tag{12.20}$$

where P is the joint probability distribution and γ denotes any permutation of the indices $\{0, \dots, N\}$. This definition can be easily extended to infinite exchangeable sequences $\{\mathbf{X}_0, \mathbf{X}_1, \dots\}$. According to *de Finetti's theorem*, independently and identically distributed variables are exchangeable due to the fact that exchangeability is a weaker condition. Before stating de Finetti's theorem and its extension, we recall the definition of *contractable sequences* from [254].

Definition 12.12 ***(Contractable sequence)*** *An infinite sequence of random elements $\xi = \{\mathbf{X}_0, \mathbf{X}_1, \ldots\}$ is contractable (spreading invariant or spreadable), if every sub-sequence has the same distribution:*

$$\{\mathbf{X}_{k_0}, \mathbf{X}_{k_1}, \ldots\} \overset{d}{=} \{\mathbf{X}_0, \mathbf{X}_1, \ldots\} \tag{12.21}$$

for any set of positive integers $k_0 < k_1 < \cdots$, where $\overset{d}{=}$ denotes equality in distribution.

Theorem 12.1 ***(de Finetti, Ryll-Nardzewski, infinite exchangeable sequences theorem)*** *Let ξ be an infinite sequence of random elements in a measurable space $(\Omega, \mathcal{F})$, then the following three conditions are equivalent, when $\mathcal{F}$ is Borel:*

(i) ξ is contractable,
(ii) ξ is exchangeable,
(iii) ξ is conditionally independent and identically distributed.

For general $\mathcal{F}$, we have (i) $\Leftrightarrow$ (ii) $\Leftarrow$ (iii).

Theorem 12.2 ***(Uniqueness theorem)*** *For an exchangeable sequence of random variables, ξ, there is an almost surely unique random probability measure P on $\mathbb{R}$ such that ξ is conditionally independent and identically distributed with respect to P. The probability measure P is referred to as the directing random measure of ξ, and ξ is said to be directed by P.*

According to de Finetti's theorem, a sequence of variables $\mathbf{X}_i$ is exchangeable, if and only if there exists a set of parameters $\boldsymbol{\theta}$ such that the elements $\mathbf{X}_i$ are independently and identically distributed samples from an unknown distribution indexed by $\boldsymbol{\theta}$. To be more precise, let us assume a prior distribution over $\boldsymbol{\theta}$ that parameterizes the underlying distribution from which $\mathbf{X}_i$ were sampled. If $\boldsymbol{\theta}$ is marginalized out, the resulting marginal distribution over $\mathbf{X}_i$ will be exchangeable. Conversely, if the sequence of $\mathbf{X}_i$ is exchangeable, then there is a random $\boldsymbol{\theta}$ such that:

$$P(\mathbf{X}_0, \mathbf{X}_1, \ldots) = \int P(\boldsymbol{\theta}) \prod_i P(\mathbf{X}_i | \boldsymbol{\theta}) d\boldsymbol{\theta}. \tag{12.22}$$

Hence, exchangeability implies the existence of a *hierarchical Bayesian model* with random latent parameters. For this reason, exchangeability is of fundamental importance in Bayesian statistics [249].

In order to build models for exchangeable data, distributions over unknown probability measures must be considered. Since the space of probability measures includes normalized non-negative functions, an infinite-dimensional $\boldsymbol{\theta}$ may be needed to index such a space, and the corresponding hierarchical Bayesian model will be nonparametric. Therefore, distributions over infinite-dimensional objects such as measures and functions are central to nonparametric Bayesian modeling [249, 250]. Section 12.5 focuses on construction of stochastic processes from their marginals.

12.5 Kolmogorov Extension Theorem

Let us consider a stochastic process with an infinite-dimensional probability measure P_X^E, whose finite-dimensional marginals form a family of probability measures P_X^I associated with the finite-dimensional subspace $\Omega_X^I \subset \Omega_X^E$.

Definition 12.13 ***(Projective family)*** *A family of probability measures is projective, if the following property holds:*

$$P_X^I = P_X^J \circ \Pi_{IJ}^{-1}, \quad I \subset J, \tag{12.23}$$

Marginals of a stochastic process measure are projective. According to *Kolmogorov extension theorem*, the converse is also true under the condition that the axes of the product space Ω_X are *Polish spaces* [255]. In order to state the Kolmogorov extension theorem, we need to recall the following definitions from topology and measure theory [256].

Definition 12.14 ***(Topology, topological space, open sets)*** *Given a set Ω, a subset of its power set 2^Ω, $\mathcal{T} \subseteq 2^\Omega$, is a topology, if and only if the following conditions hold:*

(i) $\emptyset, \Omega \in \mathcal{T}$,

(ii) $\mathcal{T}$ *is closed under finite intersections:*

$$G_1, \ldots, G_k \in \mathcal{T} \Rightarrow G_1 \cap \cdots \cap G_k \in \mathcal{T}, \tag{12.24}$$

(iii) $\mathcal{T}$ *is closed under arbitrary unions:*

$$\mathcal{T}_i \subseteq \mathcal{T} \Rightarrow \cup \mathcal{T}_i \in \mathcal{T}. \tag{12.25}$$

The pair $(\Omega, \mathcal{T})$ forms a topological space and the elements of $\mathcal{T}$ are called open sets.

Definition 12.15 ***(Dense subet)*** *Given a topological space $(\Omega, \mathcal{T})$, a subset $A \subseteq \Omega$ is dense, if and only if its closure is all of Ω:*

$$cl(A) = \Omega, \tag{12.26}$$

where $cl(A)$ consists of all points in A as well as all limit points of A.

Definition 12.16 ***(Metric space)*** *A metric space is a set Ω equipped with a distance function d:*

$$d : \Omega \times \Omega \rightarrow [0, \infty). \tag{12.27}$$

Definition 12.17 ***(Complete metric space)*** *A metric space Ω is Cauchy-complete or simply complete, if every Cauchy sequence in Ω converges in Ω. In a Cauchy sequence, as the sequence progresses, its elements become arbitrarily close to each other.*

Definition 12.18 ***(Completely metrizable space)*** *A topological space $(\Omega, \mathcal{T})$ is completely metrizable, if there exists at least one metric d on Ω such that (Ω, d) is a complete metric space and d induces the topology $\mathcal{T}$.*

Definition 12.19 ***(Polish space)*** *A Polish space is a completely metrizable topological space that has a countable base or equivalently a countable dense subset (it is separable).*

Euclidean spaces, countable discrete spaces, and countable products of Polish spaces are all Polish spaces [252].

Theorem 12.3 ***(Kolmogorov extension theorem)*** *Assume that E is an arbitrary infinite set, Ω_X is a Polish space, and $\{P_X^I | I \in \mathcal{F}(E)\}$ is a family of probability measures on the spaces $(\Omega_X^I, \mathcal{B}_X^I)$. If the family is projective, then there exists a uniquely defined probability measure P_X^E on Ω_X^E with the measures P_X^I as its marginals.*

The constructed infinite-dimensional measure P_X^E is called the *projective limit* of the family P_X^I. According to this theorem, the measure P_X^E can be completely determined by its values on a subset of events that involve a finite subset of the random variables. The values of P_X^E on such events, which can be viewed as cylinder sets with finite-dimensional bases, are in turn determined by the corresponding marginals. As a set function, the measure P_X^E is defined on $\mathcal{B}_X^E$, which is a product σ-algebra on Ω_X^E. For an uncountable index set E, this σ-algebra would be too coarse, and the measure P_X^E would not be able to assign probabilities to events of the form $\{X^E = x^E\}$ (singletons) [252].

The Kolmogorov extension theorem cannot be directly applied to extend Bayesian models, which are built on conditional probabilities as functions with two arguments. When Kolmogorov theorem is applied to each variable, due to ignoring the condition, it will not provide a proper conditional distribution on the infinite-dimensional space. Therefore, for Bayesian models, another theorem is needed to establish a correspondence between the finite-dimensional and infinite-dimensional conditional distributions [252]. Section 12.6 focuses on this issue.

12.6 Extension of Bayesian Models

A number of extension theorems for conditional distributions propose to construct sequential stochastic processes from a sequence of conditionals [257]. However, this section presents a theorem that is applicable on product spaces similar to the Kolmogorov extension theorem. As a first step towards establishing such a theorem, the projector that defines the marginals must be generalized from measures to conditionals. Let $P_X^J(\mathbf{X}^J|\mathbf{\Theta}^J)$ denote a conditional probability on the product space Ω^J, then, for $I \subset J$, the projection operator for conditionals is defined as [252]:

$$\left(\Pi_{IJ} P_X^J\right)\left(\cdot|\mathbf{\Theta}^J\right) = P_X^I\left(\Pi_{IJ}^{-1} \cdot |\mathbf{\Theta}^J\right). \tag{12.28}$$

For any specific value of the parameter, $\mathbf{\Theta}^J = \theta^J$, this definition is consistent with the projector applied to the measure

$$P_X^J\left(\cdot|\mathbf{\Theta}^J = \theta^J\right). \tag{12.29}$$

Now, the concept of regular conditional distribution is introduced, which generalizes the properties of ordinary expectations and allows for simultaneous computation of the conditional expectations of all functions of X [258].

Definition 12.20 ***(Regular conditional probability)*** *Let $(\Omega, \mathcal{F}, P)$ be a probability space, $X : (\Omega, \mathcal{F}) \to (S, \mathcal{S})$ be a measurable map, and $\mathcal{G} \subset \mathcal{F}$ be a σ-algebra. Then $\mu : \Omega \times \mathcal{S} \to [0,1]$ is a regular conditional distribution for X given $\mathcal{G}$ if*

(i) for each A, $\omega \to \mu(\omega, A)$ is a version of $P(X \in A|\mathcal{G})$, and
(ii) for almost every ω, $A \to \mu(\omega, A)$ is a probability measure on $(S, \mathcal{S})$.

If $S = \Omega$ and X is the identity map, μ is called a regular conditional probability.

Theorem 12.4 ***(Conditional expectation of functions)*** *Let $\mu(\omega, A)$ be a regular conditional distribution for X given $\mathcal{F}$. If $f : (S, \mathcal{S}) \to (\mathbb{R}, \mathcal{R})$ has $E[f(X)] < \infty$, then almost surely*

$$E\left[f(X)|\mathcal{S}\right] = \int \mu(\omega, dx) f(x). \tag{12.30}$$

In what follows, first, similar to projective families of measures, projective families of conditional probabilities are defined, and then, the counterpart of the Kolmogorov extension theorem for conditional probabilities is presented [252].

Definition 12.21 ***(Conditionally projective probability models)*** *Let $P_X^J(\mathbf{X}^J|\mathbf{\Theta}^J)$ denote a family of regular conditional probabilities on product spaces Ω^I, $\forall I \in \mathcal{F}(E)$. The family is conditionally projective, if for $I \subset J$,*

$$\left(\Pi_{IJ} P_X^J\right)\left(\cdot|\mathbf{\Theta}^J\right) =_{a.e.} P_X^I\left(\cdot|\mathbf{\Theta}^I\right), \tag{12.31}$$

where the equality needs to hold almost everywhere (a.e.), since conditional probabilities are unique almost everywhere.

Theorem 12.5 ***(Conditional probabilities extension theorem)*** *Assume that E is a countable index set and $P_X^I(\mathbf{X}^I|\boldsymbol{\Theta}^I)$ is a family of regular conditional probabilities on the product space Ω_X^I. If the family is conditionally projective, there exists a regular conditional probability $P_X^E(\mathbf{X}^E|\boldsymbol{\Theta}^E)$ on the infinite-dimensional space Ω_X^E with conditional marginals of $P_X^I(\mathbf{X}^I|\boldsymbol{\Theta}^I)$.*

Let C^E denote the following σ-algebra:

$$C^E = \sigma\left(\cup_{I\in\mathcal{F}(E)}\sigma\left(\boldsymbol{\Theta}^I\right)\right), \tag{12.32}$$

where σ refers to σ-algebra, then $P_X^E\left(X^E, C^E\right)$ is measurable. If $\Pi_{IJ}\boldsymbol{\Theta}^J = \boldsymbol{\Theta}^I$, then $P_X^E\left(X^E, C^E\right)$ can be interpreted as the conditional probability $P_X^E(\mathbf{X}^E|\boldsymbol{\Theta}^E)$ with

$$\boldsymbol{\Theta}^E = \otimes_{i\in E}\boldsymbol{\Theta}^i. \tag{12.33}$$

According to this theorem, a conditionally projective family of marginals can uniquely determine a conditional probability almost everywhere on a countably infinite dimensional product space. Hence, a nonparametric model on Ω_X^E can be built based on a family of conditionally projective parametric models on Ω_X^I for $I \in \mathcal{F}(E)$. As mentioned previously, conjugacy, which refers to closure under sampling, provides an important alternative to Bayes theorem for computing posterior distributions [248]. Section 12.7 is dedicated to this topic.

12.7 Conjugacy

Aiming at facilitating computation of the posterior, the prior and the likelihood of a Bayesian model are called conjugate if the posterior

(i) is parameterized, and
(ii) there exists a measurable mapping Γ from the data **x** and the prior parameter ψ to the posterior parameter $\psi' = \Gamma(\mathrm{x}, \psi)$.

Mapping Γ is known as the *posterior index* of the model [252].

Definition 12.22 ***(Conjugate model)*** *Let $P_\Theta(\boldsymbol{\Theta}|\mathbf{X}, \boldsymbol{\Psi})$ be the posterior of the model $P_X(\mathbf{X}|\boldsymbol{\Theta})$ under prior $P_\Theta(\boldsymbol{\Theta}|\boldsymbol{\Psi})$, where $P_X(\mathbf{X}|\boldsymbol{\Theta})$ and $P_\Theta(\boldsymbol{\Theta}|\boldsymbol{\Psi})$ are regular conditional probabilities. The model and the prior are conjugate, if there is a regular conditional probability $\kappa : \mathcal{B}_\theta \times \Omega_\Gamma \rightarrow [0,1]$, which is parameterized on a measurable Polish space $(\Omega_\Gamma, \mathcal{B}_\theta)$ and a measurable map $\Gamma : \Omega_X \times \Omega_\Psi \rightarrow \Omega_\Gamma$, such that $\forall A \in \mathcal{B}_\theta$,*

$$P_\Theta(A|\mathbf{X} = \mathbf{x}, \boldsymbol{\Psi} = \psi) = \kappa\left(A, \Gamma(\mathbf{x}, \psi)\right). \tag{12.34}$$

For $\Omega_\Gamma = \Omega_X \times \Omega_\Psi$ and Γ chosen as the identity mapping, the aforementioned definition would be trivial.

Both the Kolmogorov extension theorem and the conditional probabilities extension theorem require a projection condition on measures and models. A similar condition is defined for the posterior index mapping Γ for all $I \in \mathcal{F}(E)$ as follows [252]:

$$\left(\Pi_{IE} \circ \Gamma^{E}\right)^{-1} = \left(\Gamma^{I} \circ \Pi_{IE}\right)^{-1} \tag{12.35}$$

According to this condition, the pre-images $\left(\Gamma^{I}\right)^{-1}$ of the posterior indices Γ^{I} commute with the pre-image under the projection operator Π_{IE}. This condition holds for both Gaussian and Dirichlet processes. The following theorem explains how conjugacy can be carried over between finite-dimensional and infinite-dimensional cases by stating that [252]:

(i) Bayesian models constructed from conjugate marginals are conjugate if condition (12.35) holds.
(ii) Such conjugate models can only be constructed from conjugate marginals.

Theorem 12.6 ***(Functional conjugacy of projective limit models)*** *Let Ω_X^E and Ω_Θ^E be Polish product spaces, where E is a countable index set. Assume that there exists a Bayesian model on each finite-dimensional subspace Ω_X^I, for which the families of all priors, all observation models, and all posteriors are conditionally projective with the corresponding projective limits of $P_\Theta^E\left(\boldsymbol{\Theta}^E\right)$, $P_X^E\left(\mathbf{X}^E|\boldsymbol{\Theta}^E\right)$, and $P_\Theta^E\left(\boldsymbol{\Theta}^E|\mathbf{X}^E\right)$. Then, $P_\Theta^E\left(\boldsymbol{\Theta}^E|\mathbf{X}^E\right)$ is a posterior for the infinite-dimensional Bayesian model defined by $P_X^E\left(\mathbf{X}^E|\boldsymbol{\Theta}^E\right)$ with prior $P_\Theta^E\left(\boldsymbol{\Theta}^E\right)$ satisfying:*

(i) Assume that each finite-dimensional posterior $P_\Theta^I\left(\boldsymbol{\Theta}^I|\mathbf{X}^I\right)$ is conjugate with respect to its associated Bayesian model with posterior index Γ^I and probability kernel κ^I. If there exists a measurable mapping $\Gamma : \Omega_X^E \to \Omega_\Gamma^E$ satisfying condition (12.35), then the projective limit posterior $P_\Theta^E\left(\boldsymbol{\Theta}^E|\mathbf{X}^E\right)$ will be conjugate with posterior index Γ.

(ii) Conversely, if the infinite-dimensional posterior $P_\Theta^E\left(\boldsymbol{\Theta}^E|\mathbf{X}^E\right)$ is conjugate with posterior index Γ^E and probability kernel κ^E, then each marginal posterior $P_\Theta^I\left(\boldsymbol{\Theta}^I|\mathbf{X}^I\right)$ will be conjugate with posterior index

$$\Gamma^I = \Pi_{IE} \circ \Gamma^E \circ \Pi_{IE}^{-1} \tag{12.36}$$

and the probability kernel

$$\kappa^I\left(A^I, \Gamma^I\right) = \kappa^E\left(\Pi_{IE}^{-1} A^I, \Gamma\right), \quad \forall \Gamma \in \Pi_{IE}^{-1}\Gamma^I. \tag{12.37}$$

It is straightforward to generalize this theorem in two ways [252]:

(i) Considering the effect of hyper-parameters on posterior indices.
(ii) Considering spaces with different dimensions for each Bayesian model.

However, including additional parameters and book-keeping on dimensions will need a more complicated notation.

According to this theorem, conjugate models have conjugate marginals. In the finite-dimensional case, conjugate Bayesian models are restricted to exponential families and their natural conjugate priors. Hence, a nonparametric Bayesian model is expected to be conjugate if it is constructed from exponential family marginals based on a product space [252]. Construction of nonparametric Bayesian models is discussed in Section 12.8.

12.8 Construction of Nonparametric Bayesian Models

In the Bayesian paradigm, the concept of *sufficiency* developed by Fisher is of fundamental importance [259]. A sufficient statistic $\mathbf{S}(\mathbf{x})$ contains all the information brought by $\mathbf{x}$ about an unknown model parameter θ [44].

Definition 12.23 ***(Sufficient statistic)*** *For* $\mathbf{x} \sim f(\mathbf{x}|\theta)$, *a function* $\mathbf{S}(\mathbf{x})$, *which is known as a statistic, is sufficient, if the distribution of* $\mathbf{x}$ *conditional upon* $\mathbf{S}(\mathbf{x})$ *does not depend on* θ.

If $\mathbf{S}(\mathbf{x})$ is a sufficient statistic, then the density of $\mathbf{x}$, can be factored as the following product:

$$f(\mathbf{x}|\theta) = h\,(\mathbf{x}|\mathbf{S}(\mathbf{x}))\,g\,(\mathbf{S}(\mathbf{x})|\theta)\,, \tag{12.38}$$

where one factor, h, does not depend on θ and the other factor, g, which depends on θ, depends on $\mathbf{x}$ only through $\mathbf{S}(\mathbf{x})$.

Definition 12.24 ***(Sufficiency principle)*** *Two observations* $\mathbf{x}$ *and* $\mathbf{y}$, *which are factorized through the same value of a sufficient statistic* $\mathbf{S}(\cdot)$ *such that* $\mathbf{S}(\mathbf{x}) = \mathbf{S}(\mathbf{y})$, *lead to the same inference on* θ.

A generic model can be constructed using an exponential family and its conjugate prior. For a Bayesian model, which is built on a conjugate exponential family, the posterior index Γ is obtained from the corresponding sufficient statistic S as [252]:

$$\Gamma\,(\mathbf{x}, (\mathbf{y}, \alpha)) = (\mathbf{y} + \mathbf{S}(\mathbf{x}), \alpha + 1)\,. \tag{12.39}$$

For such a model, the projection condition (12.35) holds for the posterior indices Γ^I of a family over all dimensions $I \in \mathcal{F}\,(E)$, if and only if this condition holds for the sufficient statistics S^I of the marginals. This is due to the fact that addition and projection commute.

In Theorem 12.6, the infinite-dimensional posterior index Γ^E exists, if and only if there is a corresponding infinite-dimensional extension of the sufficient statistics

$\mathbf{S}^I$ (denoted by $\mathbf{S}^E$), for which the condition (12.35) holds. Then, the posterior index for the infinite-dimensional projective limit model will be [252]:

$$\Gamma^E\left(\mathbf{x}^E, (\mathbf{y}^E, \alpha)\right) = \left(\mathbf{y}^E + \mathbf{S}^E(\mathbf{x}^E), \alpha + 1\right). \tag{12.40}$$

For countable dimensions, Theorem 12.6 provides a construction recipe for non-parametric Bayesian models from exponential family marginals. In such cases, constructing the model reduces to checking the following two conditions:

(i) whether the models selected as finite-dimensional marginals are conditionally projective, and
(ii) whether the sufficient statistics satisfy the projection condition.

Section 12.9 covers the computability issue.

12.9 Posterior Computability

For an exchangeable sequence, $\xi = \{\mathbf{X}_0, \mathbf{X}_1, \dots\}$, the *posterior predictive*, which is denoted by $P(\mathbf{X}_{k+1}|\mathbf{X}_{0:k})$, may be uncomputable, even when the distribution of ξ is computable. In most cases, the knowledge we have about an exchangeable sequence is in the form of an algorithm, which provides the conditional distribution of an element $\mathbf{X}_{k+1}$, given the previous samples in the sequence $\mathbf{X}_{0:k} = \{\mathbf{X}_0, \dots, \mathbf{X}_k\}$. By induction, the prediction rule can be used to sample from the conditional distribution of $\mathbf{X}_{k+2}$ given the prefix $\mathbf{X}_{0:k+1}$, and this procedure can be subsequently continued to form an infinite exchangeable sequence based on the original prefix $\mathbf{X}_{0:k}$. The following theorem states the equivalence between [260]:

(i) the ability to from the true posterior predictive by consistently extending a prefix to an infinite sequence, and
(ii) the ability to compute the posterior distribution regarding the latent distribution that generates the sequence.

This equivalence relation can be mathematically expressed as [260, 261]:

$$\text{Predictive Computable} \Leftrightarrow \text{Posterior Computable} \tag{12.41}$$

Theorem 12.7 ***(Posterior computability)*** *Let $\xi = \{\mathbf{X}_0, \mathbf{X}_1, \dots\}$ be an exchangeable sequence of random variables with directing random measure μ. There exists a program that provides a representation of the sequence of posterior distributions $\{P(\mu|\mathbf{X}_{0:k})\}_{k\geq 0}$, given a representation of the sequence of posterior predictives $\{P(\mathbf{X}_{k+1}|\mathbf{X}_{0:k})\}_{k\geq 0}$, and vice-versa.*

It is worth noting that this theorem emphasizes on the directing random measure not any parametrization. The directing random measure is always identifiable and computable, although a particular parametrization may not be [260].

As shown in Section 12.8, for a nonparametric model, a closed form expression for the posterior can be obtained in terms of sufficient statistics. To do so, the nonparametric model must be constructed as the projective limit of a family of models, which admit both sufficient statistics and conjugate posteriors. In terms of computability, the posterior distribution can be computably recovered from sufficient statistics of the observations, if and only if new observations can be computably sampled, given sufficient statistics of past observations [260].

Corollary 12.1 ***(Posterior computability in terms of sufficient statistics)*** *Let $\xi = \{X_0, X_1, \ldots\}$ be an exchangeable sequence of random variables with directing random measure μ, and $\sum_{i=0}^{k} \mathbf{S}(X_i)$ with $\mathbf{S} : \mathbb{R} \to \mathbb{R}^m$ be a sufficient statistic for X_{k+1} given $X_{0:k}$. The sequence of posterior distributions $P\left(\mu | \sum_{i=0}^{k} \mathbf{S}(X_i)\right)$ for $k \geq 0$ is computable, if and only if the sequence of conditional distributions $P\left(X_{k+1} | \sum_{i=0}^{k} \mathbf{S}(X_i)\right)$ for $k \geq 0$ and the sufficient statistic $\mathbf{S}$ are computable.*

The considered form for the sufficient statistics, $\sum_{i=0}^{k} \mathbf{S}(X_i)$ with $\mathbf{S} : \mathbb{R} \to \mathbb{R}^m$ being a continuous function, covers natural exponential family likelihoods. The framework provided by Theorem 12.7 and Corollary 12.1 sheds light on the reason of success of ad-hoc methods for computing conditional distributions, even when the general task is not computable [260]. Section 12.10 studies the notion of algorithmic sufficiency.

12.10 Algorithmic Sufficiency

According to the *criterion of sufficiency*, the chosen statistic must summarize all of the relevant information provided by the sample [259]. In this regard, a *sufficient statistic* contains all of the information about the *model class* that exists in the data. Let the family of distributions $\{P_\theta\}$, which is parameterized by the set of parameters $\theta \in \Theta$, define a model class for the random variable $\mathbf{X}$ with the finite or countable alphabet $\mathcal{X}$. A *statistic* of the data in $\mathcal{X}$ is defined as any function $\mathbf{S} : \mathcal{X} \to \mathcal{S}$ that takes values in a set $\mathcal{S}$. According to Definition 12.23, $\mathbf{S}$ is a sufficient statistic, if $\forall \mathbf{s} \in \mathcal{S}$, the conditional distribution $P_\theta\,(\mathbf{X} = \cdot \mid \mathbf{S}(\mathbf{x}) = \mathbf{s})$ is invariant with respect to θ. In other words, $\mathbf{S}(\mathbf{x})$ contains all of the information about θ that exists in $\mathbf{x}$ [262].

Sufficient statistic can be expressed in terms of the probability mass function, $P_\theta(\mathbf{X} = \mathbf{x}) = f_\theta(\mathbf{x})$. Regarding the following conditional distribution:

$$f_\theta(\mathbf{x}|\mathbf{s}) = \begin{cases} f_\theta(\mathbf{x}) / \left(\sum_{\mathbf{x}\in\mathcal{X}, \mathbf{S}(\mathbf{x})=\mathbf{s}} f_\theta(\mathbf{x})\right), & \mathbf{S}(\mathbf{x}) = \mathbf{s}, \\ 0, & \mathbf{S}(\mathbf{x}) \neq \mathbf{s}, \end{cases} \tag{12.42}$$

sufficiency of $\mathbf{S}$ would be equivalent to existence of a function $g : \mathcal{X} \times \mathcal{S} \to \mathbb{R}$ that satisfies:

$$g(\mathbf{x}|\mathbf{s}) = f_\theta(\mathbf{x}|\mathbf{s}), \quad \forall\, \mathbf{x} \in \mathcal{X}, \mathbf{s} \in \mathcal{S}, \theta \in \Theta. \tag{12.43}$$

$\mathbf{S}$ is a *probabilistic sufficient statistic* for $\{f_\theta\}$, if:

$$\log \frac{1}{g(\mathbf{x}|\mathbf{s})} = \log \frac{1}{f_\theta(\mathbf{x}|\mathbf{s})}, \quad \forall\, \mathbf{x} \in \mathcal{X}, \mathbf{s} \in \mathcal{S}, \theta \in \Theta. \tag{12.44}$$

Defining an a priori distribution $p(\theta)$ over Θ, we have:

$$p(\mathbf{x}, \theta) = p(\theta)\, f_\theta(\mathbf{x}). \tag{12.45}$$

For $\mathbf{S}$ to be sufficient, the condition (12.44) must be satisfied $\forall\, \mathbf{x}$. An *expectation* version of the aforementioned definition can be obtained through the equivalence between the following two statements:

$$\sum_{\mathbf{x}} f_\theta(\mathbf{x}) \log \frac{1}{f_\theta(\mathbf{x}|\mathbf{s})} = \sum_{\mathbf{x}} f_\theta(\mathbf{x}) \log \frac{1}{g(\mathbf{x}|\mathbf{s})}, \qquad \forall\, \theta \in \Theta, \tag{12.46}$$

$$\sum_{\mathbf{x},\theta} p(\mathbf{x}, \theta) \log \frac{1}{f_\theta(\mathbf{x}|\mathbf{s})} = \sum_{\mathbf{x},\theta} p(\mathbf{x}, \theta) \log \frac{1}{g(\mathbf{x}|\mathbf{s})}, \quad \forall\, p(\theta) \text{ on } \Theta. \tag{12.47}$$

For *mutual information*, $I(\cdot\,;\cdot)$, the following inequality:

$$I(\Theta; \mathbf{X}) \geq I\left(\Theta; \mathbf{S}(\mathbf{X})\right), \quad \forall\, \mathbf{S}, \tag{12.48}$$

implies that $I(\Theta; \mathbf{X})$ cannot be enhanced through data processing. In other words, sufficiency cannot be improved. Regarding the inequality (12.48), the probabilistic sufficient statistic can be defined in terms of mutual information. To be more precise, the statistic $\mathbf{S}$ is sufficient for $\{f_\theta\}$, if and only if the mutual information is invariant under taking the statistic [262]:

$$I(\Theta; \mathbf{X}) = I\left(\Theta; \mathbf{S}(\mathbf{X})\right), \quad \forall\, p(\theta) \text{ on } \Theta. \tag{12.49}$$

A sufficient statistic may contain irrelevant information. Hence, the notion of *minimal sufficient statistic* is defined as a function of all other sufficient statistics with the following characteristics [262]:

- It does not contain any irrelevant information.
- It provides maximal compression of the information about the model class that exists in the data.

The probabilistic sufficient statistic can be viewed as extracting the relevant patterns from data that allow for determining the parameters of a model class. In a general framework, rather than focusing on a finite-dimensional parametric model class, the considered model class may include all computable distributions or all computable sets for which the collected data is an element. For such model classes, the notion of *algorithmic sufficient statistic* provides a counterpart for probabilistic sufficient statistic. The algorithmic sufficient statistic summarizes all relevant information that exists in the data, for which *Kolmogorov complexity* provides a measure of conciseness [262]. Kolmogorov complexity of an object is the length of the shortest program, which produces the object as output [263]. Kolmogorov complexity is a measure of the contained information in the object without any redundancy.

Regarding computer models, a universal computer refers to an abstract machine that can mimic the behavior of other computers. As a canonical and conceptually simple universal computer, the *universal Turing machine* can be used to compute any computable sequence. The *Turing machine* can be viewed as a map from a set of binary strings with finite lengths to a set of binary strings with finite or infinite lengths [26]:

$$f : \{0,1\}^* \to \{0,1\}^* \cup \{0,1\}^\infty. \tag{12.50}$$

If the computation does not halt, the value of the function will be undefined. The set of computable functions by Turing machines is known as the set of *partial recursive functions*. In mathematical terms, Kolmogorov complexity is defined as follows [26, 263].

Definition 12.25 ***(Kolmogorov complexity)*** *For a binary string x, Kolmogorov complexity, $K_{\mathcal{U}}(x)$, with respect to a universal computer, $\mathcal{U}$, is the length of the shortest program among all programs, p, that print x and halt:*

$$K_{\mathcal{U}}(x) = \min_p \ \{l(p) \mid \mathcal{U}(p) = x, p \in \{0,1\}^*\}, \tag{12.51}$$

where $l(p)$ denotes the number of bits in the computer program p, which can be viewed as a binary string.

In other words, $K_{\mathcal{U}}(x)$ is the shortest description length among all descriptions of x, which are interpretable by $\mathcal{U}$. The following theorem states the universality of Kolmogorov complexity [26].

Theorem 12.8 ***(Universality of Kolmogorov complexity)*** *Assume that $\mathcal{U}$ is a universal computer. Then, for any other computer $\mathcal{A}$, there exists a constant $c_{\mathcal{A}}$ independent of the string x such that:*

$$K_{\mathcal{U}}(x) \leq K_{\mathcal{A}}(x) + c_{\mathcal{A}}, \quad \forall\, x \in \{0,1\}^*. \tag{12.52}$$

Data may contain both *meaningful information* and *accidental information*. The latter refers to meaningless information such as measurement error and random noise. In statistics and machine learning, it is desired to separate these two categories of information. Therefore, Kolmogorov complexity can be considered as the shortest length program with two parts [262]:

- The first part describes a model as a Turing machine, T, for regular aspects of the data. The valuable or relevant information in the data is captured by this model.
- The second part describes irregular aspects of the data as a program, p, which is interpreted by the Turing machine, T. This part contains the useless or irrelevant information.

Although different combinations of T and p may describe the data, x, it is desired to find a Turing machine that describes the relevant information in x. The two-part nature of Kolmogorov complexity can be expressed in mathematical terms as:

$$K(x) = \min_{p,i} \left\{ K_{T_i}(x) + l(p) \mid T_i(p) = x, p \in \{0,1\}^*, i \in \{1,2,\dots\} \right\} + O(1). \tag{12.53}$$

For a finite set $S \subseteq \{0,1\}^*$ that contains x, we have:

$$\begin{aligned} K(x) &\leq K(x, S) \\ &\leq K(S) + K(x|S) + O(1) \\ &\leq K(S) + \log|S| + O(1). \end{aligned} \tag{12.54}$$

Regarding (12.54), the algorithmic sufficient statistic is defined as follows [262].

Definition 12.26 ***(Algorithmic sufficient statistic, algorithmic minimal sufficient statistic)*** *An algorithmic sufficient statistic of x is the shortest program for a set S containing x, which is optimal in the sense that:*

$$K(x) = K(S) + \log|S| + O(1). \tag{12.55}$$

An algorithmic sufficient statistic with the optimal set S is minimal, if:

$$\nexists\, S', \quad K(S') < K(S). \tag{12.56}$$

In order to extend the concepts of probabilistic and algorithmic statistics to strings with arbitrary lengths, n, the notion of *sequential statistic* has been defined [262].

Definition 12.27 ***(Sequential statistic)*** *A function $S : \{0,1\}^* \rightarrow 2^{\{0,1\}^*}$ is a sequential statistic, if $\forall n$ and $\forall x \in \{0,1\}^n$:*

- $S(x) \subseteq \{0,1\}^n$,
- $x \in S(x)$,
- *$\forall n$, the set $\{s \mid \exists x \in \{0,1\}^n, S(x) = s\}$ is a partition of $\{0,1\}^n$.*

Definition 12.28 ***(Algorithmic sufficient sequential statistic)*** *S is a sufficient sequential statistic in the algorithmic sense, if* $\forall n$ *and* $\forall x \in \{0,1\}^n$*, there exists a constant c such that the program, which generates S(x), is an algorithmic sufficient statistic for x:*

$$K(S(x)) + \log |S(x)| \leq K(x) + c. \tag{12.57}$$

The conditions in (12.44) and (12.46) do not exactly hold for algorithmic sufficient statistics, because algorithmic sufficiency holds within additive constants. Therefore, the notion of *nearly-sufficient statistic* has been defined to address this issue [262].

Definition 12.29 ***(Nearly-sufficient sequential statistic)*** *S is a nearly-sufficient sequential statistic for* $\{f_\theta\}$ *in the probabilistic-individual sense, if* $\forall \theta$, $\forall n$, *and* $\forall x \in \{0,1\}^n$*, there exist a set of functions* $g^n(\cdot)$ *and a constant c such that:*

$$\left| \log \frac{1}{f_\theta^n(x|S(x))} - \log \frac{1}{g^n(x|S(x))} \right| \leq c. \tag{12.58}$$

This inequality can be interpreted as (12.44) holds within a constant.

S is a nearly-sufficient sequential statistic for $\{f_\theta\}$ *in the probabilistic-expectation sense, if* $\forall \theta$, $\forall n$, *and* $\forall x \in \{0,1\}^n$*, there exist a set of functions* $g^n(\cdot)$ *and a constant c such that:*

$$\left| \sum_{x\in\{0,1\}^n} f_\theta^n(x) \left(\log \frac{1}{f_\theta^n(x|S(x))} - \log \frac{1}{g^n(x|S(x))} \right) \right| \leq c. \tag{12.59}$$

This inequality can be interpreted as (12.46) holds within a constant.

The following theorem relates the algorithmic sufficiency to the probabilistic sufficiency [262].

Theorem 12.9 ***(Algorithmic and probabilistic sufficient statistics are equivalent)*** *Assume that S is an algorithmic sufficient sequential statistic and g has uniform distribution. If* $\sup_{\theta\in\Theta} K(f_\theta) < \infty$*, then S will be a nearly-sufficient statistic for* $\{f_\theta\}$ *in the probabilistic-expectation sense.*

In other words, $\forall n$ *and* $\forall \theta$ *with* $K(f_\theta^n) < \infty$*, there exists a constant c such that (12.59) holds for conditional probability mass functions* $g^n(\cdot)$ *with uniform distributions:*

$$g^n(\mathbf{x}|\mathbf{s}) = \begin{cases} 1/\left|\{x \in \{0,1\}^n \mid S(x) = s\}\right|, & \mathbf{S}(\mathbf{x}) = \mathbf{s}, \\ 0, & \mathbf{S}(\mathbf{x}) \neq \mathbf{s}. \end{cases} \tag{12.60}$$

12.11 Applications

In this section, three case studies are reviewed, which are focused on multiple object tracking, probabilistic optimal flow in power systems, and single-molecule fluorescence microscopy.

12.11.1 Multiple Object Tracking

Multiple target tracking algorithms must be able to estimate the positions of an unknown and time-varying number of objects in the presence of noise and clutter. Hence, such tracking problems are usually challenging and computationally intensive. A nonparametric Bayesian method has been proposed for multiple object tracking, which is based on using a family of dependent *Pitman–Yor* processes to model the state prior [264]. In the considered scenario, it is assumed that moving objects may enter the scene, leave the scene, or remain in the scene.

12.11.2 Data-Driven Probabilistic Optimal Power Flow

To address the probabilistic optimal power flow problem, a nonparametric Bayesian framework has been proposed, which deploys the Dirichlet process mixture model and the variational Bayesian inference [265]. The constructed model takes account of uncertainties that occur in power systems due to wind generation and load power. In the proposed method, samples are drawn from the Dirichlet process mixture model using quasi-Monte Carlo sampling, where the number of components in the mixture model is determined in a data-driven manner. The effectiveness of the algorithm was tested on several IEEE benchmark power systems.

12.11.3 Analyzing Single-Molecule Tracks

The *single-molecule analysis by unsupervised Gibbs sampling* (SMAUG) algorithm has been proposed for real-time probing of nanoscale subcellular biological systems [266]. SMAUG deploys a nonparametric Bayesian framework for information extraction from single-particle trajectory datasets. In complex systems with multiple biological states that lead to multiple observed mobility states, SMAUG aims at providing the following information:

- the number of mobility states,
- the average diffusion coefficient of single molecules in each state,
- the fraction of single molecules in each state,
- the localization noise,
- the probability of transitioning between two different states.

SMAUG was evaluated using real experimental systems in both *prokaryotes* and *eukaryotes*.

12.12 Concluding Remarks

Modeling deals with *uncertainty quantification* regarding possible predictions based on the available information. In order to be effective, probabilistic models must be flexible enough to capture the relevant information about the system under study. Increasing the number of model parameters will enhance the flexibility of the model. Hence, nonparametric models with infinitely many parameters are superior to parametric models, which have a finite number of parameters. However, wide adoption of nonparametric Bayesian models heavily relies on successful applications of such models to real-world complex systems. Addressing the issues that arise in such systems calls for creating new classes of models as well as deriving general scalable inference methods [250]. In this regard, construction of nonparametric Bayesian models from parametric ones can play a critical role [252].

References

1 R. E. Kalman, "On the general theory of control systems," in *Proceedings First International Conference on Automatic Control, Moscow, USSR*, 1960, pp. 481–492.

2 M. Karl, M. Soelch, J. Bayer, and P. van der Smagt, "Deep variational Bayes filters: Unsupervised learning of state space models from raw data," in *International Conference on Learning Representations*, 2017, pp. 1–13.

3 R. G. Brown and P. Y. Hwang, *Introduction to Random Signals and Applied Kalman Filtering with MATLAB Exercises*, 4th ed. John Wiley & Sons, New York, NY, USA, 2012.

4 B. C. Smith, "The foundations of computing," in *Computationalism: New Directions*, M. Scheutz, Ed. Cambridge, MA, USA: The MIT Press, 2002, Ch. 2, pp. 23–58.

5 L. Breiman, "Statistical modeling: The two cultures (with comments and a rejoinder by the author)," *Statistical Science*, vol. 16, no. 3, pp. 199–231, 2001.

6 A. Y. Ng and M. I. Jordan, "On discriminative vs. generative classifiers: A comparison of logistic regression and naive Bayes," in *Advances in Neural Information Processing Systems*, 2002, pp. 841–848.

7 R. Shwartz-Ziv and N. Tishby, "Opening the black box of deep neural networks via information," *arXiv preprint arXiv:1703.00810*, 2017.

8 K. P. Murphy, "Dynamic Bayesian networks: Representation, inference and learning," Ph.D. dissertation, University of California, Berkeley, 2002.

9 K. R. Muske and T. F. Edgar, "Nonlinear state estimation," in *Nonlinear Process Control*, M. A. Henson and D. E. Seborg, Eds. Upper Saddle River, NJ, USA: Prentice Hall, 1997, Ch. 6, pp. 311–370.

10 N. Wiener, *Cybernetics: Or Control and Communication in the Animal and the Machine*, 2nd ed. The MIT Press, 1965.

11 J. Pearl, *Causality*, 2nd ed. Cambridge University Press, 2009.

12 A. Isidori, *Nonlinear Control Systems*, 3rd ed. Springer Science & Business Media, 2013.

Nonlinear Filters: Theory and Applications, First Edition. Peyman Setoodeh, Saeid Habibi, and Simon Haykin.

13 R. Hermann and A. Krener, "Nonlinear controllability and observability," *IEEE Transactions on Automatic Control*, vol. 22, no. 5, pp. 728–740, 1977.

14 T. Glad and L. Ljung, *Control Theory: Multivariable and Nonlinear Methods*. CRC Press, 2000.

15 G. Besançon and G. Bornard, "*On characterizing classes of observer forms for nonlinear systems*," in *1997 European Control Conference (ECC)*. IEEE, 1997, pp. 3113–3118.

16 T. Kailath, *Linear Systems*. Prentice-Hall Englewood Cliffs, NJ, USA, 1980, vol. 156.

17 S. Skogestad and I. Postlethwaite, *Multivariable Feedback Control: Analysis and Design*, 2nd ed. Wiley-Interscience, 2006.

18 K. J. Astrom and B. Wittenmark, *Computer-Controlled Systems: Theory and Design*, 3rd ed. Prentice-Hall, 1997.

19 J. M. Mendel, *Lessons in Estimation Theory for Signal Processing, Communications, and Control*, 2nd ed. Pearson Education, 2008.

20 W. Respondek, "Geometry of static and dynamic feedback," in *Lecture Notes at the Summer School on Mathematical Control Theory, Italy, September 2001 and Bedlewo-Warsaw, Poland, September 2002*, 2001.

21 A. J. Krener, "Nonlinear observers," in *Control Systems, Robotics and Automation: Nonlinear, Distributed, and Time Delay Ststems-II*, H. Unbehauen, Ed. Cambridge, MA, USA: The MIT Press, 2009, Ch. 6, pp. 153–179.

22 J.-J. E. Slotine, W. Li *et al.*, *Applied Nonlinear Control*. Prentice Hall Englewood Cliffs, NJ, USA, 1991.

23 H. K. Khalil, *Nonlinear Systems*, 3rd ed. Pearson, 2002.

24 A. J. Krener and W. Respondek, "Nonlinear observers with linearizable error dynamics," *SIAM Journal on Control and Optimization*, vol. 23, no. 2, pp. 197–216, 1985.

25 W. Lee and K. Nam, "Observer design for autonomous discrete-time nonlinear systems," *Systems & Control Letters*, vol. 17, no. 1, pp. 49–58, 1991.

26 T. M. Cover and J. A. Thomas, *Elements of Information Theory*, 2nd ed. John Wiley & Sons, 2006.

27 B. R. Frieden, *Science from Fisher Information*. Cambridge University Press, 2004.

28 A. R. Liu and R. R. Bitmead, "Stochastic observability in network state estimation and control," *Automatica*, vol. 47, no. 1, pp. 65–78, 2011.

29 R. Brown, "Not just observable, but how observable?" in *National Electronics Conference, 22 nd, Chicago, ILL*, 1966, pp. 709–714.

30 F. M. Ham and R. G. Brown, "Observability, eigenvalues, and Kalman filtering," *IEEE Transactions on Aerospace and Electronic Systems*, no. 2, pp. 269–273, 1983.

31 R. Mohler and C. Hwang, "Nonlinear data observability and information," *Journal of the Franklin Institute*, vol. 325, no. 4, pp. 443–464, 1988.

32 M. Kam, R. Cheng, and P. Kalata, "An information-theoretic interpretation of stability and observability," in *Proceedings of American Control Conference*, 1987, pp. 1957–1962.

33 M. Fatemi, P. Setoodeh, and S. Haykin, "Observability of stochastic complex networks under the supervision of cognitive dynamic systems," *Journal of Complex Networks*, vol. 5, no. 3, pp. 433–460, 2017.

34 B. Baxter, L. Graham, and S. Wright, "Invertible and non-invertible information sets in linear rational expectations models," *Journal of Economic Dynamics and Control*, vol. 35, no. 3, pp. 295–311, 2011.

35 S. Sundaram, "Fault-tolerant and secure control systems," *University of Waterloo, Lecture Notes*, 2012.

36 G. Ciccarella, M. Dalla Mora, and A. Germani, "Observers for discrete-time nonlinear systems," *Systems & Control Letters*, vol. 20, pp. 373–382, 1993.

37 I. Haskara, U. Ozguner, and V. Utkin, "On sliding mode observers via equivalent control approach," *International Journal of Control*, vol. 71, no. 6, pp. 1051–1067, 1998.

38 S. Drakunov and V. Utkin, "Sliding mode observers. Tutorial," in *Proceedings of 1995 34th IEEE Conference on Decision and Control*, vol. 4. IEEE, 1995, pp. 3376–3378.

39 S.-K. Chang, W.-T. You, and P.-L. Hsu, "Design of general structured observers for linear systems with unknown inputs," *Journal of the Franklin Institute*, vol. 334, no. 2, pp. 213–232, 1997.

40 H. K. Khalil and L. Praly, "High-gain observers in nonlinear feedback control," *International Journal of Robust and Nonlinear Control*, vol. 24, no. 6, pp. 993–1015, 2014.

41 J. Lee, J. Choi, and H. K. Khalil, "New implementation of high-gain observers in the presence of measurement noise using stochastic approximation," in *2016 European Control Conference (ECC)*. IEEE, 2016, pp. 1740–1745.

42 C. Combastel, "A state bounding observer based on zonotopes," in *European Control Conference (ECC)*. IEEE, 2003, pp. 2589–2594.

43 B. Magee, *The Story of Philosophy: A Concise Introduction to the World's Greatest Thinkers and Their Ideas*. Dorling Kindersley, 2016.

44 C. P. Robert, *The Bayesian Choice: From Decision-Theoretic Foundations to Computational Implementation*, 2nd ed. Springer, 2007.

45 G. D'Agostini, *Bayesian Reasoning in Data Analysis: A Critical Introduction*. World Scientific, 2003.

46 B. Ristic, S. Arulampalam, and N. Gordon, *Beyond the Kalman Filter: Particle Filters for Tracking Applications*. Artech House, 2004.

47 Y. Bar-Shalom, X. R. Li, and T. Kirubarajan, *Estimation with Applications to Tracking and Navigation: Theory, Algorithms, and Software*. John Wiley & Sons, 2001.

48 Y. Ho and R. Lee, "A Bayesian approach to problems in stochastic estimation and control," *IEEE Transactions on Automatic Control*, vol. 9, no. 4, pp. 333–339, 1964.

49 R. K. Boel, M. R. James, and I. R. Petersen, "Robustness and risk-sensitive filtering," *IEEE Transactions on Automatic Control*, vol. 47, no. 3, pp. 451–461, 2002.

50 W. H. Fleming, "Deterministic nonlinear filtering," *Annali Della Scuola Normale Superiore di Pisa-Classe di Scienze*, vol. 25, no. 3–4, pp. 435–454, 1997.

51 J. C. Spall, *Introduction to Stochastic Search and Optimization: Estimation, Simulation, and Control*. John Wiley & Sons, 2005.

52 H. L. van Trees, *Detection, Estimation, and Modulation Theory, Part I: Detection, Estimation, and Linear Modulation Theory*. John Wiley & Sons, 1968.

53 J. Dauwels, "Computing Bayesian Cramer-Rao bounds," in *Proceedings. International Symposium on Information Theory (ISIT)*. IEEE, 2005, pp. 425–429.

54 H. L. van Trees and K. L. Bell, *Bayesian Bounds for Parameter Estimation and Nonlinear Filtering/Tracking*. Wiley-IEEE Press, 2007.

55 P. Tichavsky, C. H. Muravchik, and A. Nehorai, "Posterior Cramér-Rao bounds for discrete-time nonlinear filtering," *IEEE Transactions on Signal Processing*, vol. 46, no. 5, pp. 1386–1396, 1998.

56 M. S. Grewal and A. P. Andrews, *Kalman Filtering: Theory and Practice with MATLAB*, 4th ed. John Wiley & Sons, 2014.

57 H. W. Sorenson, *Kalman Filtering: Theory and Application*. IEEE, 1985.

58 R. E. Kalman, "A new approach to linear filtering and prediction problems," *Transactions of the ASME. Series D, Journal of Basic Engineering*, vol. 82, pp. 35–45, 1960.

59 J. S. Meditch, *Stochastic Optimal Linear Estimation and Control*. McGraw-Hill, 1969.

60 B. D. O. Anderson and J. B. Moore, *Optimal Filtering*. Prentice-Hall, 1979.

61 P. S. Maybeck, *Stochastic Models, Estimation, and Control*. Academic Press, 1982.

62 D. Simon, *Optimal State Estimation: Kalman, H_∞, and Nonlinear Approaches*, 2nd ed. John Wiley & Sons, 2007.

63 P. Zarchan and H. Musoff, *Fundamentals of Kalman Filtering: A Practical Approach*, 4th ed. American Institute of Aeronautics and Astronautics, 2015.

64 M. Nørgaard, N. K. Poulsen, and O. Ravn, "New developments in state estimation for nonlinear systems," *Automatica*, vol. 36, no. 11, pp. 1627–1638, 2000.

65 A. Gelb, *Applied Optimal Estimation*. The MIT Press, 1974.

66 S. J. Julier and J. K. Uhlmann, "Unscented filtering and nonlinear estimation," *Proceedings of the IEEE*, vol. 92, no. 3, pp. 401–422, 2004.

67 I. Arasaratnam and S. Haykin, "Cubature Kalman filters," *IEEE Transactions on Automatic Control*, vol. 54, no. 6, pp. 1254–1269, 2009.

68 P. Kaminski, A. Bryson, and S. Schmidt, "Discrete square root filtering: a survey of current techniques," *IEEE Transactions on Automatic Control*, vol. 16, no. 6, pp. 727–736, 1971.

69 A. G. O. Mutambara, *Decentralized Estimation and Control for Multisensor Systems*. CRC Press, 1998.

70 M. Nørgaard, N. K. Poulsen, and O. Ravn, "Advances in derivative-free state estimation for nonlinear systems," Technical University of Denmark, Tech. Rep. IMM-REP-1998-15, April 2000.

71 C.-E. Fröberg *et al.*, *Introduction to Numerical Analysis*. Addison-Wesley, 1970.

72 T. S. Schei, "A finite-difference method for linearization in nonlinear estimation algorithms," *Automatica*, vol. 33, no. 11, pp. 2053–2058, 1997.

73 S. J. Julier, J. K. Uhlmann, and H. F. Durrant-Whyte, "A new approach for filtering nonlinear systems," in *Proceedings of the American Control Conference (ACC)*, vol. 3. IEEE, 1995, pp. 1628–1632.

74 S. Julier, J. Uhlmann, and H. F. Durrant-Whyte, "A new method for the nonlinear transformation of means and covariances in filters and estimators," *IEEE Transactions on Automatic Control*, vol. 45, no. 3, pp. 477–482, 2000.

75 E. A. Wan and R. van der Merwe, "The unscented Kalman filter," in *Kalman Filtering and Neural Networks*, S. Haykin, Ed. Wiley-Interscience, 2001, pp. 221–280.

76 J. Zhang, X. He, and D. Zhou, "Generalised proportional–integral–derivative filter," *IET Control Theory and Applications*, vol. 10, no. 17, pp. 2339–2347, 2016.

77 K. Murphy, *Machine Learning: A Probabilistic Perspective*. The MIT Press, 2012.

78 Y. Bar-Shalom, P. K. Willett, and X. Tian, *Tracking and Data Fusion: A Handbook of Algorithms*. YBS Publishing, 2011.

79 T. Bak, *Lecture Notes: Estimation and Sensor Information Fusion*. Aalborg University, 2000.

80 J. D. Hol, T. Schon, F. Gustafsson, and P. J. Slycke, "Sensor fusion for augmented reality," in *International Conference on Information Fusion*. IEEE, 2006, pp. 1–6.

81 M. Rostami-Shahrbabaki, A. A. Safavi, M. Papageorgiou, P. Setoodeh, and I. Papamichail, "State estimation in urban traffic networks: A two-layer approach," *Transportation Research Part C: Emerging Technologies*, vol. 115, p. 102616, 2020.

82 Z. Kazemi, A. A. Safavi, F. Naseri, L. Urbas, and P. Setoodeh, "A secure hybrid dynamic-state estimation approach for power systems under false data injection attacks," *IEEE Transactions on Industrial Informatics*, vol. 16, no. 12, pp. 7275–7286, 2020.

83 M. Jahja, D. C. Farrow, R. Rosenfeld, and R. J. Tibshirani, "Kalman filter, sensor fusion, and constrained regression: Equivalences and insights," *arXiv preprint arXiv:1905.11436*, 2019.

84 F. Arroyo-Marioli, F. Bullano, S. Kucinskas, and C. Rondón-Moreno, "Tracking R of COVID-19: A new real-time estimation using the Kalman filter," *PLoS ONE*, vol. 16, no. 1, p. e0244474, 2021.

85 X. Feng, K. A. Loparo, and Y. Fang, "Optimal state estimation for stochastic systems: An information theoretic approach," *IEEE Transactions on Automatic Control*, vol. 42, no. 6, pp. 771–785, 1997.

86 S. K. Mitter and N. J. Newton, "Information and entropy flow in the Kalman–Bucy filter," *Journal of Statistical Physics*, vol. 118, no. 1, pp. 145–176, 2005.

87 B. Sinopoli, L. Schenato, M. Franceschetti, K. Poolla, M. I. Jordan, and S. S. Sastry, "Kalman filtering with intermittent observations," *IEEE transactions on Automatic Control*, vol. 49, no. 9, pp. 1453–1464, 2004.

88 N. J. Gordon, D. J. Salmond, and A. F. Smith, "Novel approach to nonlinear/non-Gaussian Bayesian state estimation," *IEE Proceedings F (Radar and Signal Processing)* vol. 140, no. 2, pp. 107–113, 1993.

89 A. Doucet and A. M. Johansen, "A tutorial on particle filtering and smoothing: Fifteen years later," in *The Oxford Handbook of Nonlinear Filtering*, D. Crisan and B. Rozovskii, Eds. Oxford University Press, 2011, Ch. 24, pp. 656–704.

90 G. Kitagawa, "Monte Carlo filter and smoother for non-Gaussian nonlinear state space models," *Journal of Computational and Graphical Statistics*, vol. 5, no. 1, pp. 1–25, 1996.

91 C. Musso, N. Oudjane, and F. Le Gland, "Improving regularised particle filters," in *Sequential Monte Carlo Methods in Practice*, A. Doucet, N. de Freitas, and N. Gordon, Eds. Springer, 2001, pp. 247–271.

92 W. R. Gilks and C. Berzuini, "Following a moving target-Monte Carlo inference for dynamic Bayesian models," *Journal of the Royal Statistical Society: Series B (Statistical Methodology)*, vol. 63, no. 1, pp. 127–146, 2001.

93 J. D. Hol, T. B. Schon, and F. Gustafsson, "On resampling algorithms for particle filters," in *2006 IEEE Nonlinear Statistical Signal Processing Workshop*. IEEE, 2006, pp. 79–82.

94 D. Fox, "Adapting the sample size in particle filters through KLD-sampling," *The International Journal of Robotics Research*, vol. 22, no. 12, pp. 985–1003, 2003.

95 T. B. Schön, F. Gustafsson, and R. Karlsson, "The particle filter in practice," in *The Oxford Handbook of Nonlinear Filtering*, D. Crisan and B. Rozovskii, Eds. Oxford University Press, 2011, Ch. 26, pp. 741–767.

96 R. Karlsson, T. B. Schon, D. Tornqvist, G. Conte, and F. Gustafsson, "Utilizing model structure for efficient simultaneous localization and mapping for a UAV application," in *IEEE Aerospace Conference*. IEEE, 2008, pp. 1–10.

97 S. Habibi, "The smooth variable structure filter," *Proceedings of the IEEE*, vol. 95, no. 5, pp. 1026–1059, 2007.

98 R. E. Kalman and R. S. Bucy, "New results in linear filtering and prediction theory," *Transactions of the ASME. Series D, Journal of Basic Engineering*, vol. 83, pp. 95–108, 1961.

99 H. Asada and J.-J. Slotine, *Robot Analysis and Control*. John Wiley & Sons, 1986.

100 V. Utkin, "Variable structure systems with sliding modes," *IEEE Transactions on Automatic control*, vol. 22, no. 2, pp. 212–222, 1977.

101 V. I. Utkin, "Sliding modes and their applications in variable structure systems," *Mir, Moscow*, 1978.

102 V. I. Utkin, *Sliding Modes in Control and Optimization*. Springer Science & Business Media, 1992.

103 D. Milosavljevic, "General conditions for existence of a quasi-sliding mode on the switching hyperplane in discrete variable structure systems," *Automation and Remote Control*, vol. 46, pp. 307–314, 1985.

104 S. Sarpturk, Y. Istefanopulos, and O. Kaynak, "On the stability of discrete-time sliding mode control systems," *IEEE Transactions on Automatic Control*, vol. 32, no. 10, pp. 930–932, 1987.

105 S. Drakunov and R. DeCarlo, "Discrete-time/discrete-event sliding mode design via Lyapunov approach," in *Proceedings of the American Control Conference*, vol. 3. IEEE, 1997, pp. 1719–1723.

106 K. Furuta and Y. Pan, "Variable structure control with sliding sector," *Automatica*, vol. 36, no. 2, pp. 211–228, 2000.

107 S. K. Spurgeon, "Sliding mode observers: A survey," *International Journal of Systems Science*, vol. 39, no. 8, pp. 751–764, 2008.

108 Y. Shtessel, C. Edwards, L. Fridman, and A. Levant, *Sliding Mode Control and Observation*. Springer, 2014, vol. 10.

109 S. Drakunov and V. Utkin, "Discrete-event sliding mode observers for continuous-time systems," in *Proceedings of the IEEE Conference on Decision and Control*, vol. 4. IEEE, 1995, pp. 3403–3405.

110 S. Drakunov, "An adaptive quasioptimal filter with discontinuous parameters," *Automation and Remote Control*, vol. 44, no. 9, pp. 1167–1175, 1983.

111 J.-J. Slotine, J. K. Hedrick, and E. A. Misawa, "On sliding observers for nonlinear systems," *ASME Journal of Dynamic Systems, Measurement, and Control*, vol. 109, no. 3, pp. 245–252, 1987.

112 B. Walcott and S. Zak, "State observation of nonlinear uncertain dynamical systems," *IEEE Transactions on Automatic Control*, vol. 32, no. 2, pp. 166–170, 1987.

113 B. L. Walcott, M. J. Corless, and S. H. Zak, "Comparative study of non-linear state-observation techniques," *International Journal of Control*, vol. 45, no. 6, pp. 2109–2132, 1987.

114 B. L. Walcott and S. H. Zak, "Combined observer-controller synthesis for uncertain dynamical systems with applications," *IEEE Transactions on Systems, Man, and Cybernetics*, vol. 18, no. 1, pp. 88–104, 1988.

115 A. S. Zinober, *Variable Structure and Lyapunov Control*. Springer, 1994, vol. 193.

116 C. Edwards, S. K. Spurgeon, and R. J. Patton, "Sliding mode observers for fault detection and isolation," *Automatica*, vol. 36, no. 4, pp. 541–553, 2000.

117 A. J. Koshkouei and A. S. I. Zinober, "Sliding mode state observers for discrete-time linear systems," *International Journal of Systems Science*, vol. 33, no. 9, pp. 751–758, 2002.

118 E. A. Misawa, "Nonlinear state estimation using sliding observers," Ph.D. dissertation, Massachusetts Institute of Technology, 1988.

119 A. Akhenak, M. Chadli, D. Maquin, and J. Ragot, "Sliding mode multiple observer for fault detection and isolation," in *Proceedings of the IEEE International Conference on Decision and Control*, vol. 1. IEEE, 2003, pp. 953–958.

120 T. Floquet, J.-P. Barbot, W. Perruquetti, and M. Djemai, "On the robust fault detection via a sliding mode disturbance observer," *International Journal of Control*, vol. 77, no. 7, pp. 622–629, 2004.

121 J. Ackermann and V. Utkin, "Sliding mode control design based on Ackermann's formula," *IEEE Transactions on Automatic Control*, vol. 43, no. 2, pp. 234–237, 1998.

122 W. Lohmiller and J.-J. Slotine, "Contraction analysis of non-linear distributed systems," *International Journal of Control*, vol. 78, no. 9, pp. 678–688, 2005.

123 S. R. Habibi and R. Burton, "The variable structure filter," *ASME Journal of Dynamic Systems, Measurement, and Control*, vol. 125, no. 3, pp. 287–293, 2003.

124 C. M. Dorling, "The design of variable structure control systems," in *Manual for the VASSYD CAD Package*. Dept. Appl. Comput. Math., Univ. Sheffield, 1985.

125 C. M. Dorling and A. S. I. Zinober, "Robust hyperplane design in multivariable variable structure control systems," *International Journal of Control*, vol. 48, no. 5, pp. 2043–2054, 1988.

126 P. A. Cook, *Nonlinear Dynamical Systems.* Prentice Hall International (UK) Ltd., 1994.

127 E. Kreyszig, *Advanced Engineering Mathematics*, 10th edition, Wiley, 2011.

128 J.-J. E. Slotine, "Sliding controller design for non-linear systems," *International Journal of control*, vol. 40, no. 2, pp. 421–434, 1984.

129 D. G. Luenberger, *Introduction to Dynamic Systems.* Wiley, New York, 1979.

130 H. H. Afshari, S. A. Gadsden, and S. Habibi, "A nonlinear second-order filtering strategy for state estimation of uncertain systems," *Signal Processing*, vol. 155, pp. 182–192, 2019.

131 S. A. Gadsden and S. R. Habibi, "A new form of the smooth variable structure filter with a covariance derivation," in *49th IEEE Conference on Decision and Control (CDC).* IEEE, 2010, pp. 7389–7394.

132 S. A. Gadsden, M. El Sayed, and S. R. Habibi, "Derivation of an optimal boundary layer width for the smooth variable structure filter," in *Proceedings of the 2011 American Control Conference.* IEEE, 2011, pp. 4922–4927.

133 M. Avzayesh, M. Abdel-Hafez, M. AlShabi, and S. Gadsden, "The smooth variable structure filter: A comprehensive review," *Digital Signal Processing*, vol. 110, p. 102912, 2020.

134 M. Attari, Z. Luo, and S. Habibi, "An SVSF-based generalized robust strategy for target tracking in clutter," *IEEE Transactions on Intelligent Transportation Systems*, vol. 17, no. 5, pp. 1381–1392, 2015.

135 M. Messing, S. Rahimifard, T. Shoa, and S. Habibi, "Low temperature, current dependent battery state estimation using interacting multiple model strategy," *IEEE Access*, vol. 9, pp. 99876–99889, 2021.

136 J. Moor, "The Dartmouth College artificial intelligence conference: The next fifty years," *AI Magazine*, vol. 27, no. 4, p. 87, 2006.

137 R. Cordeschi, "AI turns fifty: Revisiting its origins," *Applied Artificial Intelligence*, vol. 21, no. 4–5, pp. 259–279, 2007.

138 S. Haykin, *Neural Networks and Learning Machines*, 3rd ed. Prentice Hall, 2009.

139 Y. Bengio, I. Goodfellow, and A. Courville, *Deep Learning.* The MIT Press, 2016.

140 J. Schulman, N. Heess, T. Weber, and P. Abbeel, "Gradient estimation using stochastic computation graphs," *arXiv preprint arXiv:1506.05254*, 2015.

141 M. Toussaint, "Lecture notes: Some notes on gradient descent," Technical University of Berlin, Tech. Rep., May 2012.

142 A. Devarakonda, M. Naumov, and M. Garland, "AdaBatch: Adaptive batch sizes for training deep neural networks," *arXiv preprint arXiv:1712.02029*, 2018.

143 S. L. Smith, P.-J. Kindermans, C. Ying, and Q. V. Le, "Don't decay the learning rate, increase the batch size," *arXiv preprint arXiv:1711.00489*, 2018.

144 D. P. Kingma and J. Ba, "Adam: A method for stochastic optimization," *arXiv preprint arXiv:1412.6980*, 2014.

145 S.-I. Amari, "Natural gradient works efficiently in learning," *Neural Computation*, vol. 10, no. 2, pp. 251–276, 1998.

146 R. Pascanu and Y. Bengio, "Revisiting natural gradient for deep networks," *arXiv preprint arXiv:1301.3584*, 2014.

147 A. Amini and A. Soleimany, *Introduction to Deep Learning*. Lecture Notes. MIT, 2021.

148 Y. Gal and Z. Ghahramani, "Dropout as a Bayesian approximation: Representing model uncertainty in deep learning," in *International Conference on Machine Learning*. PMLR, 2016, pp. 1050–1059.

149 X. Glorot and Y. Bengio, "Understanding the difficulty of training deep feedforward neural networks," in *Proceedings of the Thirteenth International Conference on Artificial Intelligence and Statistics*. JMLR Workshop and Conference Proceedings, 2010, pp. 249–256.

150 K. He, X. Zhang, S. Ren, and J. Sun, "Delving deep into rectifiers: Surpassing human-level performance on imagenet classification," in *Proceedings of the IEEE International Conference on Computer Vision*, 2015, pp. 1026–1034.

151 N. Buduma and N. Locascio, *Fundamentals of Deep Learning: Designing Next-Generation Machine Intelligence Algorithms*. O'Reilly Media, Inc., 2017.

152 J. Schmidhuber, "Deep learning in neural networks: an overview," *Neural Networks*, vol. 61, pp. 85–117, 2015.

153 K. Greff, R. K. Srivastava, J. Koutník, B. R. Steunebrink, and J. Schmidhuber, "LSTM: A search space odyssey," *IEEE Transactions on Neural Networks and Learning Systems*, vol. 28, no. 10, pp. 2222–2232, 2016.

154 S. Hochreiter and J. Schmidhuber, "Long short-term memory," *Neural Computation*, vol. 9, no. 8, pp. 1735–1780, 1997.

155 F. A. Gers and J. Schmidhuber, "Recurrent nets that time and count," in *Proceedings of the IEEE-INNS-ENNS International Joint Conference on Neural Networks (IJCNN) 2000. Neural Computing: New Challenges and Perspectives for the New Millennium*, vol. 3. IEEE, 2000, pp. 189–194.

156 C. Olah, "Understanding LSTM networks," blog, August 2015. [Online]. Available: https://colah.github.io/posts/2015-08-Understanding-LSTMs/.

157 K. Cho, B. Van Merriënboer, C. Gulcehre, D. Bahdanau, F. Bougares, H. Schwenk, and Y. Bengio, "Learning phrase representations using RNN encoder-decoder for statistical machine translation,"*arXiv preprint arXiv:1406.1078*, 2014.

158 D. O. Hebb, *The Organization of Behavior: A Neuropsychological Theory*. John Wiley & Sons, 1949.

159 S. Ermon, *Automated Reasoning*. Lecture Notes. Stanford University, 2016.

160 D. H. Ackley, G. E. Hinton, and T. J. Sejnowski, "A learning algorithm for Boltzmann machines," *Cognitive Science*, vol. 9, no. 1, pp. 147–169, 1985.

161 R. Salakhutdinov and G. Hinton, "Deep Boltzmann machines," in *Artificial Intelligence and Statistics*. PMLR, 2009, pp. 448–455.

162 D. Koller and N. Friedman, *Probabilistic Graphical Models: Principles and Techniques*. The MIT Press, 2009.

163 B. A. Olshausen and D. J. Field, "Sparse coding with an overcomplete basis set: A strategy employed by v1?" *Vision Research*, vol. 37, no. 23, pp. 3311–3325, 1997.

164 C. Doersch, "Tutorial on variational autoencoders," *arXiv preprint arXiv:1606.05908*, 2021.

165 A. Creswell, T. White, V. Dumoulin, K. Arulkumaran, B. Sengupta, and A. A. Bharath, "Generative adversarial networks: An overview," *IEEE Signal Processing Magazine*, vol. 35, no. 1, pp. 53–65, 2018.

166 I. J. Goodfellow, J. Pouget-Abadie, M. Mirza, B. Xu, D. Warde-Farley, S. Ozair, A. Courville, and Y. Bengio, "Generative adversarial networks," *arXiv preprint arXiv:1406.2661*, 2014.

167 A. Vaswani, N. Shazeer, N. Parmar, J. Uszkoreit, L. Jones, A. N. Gomez, L. Kaiser, and I. Polosukhin, "Attention is all you need," *arXiv preprint arXiv:1706.03762*, 2017.

168 J. Thickstun, "The transformer model in equations," University of Washington, Tech. Rep., 2021.

169 Y. Tay, M. Dehghani, D. Bahri, and D. Metzler, "Efficient transformers: A survey," *arXiv preprint arXiv:2009.06732*, 2020.

170 K. Greff, R. K. Srivastava, and J. Schmidhuber, "Highway and residual networks learn unrolled iterative estimation," *arXiv preprint arXiv:1612.07771*, 2016.

171 Y. LeCun, Y. Bengio, and G. Hinton, "Deep learning," *Nature*, vol. 521, no. 7553, pp. 436–444, 2015.

172 A. Graves, G. Wayne, M. Reynolds, T. Harley, I. Danihelka, A. Grabska-Barwińska, S. G. Colmenarejo, E. Grefenstette, T. Ramalho, J. Agapiou *et al.*, "Hybrid computing using a neural network with dynamic external memory," *Nature*, vol. 538, no. 7626, pp. 471–476, 2016.

173 S. Haykin, *Cognitive Dynamic Systems: Perception-Action Cycle, Radar, and Radio*. Cambridge University Press, 2012.

174 J. M. Fuster, *Cortex and Mind: Unifying Cognition*. Oxford University Press, 2002.

175 J. Marino, M. Cvitkovic, and Y. Yue, "A general method for amortizing variational filtering," *arXiv preprint arXiv:1811.05090*, 2018.

176 S. Särkkä, *Bayesian Filtering and Smoothing*. Cambridge University Press, 2013, no. 3.

177 M. I. Jordan, Z. Ghahramani, T. S. Jaakkola, and L. K. Saul, "An introduction to variational methods for graphical models," *Machine Learning*, vol. 37, no. 2, pp. 183–233, 1999.

178 S. Gershman and N. Goodman, "Amortized inference in probabilistic reasoning," in *Proceedings of the Annual Meeting of the Cognitive Science Society*, vol. 36, no. 36, 2014.

179 J. Marino, Y. Yue, and S. Mandt, "Iterative amortized inference," in *International Conference on Machine Learning*. PMLR, 2018, pp. 3403–3412.

180 R. Ranganath, S. Gerrish, and D. Blei, "Black box variational inference," in *Artificial intelligence and statistics*. PMLR, 2014, pp. 814–822.

181 D. P. Kingma and M. Welling, "Auto-encoding variational Bayes," in *International Conference on Learning Representations*, 2013, pp. 1–14.

182 D. P. Kingma and M. Welling, "Stochastic gradient VB and the variational auto-encoder," in *International Conference on Learning Representations*, vol. 19, 2014.

183 T. A. Le, M. Igl, T. Rainforth, T. Jin, and F. Wood, "Auto-encoding sequential Monte Carlo," *arXiv preprint arXiv:1705.10306*, 2017.

184 I. Dasgupta, E. Schulz, J. B. Tenenbaum, and S. J. Gershman, "A theory of learning to infer." *Psychological Review*, vol. 127, no. 3, p. 412, 2020.

185 M. Andrychowicz, M. Denil, S. Gomez, M. W. Hoffman, D. Pfau, T. Schaul, B. Shillingford, and N. De Freitas, "Learning to learn by gradient descent by gradient descent," *arXiv preprint arXiv:1606.04474*, 2016.

186 D. Zhao, N. Zucchet, J. Sacramento, and J. von Oswald, "Learning where to learn," in *Proceedings of the International Conference on Learning Representations (ICLR)*, vol. 3, 2021, pp. 1–7.

187 P. Kavumba, B. Heinzerling, A. Brassard, and K. Inui, "Learning to learn to be right for the right reasons," *arXiv preprint arXiv:2104.11514*, 2021.

188 R. G. Krishnan, U. Shalit, and D. Sontag, "Deep Kalman filters," *arXiv preprint arXiv:1511.05121*, 2015.

189 T. Haarnoja, A. Ajay, S. Levine, and P. Abbeel, "Backprop KF: Learning discriminative deterministic state estimators," *arXiv preprint arXiv:1605.07148*, 2017.

190 R. Jonschkowski, D. Rastogi, and O. Brock, "Differentiable particle filters: end-to-end learning with algorithmic priors," *arXiv preprint arXiv:1805.11122*, 2018.

191 R. Kurle, S. S. Rangapuram, E. de Bézenac, S. Günnemann, and J. Gasthaus, "Deep Rao-Blackwellised particle filters for time series forecasting," *Advances in Neural Information Processing Systems*, vol. 33, 2020, pp. 1–12.

192 K. Murphy and S. Russell, "Rao-Blackwellised particle filtering for dynamic Bayesian networks," in *Sequential Monte Carlo Methods in Practice*, A. Doucet, N. de Freitas, and N. Gordon, Eds. Springer, 2001, pp. 499–515.

193 P. Del Moral, *Feynman-Kac Formulae: Genealogical and Interacting Particle Systems with Applications*. Springer, 2004.

194 C. K. Sonderby, T. Raiko, L. Maaloe, S. K. Sonderby, and O. Winther, "Ladder variational autoencoders," *arXiv preprint arXiv:1602.02282*, 2016.

195 S. Roweis and Z. Ghahramani, "Learning nonlinear dynamical systems using the expectation–maximization algorithm," in *Kalman Filtering and Neural Networks*, S. Haykin, Ed. New York, NY, USA: John Wiley & Sons, 2001, Ch. 6, pp. 175–220.

196 G. E. Hinton and D. van Camp, "Keeping the neural networks simple by minimizing the description length of the weights," in *Proceedings of the 6th Annual Conference on Computational Learning Theory*, 1993, pp. 5–13.

197 N. Das, M. Karl, P. Becker-Ehmck, and P. van der Smagt, "Beta DVBF: Learning state-space models for control from high dimensional observations," *arXiv preprint arXiv:1911.00756*, 2019.

198 P. Becker-Ehmck, J. Peters, and P. van der Smagt, "Switching linear dynamics for variational Bayes filtering," in *International Conference on Machine Learning*. PMLR, 2019, pp. 553–562.

199 M. Fraccaro, S. Kamronn, U. Paquet, and O. Winther, "A disentangled recognition and nonlinear dynamics model for unsupervised learning," *arXiv preprint arXiv:1710.05741*, 2017.

200 D. J. Rezende, S. Mohamed, and D. Wierstra, "Stochastic backpropagation and approximate inference in deep generative models," *ser. Proceedings of Machine Learning Research*, E. P. Xing and T. Jebara, Eds., vol. 32, no. 2. Bejing, China: PMLR, 22–24 Jun 2014, pp. 1278–1286.

201 M. Watter, J. T. Springenberg, J. Boedecker, and M. Riedmiller, "Embed to control: A locally linear latent dynamics model for control from raw images," *arXiv preprint arXiv:1506.07365*, 2015.

202 F. Creutzig, A. Globerson, and N. Tishby, "Past-future information bottleneck in dynamical systems," *Physical Review E*, vol. 79, no. 4, p. 041925, 2009.

203 A. A. Alemi, I. Fischer, J. V. Dillon, and K. Murphy, "Deep variational information bottleneck," *arXiv preprint arXiv:1612.00410*, 2019.

204 A. Kolchinsky, B. D. Tracey, and D. H. Wolpert, "Nonlinear information bottleneck," *Entropy*, vol. 21, no. 12, p. 1181, 2019.

205 S. Shafieezadeh-Abadeh, V. A. Nguyen, D. Kuhn, and P. M. Esfahani, "Wasserstein distributionally robust Kalman filtering," *arXiv preprint arXiv:1809.08830*, 2018.

206 C. R. Givens and R. M. Shortt, "A class of Wasserstein metrics for probability distributions," *Michigan Mathematical Journal*, vol. 31, no. 2, pp. 231–240, 1984.

207 G. Detommaso, J. Kruse, L. Ardizzone, C. Rother, U. Köthe, and R. Scheichl, "HINT: Hierarchical invertible neural transport for general and sequential Bayesian inference," *Stat*, vol. 1050, p. 25, 2019.

208 L. Dinh, J. Sohl-Dickstein, and S. Bengio, "Density estimation using real NVP," *arXiv preprint arXiv:1605.08803*, 2017.

209 G. Strang, *Introduction to Linear Algebra*, 5th ed. Wellesley-Cambridge Press, 2016.

210 J. Stoer and R. Bulirsch, *Introduction to Numerical Analysis*. Springer Science & Business Media, 2013, vol. 12.

211 C. Villani, *Optimal Transport: Old and New*. Springer Science & Business Media, 2008, vol. 338.

212 Y. Marzouk, T. Moselhy, M. Parno, and A. Spantini, "Sampling via measure transport: An introduction," in Handbook of Uncertainty Quantification, R. Ghanem, D. Higdon, and H. Owhadi, Ed., Springer, 2017, pp. 787–825.

213 L. Ardizzone, C. Lüth, J. Kruse, C. Rother, and U. Köthe, "Guided image generation with conditional invertible neural networks," *arXiv preprint arXiv:1907.02392*, 2019.

214 C. Winkler, D. Worrall, E. Hoogeboom, and M. Welling, "Learning likelihoods with conditional normalizing flows," *arXiv preprint arXiv:1912.00042*, 2019.

215 A. Hyvärinen and P. Pajunen, "Nonlinear independent component analysis: Existence and uniqueness results," *Neural Networks*, vol. 12, no. 3, pp. 429–439, 1999.

216 SAE On-Road Automated Vehicle Standards Committee, "Taxonomy and definitions for terms related to on-road motor vehicle automated driving systems," *SAE Standard J*, vol. 3016, pp. 1–16, 2014.

217 D. Watzenig and M. Horn, *Automated Driving: Safer and More Efficient Future Driving*. Springer, 2016.

218 A. Geiger, P. Lenz, C. Stiller, and R. Urtasun, "The KITTI vision benchmark suite," http://www.cvlibs.net/datasets/kitti, vol. 2, 2015.

219 T. B. Schon, A. Wills, and B. Ninness, "System identification of nonlinear state-space models," *Automatica*, vol. 47, no. 1, pp. 39–49, 2011.

220 M. Klaas, M. Briers, N. De Freitas, A. Doucet, S. Maskell, and D. Lang, "Fast particle smoothing: If I had a million particles," in *Proceedings of the 23rd International Conference on Machine Learning*, 2006, pp. 481–488.

221 P. Fearnhead, D. Wyncoll, and J. Tawn, "A sequential smoothing algorithm with linear computational cost," *Biometrika*, vol. 97, no. 2, pp. 447–464, 2010.

222 K. Greff, S. Van Steenkiste, and J. Schmidhuber, "Neural expectation maximization," *arXiv preprint arXiv:1708.03498*, 2017.

223 S. Van Steenkiste, M. Chang, K. Greff, and J. Schmidhuber, "Relational neural expectation maximization: Unsupervised discovery of objects and their interactions," *arXiv preprint arXiv:1802.10353*, 2018.

224 A. Wills, T. B. Schön, and B. Ninness, "Parameter estimation for discrete-time nonlinear systems using EM," *IFAC Proceedings Volumes*, vol. 41, no. 2, pp. 4012–4017, 2008.

225 P. M. Djuric, M. Khan, and D. E. Johnston, "Particle filtering of stochastic volatility modeled with leverage," *IEEE Journal of Selected Topics in Signal Processing*, vol. 6, no. 4, pp. 327–336, 2012.

226 M. G. Bellemare, Y. Naddaf, J. Veness, and M. Bowling, "The Arcade learning environment: An evaluation platform for general agents," *Journal of Artificial Intelligence Research*, vol. 47, pp. 253–279, 2013.

227 J. S. Garofolo, L. F. Lamel, W. M. Fisher, J. G. Fiscus, D. S. Pallett, and N. L. Dahlgren, "DARPA TIMIT acoustic-phonetic continuous speech corpus," 1993.

228 N. Boulanger-Lewandowski, Y. Bengio, and P. Vincent, "Modeling temporal dependencies in high-dimensional sequences: Application to polyphonic music generation and transcription," *arXiv preprint arXiv:1206.6392*, 2012.

229 C. Schuldt, I. Laptev, and B. Caputo, "Recognizing human actions: A local SVM approach," in *Proceedings of the 17th International Conference on Pattern Recognition, 2004. ICPR 2004*, vol. 3. IEEE, 2004, pp. 32–36.

230 T. D. Ullman, E. Spelke, P. Battaglia, and J. B. Tenenbaum, "Mind games: Game engines as an architecture for intuitive physics," *Trends in Cognitive Sciences*, vol. 21, no. 9, pp. 649–665, 2017.

231 B. M. Lake, T. D. Ullman, J. B. Tenenbaum, and S. J. Gershman, "Building machines that learn and think like people," *Behavioral and Brain Sciences*, vol. 40, 2017, 1–72.

232 S. Chatterjee, O. Romero, and S. Pequito, "Analysis of generalized expectation-maximization for Gaussian mixture models: A control systems perspective," *arXiv preprint arXiv:1903.00979*, 2019.

233 R. S. Sutton and R. G. Barto, *Reinforcement Learning: An Introduction*, 2nd ed. The MIT Press, Cambridge, MA, USA, 2018.

234 D. P. Bertsekas and J. N. Tsitsiklis, *Neuro-Dynamic Programming*. Athena Scientific, Belmont, MA, USA, 1996.

235 S. K. Mitter and N. J. Newton, "The duality between estimation and control," in *Optimal Control and Partial Differential Equations: In Honour of Professor Alain Bensoussan's 60th Birthday*, J. L. Menaldi, Ed. IOS Press, 2001, pp. 47–58.

236 T. Weber, N. Heess, S. M. A. Eslami, J. Schulman, D. Wingate, and D. Silver, "Reinforced variational inference," in *Proceedings of Neural Information Processing Systems (NIPS)*, 2015, pp. 1–9.

237 D. Silver, *Introduction to Reinforcement Learning*. Lecture Notes. University College London, 2015.

238 I. Grondman, L. Busoniu, G. A. D. Lopes, and R. Babuska, "A survey of actor-critic reinforcement learning: Standard and natural policy gradients," *IEEE Transactions on Systems, Man, and Cybernetics Part C: Applications and Reviews*, vol. 42, no. 6, pp. 1291–1307, 2012.

239 R. J. Williams, "Simple statistical gradient-following algorithms for connectionist reinforcement learning," *Machine Learning*, vol. 8, no. 3–4, pp. 229–256, 1992.

240 C. J. C. H. Watkins, "Learning from delayed rewards," Ph.D. dissertation, King's College, Cambridge, United Kingdom, 1989.

241 C. J. Watkins and P. Dayan, "Q-learning," *Machine Learning*, vol. 8, no. 3–4, pp. 279–292, 1992.

242 G. A. Rummery and M. Niranjan, "On-line Q-learning using connectionist systems," University of Cambridge, United Kingdom, Tech. Rep., 1994.

243 H. R. Berenji and D. Vengerov, "A convergent actor-critic-based FRL algorithm with application to power management of wireless transmitters," *IEEE Transactions on Fuzzy Systems*, vol. 11, no. 4, pp. 478–485, 2003.

244 V. R. Konda and J. N. Tsitsiklis, "On actor-critic algorithms," *SIAM Journal on Control and Optimization*, vol. 42, no. 4, pp. 1143–1166, 2003.

245 R. S. Sutton, "Dyna, an integrated architecture for learning, planning, and reacting," *ACM Sigart Bulletin*, vol. 2, no. 4, pp. 160–163, 1991.

246 H. H. Afshari, M. Attari, R. Ahmed, M. Farag, and S. Habibi, "Modeling, parameterization, and state of charge estimation of Li-Ion cells using a circuit model," in *2016 IEEE Transportation Electrification Conference and Expo (ITEC)*. IEEE, 2016, pp. 1–6.

247 M. Kim, K. Kim, J. Kim, J. Yu, and S. Han, "State of charge estimation for lithium ion battery based on reinforcement learning," *IFAC-PapersOnLine*, vol. 51, no. 28, pp. 404–408, 2018.

248 P. Orbanz, *Lecture Notes on Bayesian Nonparametrics*. University College London, 2014.

249 P. Orbanz and Y. W. Teh, "Bayesian nonparametric models," in *in Encyclopedia of Machine Learning*, C. Sammut and G. I. Webb, Ed. Springer, pp. 81–88, 2011.

250 Z. Ghahramani, "Bayesian non-parametrics and the probabilistic approach to modelling," *Philisophical Transactions of the Royal Society A*, vol. 371, no. 1984, pp. 1–20, 2013.

251 D. M. Roy, "Computability, inference and modeling in probabilistic programming," Ph.D. dissertation, Massachusetts Institute of Technology, 2011.

252 P. Orbanz, "Construction of nonparametric Bayesian models from parametric Bayes equations," in *Proceedings of Neural Information Processing Systems (NIPS)*, 2009, pp. 1–9.

253 L. Aggoun and R. J. Elliott, *Measure Theory and Filtering: Introduction and Applications*. Cambridge University Press, 2004.

254 O. Kallenberg, *Probabilistic Symmetries and Invariance Principles*. Springer, 2005.

255 H. Bauer, *Probability Theory*. de Gruyter, 1996.

256 E. E. Doberkat, *Special Topics in Mathematics for Computer Scientists: Sets, Categories, Topologies and Measures*. Springer, 2015.

257 M. M. Rao, *Conditional Measures and Applications*, 2nd ed. Chapman and Hall, 2005.

258 R. Durrett, *Probability: Theory and Examples*, 5th ed. Cambridge University Press, 2019.

259 R. A. Fisher, "On the mathematical foundations of theoretical statistics," *Philosophical Transactions of the Royal Society of London. Series A, Containing Papers of a Mathematical or Physical Character*, vol. 222, no. 594-604, pp. 309–368, 1922.

260 C. E. Freer and D. M. Roy, "Posterior distributions are computable from predictive distributions," in *Proceedings of AISTATS*, 2010, pp. 233–240.

261 C. E. Freer and D. M. Roy, "Predictive computable $\Leftrightarrow$ posterior computable," in *Proceedings of Nonparametric Bayes Workshop*, 2009, pp. 1–3.

262 P. Grünwald and P. Vitányi, "Shannon information and Kolmogorov complexity," *arXiv:cs/041000*, pp. 1–54, 2010.

263 M. Li, P. Vitányi *et al.*, *An Introduction to Kolmogorov Complexity and Its Applications*, 4th ed. Springer, 2019.

264 B. Moraffah, A. Papandreou-Suppappola, and M. Rangaswamy, "Nonparametric Bayesian methods and the dependent Pitman-Yor process for modeling evolution in multiple object tracking," in *2019 22th International Conference on Information Fusion (FUSION)*. IEEE, 2019, pp. 1–6.

265 W. Sun, M. Zamani, M. R. Hesamzadeh, and H.-T. Zhang, "Data-driven probabilistic optimal power flow with nonparametric Bayesian modeling and inference," *IEEE Transactions on Smart Grid*, vol. 11, no. 2, pp. 1077–1090, 2019.

266 J. D. Karslake, E. D. Donarski, S. A. Shelby, L. M. Demey, V. J. DiRita, S. L. Veatch, and J. S. Biteen, "SMAUG: Analyzing single-molecule tracks with nonparametric Bayesian statistics," *Methods*, vol. 193, pp. 16–26, 2021.

Index

Nonlinear Filters: Theory and Applications, First Edition. Peyman Setoodeh, Saeid Habibi, and Simon Haykin.

e

f

g

Printed and bound by CPI Group (UK) Ltd, Croydon, CR0 4YY